Slosh Dynamics

Slosh Dynamics

Theory and Practice

Musaddiq Al Ali

Associate Professor, Graduate School of Engineering,
The University of Osaka, Suita City, Osaka Prefecture, Japan

ELSEVIER

Elsevier
Radarweg 29, PO Box 211, 1000 AE Amsterdam, Netherlands
125 London Wall, London EC2Y 5AS, United Kingdom
50 Hampshire Street, 5th Floor, Cambridge, MA 02139, United States

ISBN: 978-0-443-44330-5

For Information on all Elsevier publications
visit our website at https://www.elsevier.com/books-and-journals

Publisher: Matthew Deans
Acquisitions Editor: Dennis McGonagle
Editorial Project Manager: Tripti Rawat
Production Project Manager: Maria Bernard
Cover Designer: Raman Kumar

Typeset by MPS Limited, Chennai, India

Dedication

To my mother, the enduring source of patience, strength, and continuity.

Contents

The link for the companion site is:
https://shop.elsevier.com/books/book-companion/9780443443305.

About the author

Musaddiq Al Ali is an associate professor in the Graduate School of Engineering at The University of Osaka. He received his PhD in transportation and environmental systems from Hiroshima University. His research focuses on fluid–structure interaction, additive manufacturing, multiphysics optimisation, and computational mechanics.

Prof. Al Ali has extensive experience collaborating with industry and has served as a consultant to major Japanese corporations in the automotive, electronics, and heavy engineering sectors. These industrial engagements have strongly influenced his physics-first, method-independent approach to modelling and simulation. Prior to joining The University of Osaka, he was affiliated with Toyota Technological Institute, where he contributed to research, education, and industry partnership initiatives.

Preface

Engineering problems rarely announce themselves as equations. They appear first as small, almost trivial disturbances in everyday systems: a spill that should not occur, a vibration that should not exist, a sound that engineers struggle to justify, or a failure that cannot be traced to an obvious design flaw. Sloshing dynamics belongs precisely to this class of phenomena. It is often dismissed as secondary, treated as a correction, or assumed to be automatically handled by numerical software. Yet, in practice, sloshing repeatedly emerges as a governing physical mechanism that determines safety, performance, noise, durability, and even public confidence in engineered systems.

The motivation for writing this book did not originate from a laboratory experiment or a theoretical curiosity, but from an ordinary moment in a highly engineered environment. While dining at a conveyor-belt restaurant in Japan, a glass of liquid was delivered via a moving belt system. When the glass was lifted from the belt, a noticeable amount of liquid had spilled onto the belt surface, leaving the exterior of the glass wet. From a design perspective, this should not have happened. The transport speed, acceleration profile, and stopping mechanism of the belt had been optimised for efficiency and throughput, yet the liquid inside the glass behaved independently of those optimisations. The system had been designed to move solid objects, not free-surface fluids.

This simple observation is a direct manifestation of sloshing dynamics. The belt velocity and acceleration were specified without considering the inertial response of the liquid, the phase lag between container motion and free-surface deformation, or the transient resonance induced during acceleration and deceleration. The liquid was not defective, the container was not poorly manufactured, and the belt mechanism was not malfunctioning. The issue lay in the absence of sloshing physics from the design loop. This type of oversight is far more common than engineers may wish to admit, and it forms the conceptual entry point of this book.

Japan provides a particularly instructive context for understanding the importance of sloshing dynamics. The country possesses one of the most advanced chemical industries in the world, supplying ultra-high-purity chemicals essential for semiconductor manufacturing. These chemicals are stored, transported, and processed in tanks of various geometries, often under stringent safety requirements. At the same time, Japan is a seismically active region, where earthquakes are not exceptional events but expected environmental conditions.

When seismic excitation occurs, the liquid inside storage tanks does not move rigidly with the container. Instead, it undergoes sloshing motion governed by its own natural frequencies, mode shapes, and nonlinear interactions. These motions impose dynamic pressures on tank walls, alter load paths, and interact with structural supports and foundations. In many cases, the integrity of the tank is not determined by static stress limits but by resonance between ground motion and sloshing modes. A tank that satisfies all conventional structural design codes may still fail if sloshing dynamics is inadequately understood or improperly modelled.

The challenge becomes even more complex when the mechanical support system of the tank is considered. Sloshing does not act independently of the structure; it couples with structural vibration, modifies effective mass and damping, and can either amplify or attenuate dynamic response depending on frequency alignment. This fluid–structure interaction lies at the heart of safe tank design in seismic regions, yet it is often treated as a specialised topic rather than a fundamental design consideration.

Further motivation arises from the rapid transformation of the automotive industry. The widespread adoption of hybrid and electric vehicles has shifted engineering attention from engine noise to previously masked phenomena. Liquid-borne noise, vibration, and harshness have become critical design issues. Coolant systems for batteries and power electronics, washer fluid containers, and residual fuel tanks in hybrid vehicles all contain liquids that experience acceleration, braking, and cornering. Unlike conventional vehicles, where engine noise dominated the acoustic environment, modern vehicles make sloshing-induced noise perceptible and undesirable.

In this context, sloshing is no longer merely a structural concern but an acoustic and comfort issue. The motion of liquid free surfaces generates pressure fluctuations, impacts with container walls, and flow-induced vibrations that propagate through vehicle structures. Addressing these issues requires more than empirical damping treatments; it requires an understanding of the underlying physics of sloshing and its interaction with container geometry, excitation spectra, and boundary conditions.

From both industrial practice and academic teaching, a consistent pattern emerges in the existing literature. Many references focus heavily on mathematical formulations, exact solutions for idealised geometries, or highly specialised nonlinear analyses. While these contributions are scientifically valuable, they often assume that the reader already possesses a strong intuition for sloshing physics and computational modelling. At the same time, a large body of applied work relies almost exclusively on a single discretisation framework, most commonly the finite volume method, presented as a default solution rather than a modelling choice.

In many cases, the numerical method is described in detail, while the physical interpretation of the results receives comparatively little attention. Mesh refinement, time stepping, turbulence models, and solver settings are discussed extensively, yet the fundamental question of why the liquid behaves as it does under a given excitation is not always addressed. This imbalance creates a gap between simulation output and engineering understanding.

Practising engineers, particularly those working in design firms, often encounter a different and more abrupt difficulty. When faced with sloshing problems, they are confronted not by equations but by practical challenges: mesh generation for moving free surfaces, cleaning complex geometries, defining boundary conditions that reflect real operating scenarios, and selecting numerical approaches that remain stable under large deformations. These tasks are rarely covered in textbooks, yet they dominate real engineering workflows.

Furthermore, many industrial environments rely on specialised computational fluid dynamics technicians whose role is to operate specific software packages efficiently. While their expertise is invaluable, the organisational structure often leads to dependence on a single numerical method and a single software ecosystem. Transitioning to alternative formulations or complementary approaches can be perceived as unnecessary risk, professional discomfort, or even technical defection. This situation is not a consequence of resistance to innovation, but rather of the absence of accessible literature that explains sloshing dynamics in a method-independent, physics-first manner.

This book is written in response to these observations. It is shaped by experience in both industry and academia, and it aims to bridge the conceptual and practical gaps that currently characterise the field of sloshing dynamics. The intent is not to replace existing specialised references, nor to present a definitive or exhaustive treatment of every sloshing phenomenon. Instead, the goal is to provide a coherent foundation that connects physical intuition, mathematical description, and computational implementation.

The emphasis throughout the book is on understanding sloshing as a physical process before treating it as a numerical problem. Mathematical formulations are introduced to clarify mechanisms, not to overwhelm the reader. Computational methods are discussed as tools with specific strengths and limitations, rather than as universal solutions. Where possible, practical examples drawn from industrial applications are used to illustrate how theoretical concepts manifest in real systems.

This book is designed to serve multiple audiences. Graduate students and advanced undergraduates will find a structured entry point into sloshing dynamics that does not assume prior specialisation. Engineers and technicians working in industry will find guidance that reflects the realities of modelling, simulation, and design decision-making. Researchers will find a unified perspective that situates advanced analyses within a broader physical context.

It is important to state explicitly that this book does not aim to reduce sloshing dynamics to a single methodology or viewpoint. Sloshing is inherently multidisciplinary, involving fluid mechanics, structural dynamics, acoustics, numerical analysis, and experimental observation. No single framework can capture all aspects equally well. Accordingly, this book introduces multiple perspectives while maintaining a consistent physical narrative.

Ultimately, sloshing dynamics is not a niche topic reserved for specialised applications. It is a fundamental consequence of placing a liquid inside a container and subjecting that container to motion. As engineering systems become lighter, quieter, faster, and more sensitive to dynamic effects, the importance of understanding sloshing will continue to grow. The hope is that this book contributes to that understanding in a way that is rigorous, practical, and accessible, enabling engineers and researchers to address sloshing not as an afterthought, but as an integral part of system design.

Acknowledgements

This book is the outcome of many years of sustained reflection, research, and collaboration, and it has been shaped as much by human support as by technical inquiry. First and foremost, I wish to express my deepest gratitude to my family for their patience, endurance, and understanding throughout the long periods of time absorbed by the preparation of this work. Their quiet strength and unwavering support provided me with the stability necessary to pursue this project with focus and determination.

I am equally grateful to my students, whose excellence, intellectual curiosity, and sincere engagement in learning have given me constant reassurance and peace of mind. Their dedication has been a powerful motivation to continue working on this book and to refine its ideas with clarity and purpose.

I would like to express my sincere appreciation to Professor Masatoshi Shimoda and Professor Tsunehiro Takeuchi of my former faculty at Toyota Technological Institute, as well as to the broader academic community there, which I continue to regard as an academic family. The support, encouragement, and professional trust I received during my time at Toyota Technological Institute played a decisive role in shaping my academic direction and engineering perspective. I also gratefully acknowledge my faculty and academic home, the Graduate School of Engineering at The University of Osaka, whose intellectually vibrant and collaborative environment has provided an ideal setting for completing this book.

It is important for me to acknowledge Associate Professor Kentaro Yaji, to whom I owe a great deal both professionally and personally for the trust, responsibility, and academic freedom extended to me, as well as for his constructive guidance in shaping my role within the faculty. His commitment to education and research excellence has created a working environment in which meaningful scholarly contributions can flourish, and I am sincerely thankful for his support.

I would also like to thank the outstanding engineers with whom I have worked on sloshing dynamics across a wide range of industrial products and applications. While confidentiality prevents me from naming individuals or companies, their technical expertise, practical insight, and professionalism have had a significant influence on the perspectives presented in this book. These collaborations provided invaluable exposure to real-world constraints and engineering judgement that cannot be gained through theory alone.

Finally, I sincerely acknowledge Siemens Corporation for their support during the preparation of this book, particularly through the provision and use of STAR-CCM+, which remains one of the most reliable and advanced platforms for complex multiphysics analysis, including high-fidelity simulations of sloshing dynamics.

Their software tools and technical guidance contributed substantially to the computational rigour and practical relevance of this work.

To all those who have supported this journey, directly or indirectly, I offer my sincere gratitude.

Introduction to slosh dynamics

1

1.1 Introduction

At its core, sloshing dynamics is defined as the movement of a liquid's free surface within a partially filled container, triggered by external motion or changes in gravity. Rather than being a simple displacement of mass, it is a coupled phenomenon where the fluid's momentum exerts forces (referred to as slosh loads) back onto the walls of its container. Before exploring the technical foundations of sloshing dynamics, it is important to pause and ask a simple but important question: why does this subject deserve its own dedicated field of study? At first, it may seem natural to place the behaviour of liquids in partially filled containers within the wider world of classical fluid mechanics. Yet sloshing has developed into its own area because the phenomena involved, although familiar in daily life, show a level of complexity that ordinary fluid mechanics does not fully explain.

Sloshing is something almost everyone has experienced. It appears when a bartender shakes a drink, when we clean a bottle by swirling water and soap inside it, or when we are careful not to open a fizzy drink immediately after it has been shaken. We all know that the mixture of liquid and gas can suddenly surge upward and spill. A simple experiment makes the idea clearer: shake a bottle that is completely filled with water, then empty half of it and shake it again. Although the first case contains more mass, the second creates a much stronger force on the walls. This happens not because of the total weight but because the free surface of the liquid is now able to move and respond to gravity. Even the thin layer of gas above the liquid plays a role in how the motion develops.

These everyday moments point to a deeper scientific idea. Once a free surface is allowed to move, the behaviour of the liquid becomes highly sensitive to motion, container shape, and the surrounding boundaries. The fluid no longer behaves like a smooth and predictable flow. Instead, it develops a variety of oscillations, wave patterns, and strong interactions with the walls. Because of these features, the subject requires a specific term. Sloshing, which can be described as the oscillatory motion of a liquid surface inside a partially filled container, represents a rich set of interactions between fluid motion and the structure that contains it. It cannot be treated properly as a small corner of fluid mechanics.

As such, sloshing dynamics bring together physical concepts from several fields. These include fluid mechanics, structural vibration, nonlinear behaviour, acoustics, and parts of applied mathematics that study resonance and stability. It is this combination that gives sloshing both its scientific depth and its practical importance, whether in large industrial tanks under earthquake influence, rockets and spacecraft that carry liquid fuel, coastal engineering, or something as simple as trying to carry a cup of tea without spilling it. As such, slosh dynamics involves the motion of a liquid

Slosh Dynamics. DOI: https://doi.org/10.1016/B978-0-443-44330-5.00014-1

with a free surface under imposed accelerations – be they translational, rotational, or vibratory. The interface between the liquid and the overlying other fluid phase (typically gas) deforms in response to boundary motion, giving rise to time-varying hydrodynamic pressures and internal fluid resonances. These effects can result in splashing, wave breaking, or even structural compromise of the containment vessel.

The story of sloshing dynamics stretches far deeper into human history than one might expect. Long before the subject was formalised within the language of fluid mechanics, people had already learned to use sloshing for practical purposes. One of the earliest known examples comes from ancient Sumer, in the fertile crescent of what is now southern Iraq, where the technique of butter-making emerged more than 4000 years ago.

Archaeological evidence and cultural studies suggest that the Sumerians relied on simple containers made from animal skins. Vessels of this kind are still used in parts of the same region today and are commonly known as ***Shi-chu-wa***, a term meaning wineskin (as depicted in Fig. 1.1). These flexible containers were half-filled with fresh milk and then shaken to produce butter. These vessels were then shaken by hand or placed on animals and allowed to sway rhythmically as they moved. Although the method was practical and required little technology, it relied on a surprisingly sophisticated physical process. By shaking a partially filled flexible container, the milk inside underwent a strong oscillatory motion. This movement created rapid shear and stretching of the liquid, particularly at the free surface and at the deforming walls of the skin container.

Milk is not a uniform liquid but a complex multiphase mixture. It contains tiny droplets of fat suspended within water; each droplet is protected by delicate

Figure 1.1 Traditional Sumerian butter-making was an early application of sloshing dynamics.

molecular layers that hold the emulsion together. During vigorous sloshing, the internal motion produces shear rates high enough to break these protective layers. Once the fat droplets lose their stability, they begin to merge, forming larger clusters. This process continues until the mixture gradually transforms from an oil-in-water emulsion, which we recognise as milk, into a water-in-oil emulsion, which we recognise as butter.

For thousands of years, people had used the motion of liquids for practical tasks, but the underlying behaviour remained almost entirely unexamined. The complex motion of a free surface, though familiar in daily life, was not studied in a structured or mathematical way, and its governing principles were left unrecognised. This changed with the remarkable investigations of Michael Faraday in the early nineteenth century (Fig. 1.2).

His work, which we now recognise as the first formal study of parametric sloshing, provided the earliest scientific insight into how a liquid behaves when the container that holds it is made to oscillate in a repeated manner.

Figure 1.2 Michael Faraday.

Faraday placed a shallow layer of liquid in a cup and vibrated the cup vertically at a steady frequency. Instead of simply following the motion of the vessel, the liquid surface began to lift and fall in organised patterns.

Faraday noted that the waves did not match the frequency of the imposed motion. When the dish moved at a frequency Ω, the liquid responded at a frequency almost exactly $\Omega/2$, forming standing patterns that remained fixed in space even though the surface itself was in constant motion. These patterns could take the form of straight lines, squares, or more intricate arrangements, depending on the forcing conditions, depth of the liquid, and nature of the vessel. This subharmonic response was entirely unexpected at the time, and it remains the hallmark of what we now call Faraday waves.

Modern analysis provides a clear mathematical structure to what Faraday observed. For small waves in an incompressible and irrotational fluid, the free surface elevation $\eta(x, t)$ satisfies a linearised equation in which the effective gravity is periodically altered by the vertical motion of the container. When the equations are reduced to a single horizontal wave mode, the motion takes the form of a Mathieu-type differential equation,

$$\ddot{\eta} + 2\gamma\dot{\eta} + (\omega^2 + a\cos(\Omega t))\eta = 0 \tag{1.1}$$

where ω_0 is the natural frequency of the gravity capillary wave, γ represents viscous damping, and α encodes the strength of the periodic forcing through the vibration amplitude. The instability of this equation explains the appearance of Faraday's subharmonic waves: when the forcing frequency Ω is close to twice the natural frequency $2\omega_0$, the surface becomes unstable and grows into standing waves that oscillate at exactly half the driving frequency.

Faraday worked without the mathematical framework that underpins modern fluid science, yet his careful experiments revealed the essential features of the phenomenon with remarkable precision. The study of Faraday waves continues to evolve, especially as numerical tools allow us to explore more complex regimes, but a full treatment of these methods will be given in later chapters.

Sloshing dynamics, therefore stands as a subject shaped by both long human experience and careful scientific discovery. From the intuitive practices of ancient cultures to the first structured experiments of Faraday, the motion of a free liquid surface has revealed behaviours that are far richer than simple intuition suggests. Sloshing is not merely a secondary effect of moving fluids. It is a phenomenon governed by its own physical laws, marked by sensitive dependence on geometry, forcing, fluid properties, and resonant internal modes. Even in its simplest forms, sloshing can generate amplified forces, complex wave motion, and patterns that emerge from the interaction between the container and the liquid it holds.

These characteristics are what make sloshing dynamics a distinct and essential part of fluid science. It involves interactions that cannot be understood by looking at ordinary fluid motion alone. The free surface introduces new pathways of instability, resonance, energy exchange, and structural loading, many of which play decisive roles in engineering and in natural systems. As such, a dedicated study of sloshing provides the tools needed to understand behaviour that ranges from gentle surface ripples to violent impacts capable of damaging vessels, fuel tanks, and large engineered structures.

Following the background, we will examine some applications that make sloshing an important area of contemporary research. From liquid storage tanks and coastal structures to spacecraft propellant management and everyday containers, sloshing influences both performance and safety across a wide range of technologies. Understanding its behaviour is therefore not only a matter of scientific interest, but also a practical necessity in modern engineering.

1.2 Essence of studying sloshing dynamics

Understanding the motion of a liquid inside a rigid container begins with establishing the equations that describe the fluid field and the behaviour of the free surface. These equations provide the foundation for predicting internal pressure distributions, surface oscillations, and the hydrodynamic forces that develop when a tank is accelerated or disturbed. Although the physical setting appears familiar, the motion of the free surface introduces a level of complexity that demands careful mathematical treatment.

Exact analytical solutions to the governing equations exist only for a few ideal container geometries, such as vertical cylinders or rectangular tanks. In most practical situations, engineers rely on modal analysis to characterise the response of the fluid. This approach identifies the natural frequencies at which the liquid surface prefers to oscillate and the shapes of the associated modes. These quantities are central in many applications, since resonance between the sloshing modes and external excitation can greatly amplify the internal liquid motion.

The significance of natural frequencies is evident across diverse fields. In aerospace systems, the oscillation of liquid propellant may interact with the attitude control system or structural vibration modes. In large marine tanks, fluid resonances influence both structural stresses and operational safety. Civil engineering structures such as water towers or elevated tanks must also avoid matching their dynamic response with the natural frequencies of the contained fluid. Across these disciplines, accurate prediction of the modal characteristics of a liquid-filled container is essential to the design process.

To describe the behaviour of the liquid mathematically, one must specify boundary conditions on the container walls and on the free surface. For containers that only partially confine the fluid, these conditions determine how the velocity potential or pressure field behaves at the boundaries. They may fix the value of the field variable itself, prescribe the value of its normal derivative, or impose a specific relationship between the two. These belong to three classical categories: Dirichlet conditions, Neumann conditions, and Cauchy conditions. The choice depends on the physical interpretation of the boundary, such as whether the wall is rigid, whether the surface is free to move, or whether the pressure is prescribed.

Modern derivations of the governing equations often employ variational principles. Hamilton's principle, in particular, offers a compact and elegant route to obtaining both the field equations and the associated boundary conditions from energy arguments. This formulation has been refined by many researchers and provides a rigorous framework for analysing free surface oscillations in rigid tanks.

The material that follows develops the governing equations, introduces the boundary conditions appropriate to free surface flow, and presents the modal analysis framework used to describe sloshing behaviour. These foundations will support the later chapters of this book, where we explore the dynamic response of real containers, the role of external excitation, and advanced methods for controlling or exploiting sloshing phenomena.

The study of liquid sloshing in rigid containers has evolved through several distinct stages, each driven by the need to understand how a free surface responds to external motion. Early investigations focused on obtaining the natural frequencies of simple container shapes, since these frequencies determine the modes at which a liquid surface oscillates and the conditions under which resonance may occur. Analytical progress was greatest for geometries such as upright cylindrical tanks, rectangular vessels, and containers with planar walls, where the separation of variables could be employed to reduce the governing Laplace equation into manageable eigenvalue problems.

One of the major insights that emerged during this period was the central role of Bessel functions in describing the wave structure of liquids in circular and axisymmetric containers. For such systems, the free surface motion could be expressed in terms of radial and azimuthal modes, each associated with its own natural frequency. Similar analytical strategies were later applied to spherical tanks, conical vessels, and containers with curved walls, although the complexity of the solutions increased significantly as geometry departed from simple symmetries.

As the field progressed, researchers recognised that many real containers possess nonuniform depth, internal structures, or geometrical irregularities that violate the assumptions of classical separation. To address these challenges, new mathematical approaches were introduced. Methods such as the Ritz procedure, integral equation formulations, and matrix discretisation enabled the study of more complicated shapes, including vessels with sloping bottoms, quarter tanks, or annular cross sections. These techniques provided approximate but accurate solutions for natural frequencies and mode shapes even when analytical solutions were no longer possible.

The dynamics of propellant tanks in rockets marked a significant step forward in the understanding of sloshing behaviour. Studies in this area revealed that the interaction between liquid motion and vehicle acceleration could amplify sloshing and that even small oscillations in a microgravity environment could produce persistent fluid movement. These findings motivated the development of combined analytical and experimental techniques capable of capturing both linear and nonlinear effects. They also encouraged the exploration of internal structures such as baffles, which act to disrupt wave formation and dampen motion.

The study of sloshing dynamics has developed over many decades, starting with early work on how fluids move and lose energy. Stokes was among the first to explain how friction inside a fluid reduces wave motion (Stokes et al., 1851). His ideas helped shape the understanding of how liquids naturally slow down when they move inside containers. His model later expanded to include describing how small waves travel and carry energy, forming a basic theory still used in sloshing and wave-motion studies.

In the early 20th century, experiments began to provide real measurements. Honda and Matsushita performed some of the first controlled tests on water motion in rectangular tanks (Honda & Matsushita, 1913) (Fig. 1.3).

They measured natural frequencies, damping, and wave shapes, giving engineers useful data to compare with mathematical models. Jeffries then investigated water movement in elliptical basins, showing that the shape of the container strongly influences sloshing patterns. This insight is still important when designing tanks for buildings, ships, and industrial systems (Cooper, 1960; Jeffreys, 1925).

Figure 1.3 Kotarō Honda.

A major step forward came from Westergaard (Cooper, 1960), who proposed a simple but practical method for calculating the pressure that water exerts on dams during earthquakes. He divided the effect into two parts – an impulsive part and a convective part – making it easier to estimate sloshing forces during seismic events. This approach remains widely used in modern design standards for tanks and reservoirs (Fig. 1.4).

Work by Zhukovskii helped explain how the movement of fluid inside a structure changes its overall motion (Fig. 1.5).

His studies showed that internal liquids can shift the effective mass and alter the stability of the surrounding structure. These ideas later became very important for spacecraft, where sloshing fuel can affect control and stability (Zhukovskii, 1945).

Wedemeyer and Reese measured how liquids behave in pitching tanks, helping early aerospace engineers understand how fuel moves during flight (Widmayer & Reese, 1953). Shved examined how acceleration changes the natural sloshing

Figure 1.4 Harold Malcolm Westergaard.

Figure 1.5 Nikolai Egorovich Zhukovskii.

frequencies inside a container, showing how vibration and motion influence free-surface behaviour (Shved, 1959).

Later studies expanded these ideas into more specialised areas. Schoenhals et al. (1967) explored how vibrations affect liquid discharge from propellant tanks, helping improve fuel-emptying performance in rockets. Siekmann and Chang (1967) examined sloshing under low-gravity and strong capillary forces, showing how container flexibility and surface tension change wave motion in weak-gravity conditions. Shepherd (1969) studied sloshing and hydrodynamic pressures in petroleum tanks during earthquakes, improving the safety of industrial storage systems. He later introduced a two-mass model (Shepherd, 1972) that separates rigid-body motion from sloshing motion in elevated tanks, making dynamic analysis easier. Schwabe et al. (1978) then investigated surface-tension-driven flows during crystal-growth processes, revealing important free-surface behaviours in thermally controlled fluid systems.

1.3 Application of sloshing dynamics

Sloshing dynamics is not merely a theoretical subject but a practical one with direct impact across many modern technologies. Whenever a liquid is carried inside a moving container, the motion of its free surface can influence performance, safety, comfort, and even the structural integrity of the system. From large storage tanks and marine vessels to spacecraft, medical devices, and everyday transport, sloshing plays a quiet but significant role. The following sections explore several key applications where understanding sloshing is essential for design and operation.

1.3.1 Automotive engineering: fuel tanks, coolant systems, ride comfort, and acoustic control

Within automotive engineering, sloshing dynamics plays a far more influential role than many drivers realise. Modern vehicles, whether powered by internal combustion engines, hybrid powertrains, or even fuel cell systems, all carry significant volumes of liquid. These include fuel, coolant, lubricants, washer fluid, and in some designs even hydraulic media. Whenever the vehicle accelerates, brakes, climbs a slope, or turns a corner, these liquids respond with their own internal motion. The most critical examples occur inside the fuel tank and the coolant reservoir.

In a typical fuel tank, the liquid fuel does not remain still. Even routine driving induces oscillatory waves that travel across the free surface, strike the tank walls, and reflect back into the interior. These waves generate time-varying pressure loads that act on the tank structure. Over time, repeated exposure to these loads can cause low-cycle fatigue, vibration-induced wear, or deformation of thin-walled components. In addition to mechanical concerns, sloshing motion often produces noise. As the waves grow and break, they cause a distinct knocking or rumbling sound that can travel into the cabin through the chassis, giving passengers a sense that something is vibrating or loose within the vehicle (Fig. 1.6).

This issue becomes even more pronounced in hybrid and electric-assisted vehicles. Since the engine is often quieter or operating intermittently, the cabin environment has lower background noise. Sounds that would normally be masked by engine

Figure 1.6 Fuel tank location and integration within a passenger vehicle.

operation now become noticeable. What was once a minor sloshing vibration can become a distinct acoustic signature that reduces perceived comfort and quality. For premium and luxury vehicles, this is a design challenge of considerable importance, as manufacturers place strong emphasis on quiet ride conditions and refined interior experience.

To manage the effects of sloshing, automotive engineers use a combination of structural design, internal components, and advanced simulation techniques. One of the most effective design features is the use of internal baffles. These partitions, often made of lightweight polymer or metal sheets, are placed strategically inside the tank to interrupt the movement of the liquid. Instead of allowing a large wave to travel freely across the tank, the baffles force the flow to split into smaller regions. This reduces the effective wavelength, dissipates energy, and prevents the formation of strong resonant modes. In practice, this means lower peak pressures on the tank walls, reduced vibration transmission, and a significant reduction in noise.

In coolant reservoirs, sloshing control is equally important. Coolant plays a central role in temperature regulation, and any interruption in its delivery can cause thermal hotspots in the engine. Under sharp manoeuvres such as rapid acceleration, high-speed cornering, or steep downhill driving, the coolant can slosh strongly enough to expose the pump inlet, allowing air to enter the circulation system. This can lead to loss of cooling efficiency and local overheating. Engineers therefore design coolant tanks with particular geometries that stabilise the free surface, maintain pump immersion, and reduce the likelihood of sloshinginduced aeration.

The design process is supported by numerical tools, including computational fluid dynamics (CFD) and fluid structure interaction modelling, which allow engineers to predict sloshing behaviour under a wide variety of driving scenarios. These simulations help determine the optimal placement and number of baffles, tank geometry, fill level effects, and dynamic response. Combined with experimental testing such as shaker rigs and road simulation benches, engineers can evaluate sloshing performance long before a vehicle reaches production.

1.3.2 Sloshing dynamics in food processing and packaging

In the food industry, the behaviour of liquids during handling is of central importance, particularly in processes that involve filling, transporting, and sealing containers. Milk, fruit juices, broths, and many other products are packaged in large volumes everyday, often using containers that are flexible or semirigid. In such systems, the motion of the free surface becomes a critical factor in maintaining product quality and operational safety.

During the filling stage, the liquid enters the container at high speed, striking the surface and generating transient waves. These waves can rise unexpectedly and approach the container opening, increasing the risk of overflow. Even a small spill during filling can compromise hygiene standards, lead to product loss, or interrupt the automated process. The situation becomes more delicate when the container is made from thin polymer films, tetra pack, or lightweight bottles. These structures deform easily, and their geometry can change slightly as the liquid enters them. This deformation in turn influences the sloshing motion, sometimes amplifying it if the container walls flex inward or outward (Fig. 1.7).

Once the container is partially filled, the remaining free surface can move strongly when the line conveyor accelerates or stops. A sudden change in speed can push the liquid forward, creating a surge that impacts the walls and causes temporary bulging in flexible packaging. If the surge coincides with a natural resonance of the container shape, the internal wave can grow large enough to cause spillage or disrupt the sealing stage. For nonrigid containers such as milk cartons or sachets, this interaction between fluid motion and wall deformation requires careful design to maintain dimensional stability throughout the filling cycle.

In bottled products, sloshing plays an equally important role. When the bottle is nearly full, the small air pocket at the top compresses and expands as the liquid oscillates, generating pressure fluctuations that can affect the cap or sealing closure. If the bottle is subjected to vibration or rapid movement during packaging or transport, sloshing may create repeated impacts against the walls, leading to internal foaming, loss of carbonation in carbonated drinks, or even deformation of thin-walled bottles. These effects can reduce shelf stability or change the perceived quality of the product.

To manage these challenges, the food industry relies on a combination of container design, filling protocol, and numerical simulation. Baffles or internal shaping features

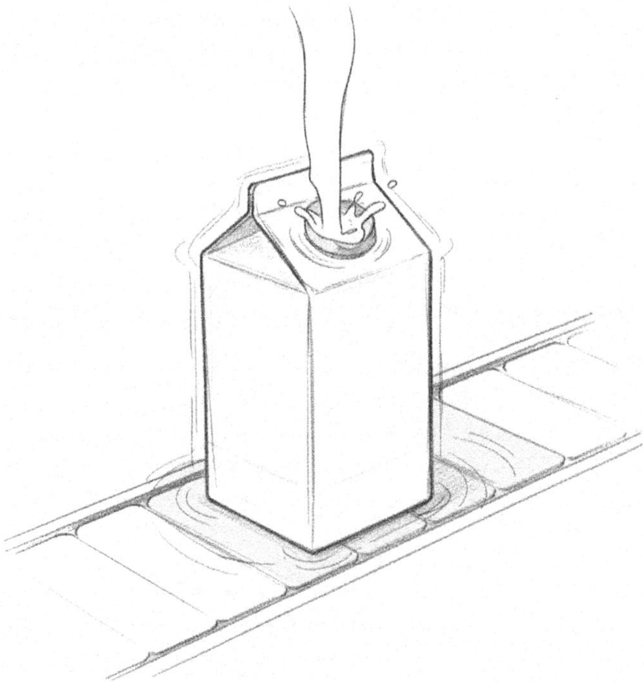

Figure 1.7 Liquid sloshing during filling of a Tetra Pak carton on a conveyor line.

may be incorporated to guide the flow and minimise wave growth. Filling nozzles are designed to reduce impact energy and allow a smooth level rise. Conveyor systems are programmed with controlled acceleration profiles that limit sudden sloshing. Computational modelling helps predict free surface motion in flexible packages, ensuring that the walls remain stable and the liquid motion stays within safe limits.

1.3.3 Sloshing dynamics in space: liquid propellant behaviour in rockets and spacecraft

Liquid propellants play a central role in modern rocket engines, and understanding their behaviour is essential for safe and efficient space flight. The choice of liquid propellants arose from a combination of practical, energetic, and engineering reasons. Liquids such as liquid oxygen, liquid hydrogen, kerosene, hydrazine, and their derivatives offer very high energy density in compact volumes, allowing rockets to carry far more chemical energy than would be possible with solid propellants of similar mass. Liquids can also be throttled, pumped, mixed, and controlled with high precision, enabling variable thrust, staged combustion, and efficient engine cycles. Their ability to flow makes them suitable for complex engine designs that require controlled injection, cooling, and regenerative heat transfer. For these reasons, the tanks that feed rocket engines almost always carry large quantities of liquid propellant.

However, the very property that makes liquid propellants useful – their ability to move freely – introduces one of the most challenging dynamical problems in space engineering: sloshing. Inside a rocket tank, the liquid is not fixed or static. It responds to every change in acceleration as the vehicle ignites, climbs, rolls, or stages. During launch, the intense acceleration forces the liquid towards the bottom of the tank, forming a well-defined free surface. As the vehicle vibrates under engine thrust or structural oscillations, the liquid generates standing waves, transient surges, and pressure pulses that travel through the tank. These internal motions can couple with the structure of the rocket, producing oscillatory forces that destabilise the vehicle. In extreme cases, this coupling can excite the natural frequencies of the rocket body, contributing to phenomena such as pogo oscillations, where propulsion and structural vibrations interact in a reinforcing cycle.

As the rocket leaves the atmosphere and approaches orbit, the gravitational environment changes dramatically. Under reduced gravity, the free surface no longer settles at the bottom of the tank. Instead, it becomes highly sensitive to even small disturbances. The liquid may drift, form irregular shapes, or attach itself to the tank walls due to surface tension. Pumps that were designed to draw propellant from a clearly defined region may instead encounter vapour, leading to ingestion of gas, unstable combustion, or engine shutdown. Sloshing behaviour therefore becomes central to ensuring that the propellant is delivered smoothly and continuously to the engine, regardless of the gravitational environment.

Once in orbit, in true microgravity, sloshing becomes even more complex (Bauer, 1991; Dodge et al., 1991; Kana, 1987). Without a dominant gravitational field, the motion of the spacecraft itself becomes the primary driver of internal fluid behaviour

(Bauer, 1982; Hill et al., 1988). Small nudges from attitude control thrusters, thermal gradients, docking impacts, or even crew movement can cause large-scale oscillations in the liquid. The fluid may form bubbles, large free surface waves, or drifting masses that periodically strike the tank walls. These impacts generate forces that can alter attitude control, disturb pointing accuracy, and degrade the stability of delicate instruments. Engineers therefore incorporate devices such as vanes, screens, diaphragms, and surface tension-based propellant management systems to control the position and motion of the liquid. These components guide the fluid using geometric confinement and capillary forces, ensuring that vital systems draw liquid rather than vapour.

In the transition from Earth's gravity to the microgravity of orbit, fuel sloshing ceases to be a secondary concern and becomes a central physical phenomenon that directly affects guidance, navigation, stability, fuel utilisation, and ultimately the operational lifetime of a spacecraft (Chen et al., 1992). In such environments, the motion of liquids is exceptionally difficult to predict, which is why researchers rely on a combination of numerical simulations and carefully designed experiments in microgravity, including parabolic flights and drop tower tests, to observe fluid behaviour under changing forces and stresses. The real strength, and indeed the quiet elegance, of fluid mechanics lies in its ability to reduce this apparent complexity by separating the problem into well-defined and manageable components through dimensional and nondimensional analysis. By doing so, scientists can construct simplified yet physically faithful models that capture the essential sloshing dynamics with remarkable accuracy, often approaching that of large-scale computational simulations while requiring far less computational effort. As a result, every space mission, from small satellites to deep space explorers, must incorporate an informed understanding of sloshing dynamics to ensure reliability, safety, and long-term performance.

1.3.4 Sloshing dynamics in satellites and deep space systems

As human activity in space continues to expand, satellites are being asked to perform functions that extend far beyond their original roles in communication and navigation (Himeno et al., 2007). Contemporary spacecraft concepts now encompass deep space exploration platforms, large-scale solar energy harvesting systems capable of exploiting uninterrupted solar radiation, and early visions of orbital computing infrastructures sustained entirely by space-based power generation. Such missions impose exceptionally strict requirements on stability, precision, and long-term operational reliability, elevating the dynamic behaviour of spacecraft subsystems to a critical design concern (Sah et al., 2024). Within this framework, the motion of liquids carried onboard, whether as propellants or thermal control media, emerges as a fundamentally important physical problem rather than a secondary engineering detail (Amin & Perrine, 2025; Jang, 2013).

In the microgravity environment of orbit, liquids no longer behave in an intuitively familiar manner. The absence of a dominant gravitational field prevents fluids from settling at the bottom of a tank, leaving the free surface governed instead by surface tension, inertia, and weak residual accelerations (Zhang et al., 2018). Under these conditions, even modest disturbances such as attitude corrections, reaction wheel desaturation, thermal expansion, docking events, or orbital perturbations can induce significant fluid motion. The liquid may drift, oscillate, or reorganise into large moving masses, generating forces and moments that couple directly to the spacecraft body. Because modern satellites require extremely accurate attitude control to maintain antenna alignment, sensor pointing, solar array orientation, or instrument calibration, even small sloshing-induced torques can degrade mission performance (Cui et al., 2014).

The interaction between the moving liquid and the spacecraft structure represents a central challenge in spacecraft dynamics. As the fluid redistributes itself within a tank, it alters the effective centre of mass and the inertia properties of the vehicle. These changes introduce unanticipated rotational responses that must be corrected by the attitude control system, often under tight control margins. When the natural frequencies of the fluid motion interact with structural modes or coincide with repeated manoeuvres, sustained oscillations may develop. Such behaviour increases actuator workload, accelerates propellant consumption, and can compromise pointing stability. In extreme situations, poorly controlled sloshing can lead to loss of attitude control, mission interruptions, or structural loading conditions that threaten vehicle integrity.

These effects become particularly pronounced in large spacecraft and deep space missions that carry substantial quantities of liquid propellant or thermal management fluids. Platforms designed for long-duration energy harvesting, precision solar array tracking, or autonomous orbital computing demand prolonged periods of highly stable orientation, often with exceptionally narrow pointing tolerances. Any internal fluid motion that perturbs this stability directly challenges mission objectives, making sloshing dynamics a core consideration in both system architecture and control strategy.

Smaller spacecraft, such as CubeSats and nanosatellites, have historically faced fewer sloshing-related issues due to their limited size and reliance on solid propulsion, cold gas thrusters, or electric propulsion systems with minimal liquid mass. However, as these platforms evolve towards higher capability missions involving orbit raising, station keeping, and increased payload complexity, the inclusion of liquid propellants is becoming more common. As a result, sloshing effects are increasingly relevant even at small scales, particularly when mission profiles involve frequent manoeuvres or extended operational lifetimes.

To mitigate the adverse effects of liquid motion, spacecraft tanks are often equipped with specialised internal features such as vanes, screens, porous media, and capillary-based devices that exploit surface tension to control fluid positioning. These propellant management systems are designed to suppress large-scale fluid movement, ensure reliable propellant delivery to thrusters, and stabilise the free surface during dynamic events. Their design is supported by advanced numerical modelling approaches, including fluid structure interaction analysis and microgravity free surface simulations, which aim to capture the coupled dynamics of liquid motion and spacecraft response. Together, these methods enable the development of tank and control system designs capable of maintaining stability and precision across the complex mechanical, thermal, and gravitational environments encountered in space.

1.3.5 Sloshing dynamics in maritime engineering

In maritime engineering, the behaviour of liquids inside large moving vessels is one of the most complex and consequential challenges in modern design. Ships that transport oil, refined fuels, LNG, chemicals, and other industrial liquids rely on massive cargo tanks that move with the vessel as it travels through the sea. Unlike rigid cargo, liquid responds continuously to every external disturbance. When a ship encounters waves, wind, or manoeuvres through tight channels, the liquid inside its tanks shifts with its own inertia. This motion produces free surface waves that travel across the tank, reflect from its boundaries, and may rise to significant heights depending on the sea state and tank geometry.

The forces generated by internal fluid motion can be surprisingly intense. As the vessel rolls, pitches, or heaves, the liquid surges toward the tank walls and impacts them with powerful pressure pulses. These impacts act over short intervals but with magnitudes far exceeding static loads. Localised pressure spikes can introduce repeated stress cycles that weaken the structure over time. The result may be fatigue damage, deformation of internal components, or, in severe conditions, structural failure. Historically, several maritime accidents involving product carriers and chemical tankers have been linked to violent sloshing events that were not adequately predicted during design.

The challenge becomes even more demanding in liquefied natural gas (LNG) carriers. LNG tanks operate at cryogenic temperatures and use advanced containment systems made from special alloys and layered insulation. These systems must remain both strong and flexible while holding extremely cold liquid. Sloshing in LNG tanks can create high-impact loads that damage insulation panels, compromise membrane structures, or cause cracks in the supporting framework. As the global fleet of LNG carriers has increased in both size and capacity, engineers have developed sophisticated numerical models, wave basin experiments, and operational guidelines to anticipate extreme sloshing conditions. Classification societies now require detailed sloshing assessments before new containment designs are approved.

Sloshing dynamics also play an important role in other maritime applications. Floating production and storage platforms in offshore energy fields manage large tanks of oil, water, or ballast fluid that move with the platform under wind and wave loads. In coastal engineering, wave energy devices rely on controlled sloshing within chambers to extract mechanical or electrical power from the motion of the sea. Firefighting vessels and naval ships must account for the effect of sloshing in onboard water tanks, since uncontrolled internal fluid motion can alter the stability and manoeuvrability of the vessel during emergency operations. Even in fishing vessels and research ships, sloshing in bait tanks, live wells, or scientific sampling containers can affect onboard stability.

To mitigate these risks, maritime engineers employ a combination of geometric design, internal features, and operational practices. Tank shapes are chosen to limit resonant wave patterns that amplify sloshing. Internal structures such as swash bulkheads, baffles, and partial partitions break up long wave paths and dissipate energy. Safe filling guidelines specify preferred load levels that avoid dangerous resonance between the fluid and the ship's motion. Modern vessels incorporate sensors and monitoring systems that detect unusual tank behaviour during operation, allowing the crew to adjust heading or reduce speed to limit sloshing severity.

Advanced simulation tools now play a central role in maritime design. CFD, fluid structure interaction models, and large-scale wave tank experiments allow engineers to capture the combined behaviour of the ship and its internal fluids under realistic sea conditions. These tools help predict sloshing loads, optimise structural reinforcement, and evaluate new containment technologies long before the vessel enters service.

1.4 Constructive uses of sloshing dynamics

Although much of engineering seeks to suppress or control sloshing, the phenomenon also offers opportunities for constructive use when its motion is shaped with intention. In several industrial processes, the natural movement of a liquid becomes an effective tool rather than a hazard. One example is the cleaning of storage tanks used in food processing, biotechnology, and chemical manufacturing. By introducing controlled fluid motion at particular frequencies, the free surface can sweep across the internal walls of the vessel, loosening sediment, disrupting bacterial films, and transporting impurities toward drainage points. This approach reduces the need for mechanical scrubbing, lowers water consumption, and helps maintain hygienic conditions in large-scale production facilities.

A more emerging direction lies in technologies that operate on much smaller scales, where the internal liquid plays an active role in shaping the behaviour of the system. Soft robotics is a notable illustration. Many soft robotic structures contain internal channels filled with liquids or gels that shift as the structure bends or stretches. The resulting sloshing produces internal pressure gradients and distributed forces that assist movement, enhance flexibility, or modulate the stiffness of the robot. In this context, sloshing becomes an intentional feature that allows the device to adapt to external loads in a way that rigid mechanical systems cannot easily replicate.

The same principle is being explored in self-healing materials. Certain experimental systems embed micro-reservoirs of liquid healing agents within a solid matrix. When the material sustains damage, the mechanical disturbance initiates small-scale sloshing within these reservoirs, enabling the fluid to move through narrow channels and reach the damaged region. Once delivered, the healing agent can bond, polymerise, or otherwise restore the mechanical properties of the structure. Here, sloshing serves as a transport mechanism, driven not by external actuation but by the deformation of the material itself.

An example of engineered utilisation of sloshing can be found in the preparation of dental amalgam. For many decades, dentists relied on a small sealed capsule that contained two separate components. One compartment held a fine metal alloy powder, typically composed of silver, tin, copper, and other elements. The second compartment contained a small membrane or sachet filled with liquid mercury. These two components had to be mixed rapidly and uniformly to produce the amalgam used for dental restorations.

The mixing process did not rely on manual stirring or mechanical grinding. Instead, the capsule was placed inside a specialised high-speed shaking device known as an amalgamator. Once activated, the device subjected the capsule to intense oscillatory motion, producing rapid sloshing of its internal contents. The inertia of the liquid mercury caused the small membrane containing it to rupture within seconds. As the capsule continued to vibrate, the now free mercury spread throughout the metal powder and began to wet and bind the particles.

The effectiveness of this procedure comes directly from the physics of sloshing. The repeated acceleration and deceleration inside the capsule generate strong internal shear forces that break apart powder agglomerates, promote homogeneous mixing, and accelerate the chemical process that forms the amalgam paste. The highly energetic motion also enhances contact between the mercury and alloy particles, allowing the reaction to proceed efficiently in a very short time.

These examples reflect a growing recognition that the internal motion of liquids can be directed to achieve specific functions. Rather than being viewed solely as a source of instability or wear, sloshing becomes part of a system's active behaviour, contributing to cleaning, movement, repair, or adaptive response. As novel materials, flexible devices, and microfluidic systems continue to develop, the controlled use of sloshing may offer new pathways for engineering structures that respond dynamically to their environment.

1.5 Book overview

Sloshing dynamics occupies a distinctive position in fluid mechanics and engineering science. It is at once a familiar everyday phenomenon and a source of some of the most challenging problems in modern engineering practice. The motion of a liquid free surface inside a container appears deceptively simple, yet its consequences span an extraordinary range of physical scales, industries, and disciplines. The aim of this book is to provide a unified, physically grounded, and computationally rigorous treatment of sloshing dynamics, extending from first principles to advanced modelling and real-world applications.

This text is written with the conviction that sloshing cannot be understood in isolation as a purely fluid-mechanical effect. Rather, it is a **coupled phenomenon**, governed simultaneously by fluid properties, interfacial physics, thermal effects, structural dynamics, and external forcing. In many practical systems, sloshing is neither an accidental side effect nor a minor disturbance. It is a dominant mechanism that influences stability, durability, safety, and performance. In some cases, it is an unwanted source of load amplification and noise; in others, it is deliberately exploited as a functional mechanism. The book therefore adopts a broad perspective, treating sloshing as a dynamic interaction between fluids, structures, and energy transfer pathways.

The structure of the book reflects this philosophy. Rather than beginning immediately with sloshing models, the text first develops the physical and mathematical foundations required to understand why sloshing behaves as it does, why it becomes unstable or resonant, and why it couples so strongly with its container. Each chapter is designed to build conceptual depth while maintaining continuity with engineering practice.

Chapter 2 establishes the physical basis upon which all later discussions of sloshing are built. The behaviour of a sloshing liquid is inseparable from its material properties, its thermodynamic state, and the presence of interfaces. For this reason, the chapter begins with a systematic discussion of fundamental fluid properties, including density, viscosity, compressibility, specific heat, thermal conductivity, and bulk modulus. These properties determine how momentum, energy, and pressure propagate through a liquid and how the free surface responds to imposed motion.

Special emphasis is placed on properties that play a decisive role in sloshing but are often underestimated in simplified analyses. Surface tension, for example, governs the stability of short-wavelength waves, capillary effects in small containers, and free-surface behaviour in reduced-gravity environments. Viscosity controls dissipation, wave damping, and the transition between linear and nonlinear regimes. Compressibility and bulk elasticity, although often neglected in low-speed flows, become important in confined volumes and during rapid excitation.

The chapter then expands naturally into thermal effects and energy scales. Temperature-dependent variations in viscosity and density are examined, together with thermal expansion and its influence on pressure and free-surface behaviour. These effects are particularly important in systems such as fuel tanks, cryogenic vessels, and industrial storage, where thermal gradients coexist with dynamic motion.

The treatment of multiphase aspects begins with the vapour phase above liquids and proceeds to cavitation phenomena within liquids. These topics are essential for understanding real sloshing systems, where phase change, vapour pockets, and nucleation processes can dramatically alter pressure loads and stability. Cavitation-induced sloshing and vapour collapse introduce nonlinearities that cannot be captured by single-phase models.

The chapter then addresses foams and emulsions, which represent extreme manifestations of interfacial dynamics. Foam formation illustrates how sloshing can actively create new interfacial area, storing energy in surface tension and altering macroscopic flow behaviour. Emulsions reveal how sloshing can drive microstructural transformation, breaking droplets, redistributing phases, and changing rheological properties. These discussions broaden the scope of sloshing beyond traditional tank problems, showing its relevance in food processing, chemical engineering, and materials science.

Finally, the introduction of dimensionless numbers provides a unifying framework for scaling and comparison. Froude, Reynolds, Weber, Bond, and capillary numbers are not presented merely as mathematical abstractions, but as physical indicators that determine which mechanisms dominate a given sloshing regime. This nondimensional perspective is essential for interpreting experiments, simulations, and real systems on equal footing.

While Chapter 2 establishes how liquids behave, Chapter 3 addresses how structures respond. Sloshing is fundamentally a problem of **interaction**, and no analysis is complete without a clear understanding of the mechanics of the container itself. This chapter develops the essential tools of solid mechanics needed to describe deformation, stress, vibration, and energy exchange.

Beginning with stress–strain relationships and constitutive equations, the chapter progresses through equilibrium and compatibility conditions to variational formulations such as virtual work and Hamilton's principle. These principles provide a powerful and elegant framework for coupling fluid motion to structural response, especially in dynamic systems.

Vibration theory plays a central role. Sloshing often becomes dangerous or beneficial precisely because it resonates with structural modes. The chapter therefore introduces vibration of discrete systems and plates, laying the groundwork for understanding how sloshing frequencies interact with container dynamics. This foundation is critical for applications ranging from spacecraft attitude control to seismic response of storage tanks.

With the physical and structural foundations in place, Chapter 4 presents the core theory of sloshing dynamics. Fundamental categories of sloshing are defined, including planar, swirling, and three-dimensional modes. The concept of natural sloshing frequency is developed in detail, highlighting its dependence on container geometry, fill level, gravity, and fluid properties.

The chapter explores both linear and nonlinear sloshing regimes. While linear theory provides valuable insight into mode shapes and resonance conditions, real systems frequently operate beyond this regime. Nonlinear effects such as wave steepening, breaking, and modal coupling are examined, together with their physical origins and consequences.

Mechanical analogues, including mass–spring and pendulum models, are introduced not as crude approximations but as reduced-order representations that capture essential dynamics. These models are particularly valuable in control applications and preliminary design, where full CFD simulations may be impractical.

Special attention is given to sloshing under reduced gravity, a regime of paramount importance in space systems. In such environments, surface tension and inertia replace gravity as dominant forces, fundamentally altering free-surface behaviour.

Damping mechanisms are treated comprehensively, including viscous dissipation, wave breaking, and energy loss through baffles. Both linear and nonlinear damping representations are discussed, along with their limitations. The chapter concludes with a nondimensional classification of sloshing regimes, providing a conceptual map of the vast parameter space encountered in practice.

Chapter 5 addresses numerical simulation, recognising that modern sloshing analysis relies heavily on computational methods. Rather than presenting numerical tools as interchangeable black boxes, the chapter develops each method from its governing principles.

The Finite Element Method is introduced through weak formulations and Galerkin discretisation, with careful attention to pressure – velocity coupling, stabilisation, and low-Mach-number behaviour. The Finite Volume Method is then presented from a conservation-based perspective, emphasising flux evaluation, boundedness, and robustness. Smoothed Particle Hydrodynamics is discussed as a mesh-free alternative, particularly well suited to large deformations and free-surface fragmentation.

Throughout the chapter, the strengths and limitations of each method are critically assessed in the context of sloshing. Issues such as numerical diffusion, interface tracking, stability, and computational cost are examined in detail.

Chapters 6 and 7 translate numerical theory into computational practice. Solver selection, time integration, turbulence modelling, convergence assessment, and validation strategies are discussed with an emphasis on reliability and physical consistency.

Mesh preparation receives dedicated treatment, reflecting its critical importance in free-surface problems. Different mesh types, refinement strategies, and adaptive techniques are evaluated, with practical guidance on avoiding common pitfalls in sloshing simulations.

Chapter 8 presents complete sloshing simulations through representative case studies. These examples demonstrate how theory, numerics, and computation are integrated into coherent workflows.

Chapter 9 explores advanced and emerging topics, including sloshing mitigation, multiphase effects, non-Newtonian fluids, and machine-learning-assisted modelling. These discussions highlight how sloshing research continues to evolve, driven by new materials, new technologies, and increasing computational power.

References

Amin, H., & Perrine, C. (2025). Simulation of Fluid Sloshing in High-Acceleration Environments for Aerospace Applications. In *2025 Regional Student Conferences, AIAA 2025-99458*. Reston, VA: American Institute of Aeronautics and Astronautics. https://doi.org/10.2514/6.2025-99458.

Bauer, H. F. (1982). Sloshing in conical tanks. *Acta Mechanica, 43*, 185–200.

Bauer, H. F. (1991). Liquid sloshing response in spin-stabilized missiles or satellites due to axial excitation. *Zeitschrift fur Flugwissenschaften und Weltraumforschung, 15*, 252–256.

Chen, K.-H., Kelecy, F. J., & Pletcher, R. H. (1992). A numerical and experimental study of three-dimensional liquid sloshing in a rotating spherical container. In *30th Aerospace Sciences Meeting and Exhibit: Aerospace Sciences Meetings*. Reston, VA: American Institute of Aeronautics and Astronautics, 829. https://doi.org/10.2514/6.1992-829.

Cooper, R. M. (1960). Dynamics of liquids in moving containers. *ARS Journal, 30*(8), 725–729. https://doi.org/10.2514/8.5212.

Cui, D.-L., Yan, S.-Z., Guo, X.-S., & Gao, R. X. (2014). Parametric resonance of liquid sloshing in partially filled spacecraft tanks during the powered-flight phase of rocket. *Aerospace Science and Technology, 35*, 93–105. https://doi.org/10.1016/j.ast.2014.03.006.

Dodge, F. T., Green, S. T., & Kana, D. D. (1991). *Fluid management technology: Liquid slosh dynamics and control*. Cleveland, OH: National Aeronautics and Space Administration. https://ntrs.nasa.gov/api/citations/19920005102/downloads/19920005102.pdf.

Hill, D. E., Baumgarten, J. R., & Miller, J. T. (1988). Dynamic simulation of spin-stabilized spacecraft with sloshing fluid stores. *Journal of Guidance, Control, and Dynamics: a*

Publication of the American Institute of Aeronautics and Astronautics Devoted to the Technology of Dynamics and Control, 11, 597–599.

Himeno, T., Watanabe, T., Nonaka, S., Naruo, Y., Inatani, Y., & Aoki, H. (2012). Numerical and experimental investigation on sloshing in rocket tanks with damping devices. In *43rd AIAA/ASME/SAE/ASEE Joint Propulsion Conference & Exhibit*. Reston, VA: American Institute of Aeronautics and Astronautics, 5557. https://doi.org/10.2514/6.2007-5557.

Honda, K., & Matsushita, T. (1913). An investigation of the oscillations of tank water. *Scientific Reports Tohoku, Imperial University, First Series, 21*, 131–148.

Jang, J. (2013). Mechanical slosh models for rocket-propelled spacecrafts. In *AIAA Guidance, Navigation, and Control (GNC) Conference*. Reston, VA: American Institute of Aeronautics and Astronautics, 4651. https://doi.org/10.2514/6.2013-4651.

Jeffreys, H. (1925). The free oscillations of water in an elliptical lake. In: *Proceedings of the London Mathematical Society, s2-23*, 455–476. https://doi.org/10.1112/plms/s2-23.1.455.

Kana, D. D. (1987). A model for nonlinear rotary slosh in propellant tanks. *Journal of Spacecraft and Rockets, 24*, 169–177.

Sah, R., Srivastava, R., Lima, R., Vatsal, V., & Das, K. (2024). Numerical modeling of liquid propellant mass transfer with sloshing during on-orbit refueling. In *2024 IEEE Aerospace Conference, Big Sky, MT, USA*. IEEE, 1–14. https://doi.org/10.1109/AERO58975.2024.10521091.

Schoenhals, R. J., Winter, E. R. F., & Griggs, E. I. (1967). Effects of longitudinal vibration on discharge of liquids from propellant tanks. In: In *Proceedings of the 1967 Heat Transfer and Fluid Mechanics Institute (HTFMI)*. San Diego, CA: Stanford University Press, 277–297.

Schwabe, D., Scharmann, A., Preisser, F., & Oeder, R. (1978). Experiments on surface tension driven flow in floating zone melting. *Journal of Crystal Growth, 43*, 305–312.

Shepherd, R. (1969). Earthquake resistant design of petroleum storage tanks. In: In *Proceedings of the 20th Australasian Conference on the Mechanics of Structures and Materials, Adelaide, Australia*1–11.

Shepherd, R. (1972). The two-mass representation of a water tower structure. *Journal of Sound and Vibration, 23*, 391–396.

Shved, G. L. (1959). Small oscillations of the free surface of an ideal fluid in a plane moving vessel with smooth vertical walls. *Tr. Odess. Tekhnologicheskogo Instituta Pishchevoi i Kholodilnoi Promyshlennosti, 8*, 150–160.

Siekmann, J., & Chang, S. C. (1967). On liquid sloshing in a cylindrical tank with a flexible bottom under strong capillary and weak gravity conditions. *The Journal of the Astronautical Sciences, 14*, 167–172.

Stokes, G. G., et al. (1851). On the effect of the internal friction of fluids on the motion of pendulums. *Transactions of the Cambridge Philosophical Society, 9*, 1880 1095.

Widmayer, E., & Reese, J. R. (1953). *Moment of inertia and damping of fluid in tanks undergoing pitching oscillations*. National Advisory Committee for Aeronautics.

Zhang, D., Qi, H., Xu, J., Yang, Y., & Fu, B. (2018). Modeling and analysis of aeroelasticity and sloshing for liquid rocket. *IEEE Access, 7*, 4453–4465.

Zhukovskii, N. E. (1945). On the motion of a rigid body having cavities filled with a homogeneous liquid. *Russian Journal of Physical Chemistry A, 17*, 231–280 81-113,145-199.

Fundamentals of fluid dynamics

2

2.1 Introduction

Fluid dynamics provides the physical framework through which the motion, deformation, and interaction of liquids and gases are understood across scales ranging from molecular transport to macroscopic flow and structural loading. In many engineering systems, fluids do not exist in isolation but interact continuously with solid boundaries, thermal fields, and additional fluid phases. These interactions give rise to complex behaviours that cannot be captured by kinematics alone, but instead emerge from the coupled action of inertia, viscosity, compressibility, interfacial forces, and thermodynamic driving mechanisms.

In this context, sloshing dynamics represents a particularly rich and practically important class of fluid motion. Sloshing refers to the unsteady, often oscillatory movement of a fluid free-surface within a partially filled container subjected to external excitation such as acceleration, vibration, or rotation. While the visible manifestation of sloshing is surface wave motion, the underlying physics is far more intricate. Internal velocity fields, pressure fluctuations, phase interactions, and energy dissipation processes evolve simultaneously, producing time-dependent forces on container walls and altering the effective mechanical response of the system. These effects are central to applications ranging from fuel tanks in aerospace and automotive systems to marine transport, industrial reactors, and energy storage devices.

A defining feature of realistic sloshing problems is the coexistence and interaction of multiple phases. The liquid phase is commonly accompanied by a gas phase above the free surface, and under sufficiently strong excitation, gas may also become entrained within the liquid in the form of bubbles. In the presence of surface-active species or multiple immiscible liquids, sloshing can further induce the formation of foams and emulsions, transforming an initially simple single-phase liquid into a complex, evolving multiphase material. These transformations fundamentally modify effective density, viscosity, compressibility, and energy dissipation pathways, thereby feeding back into the sloshing dynamics itself. Consequently, an accurate physical description of sloshing must extend beyond classical free-surface hydrodynamics to include interfacial thermodynamics and multiphase transport phenomena.

The purpose of this chapter is to establish the physical foundations required to analyse such coupled behaviour. Rather than treating sloshing as an isolated phenomenon, the chapter situates it within the broader framework of fluid physics, beginning with the microscopic origins and macroscopic consequences of fluid properties. Fundamental thermophysical quantities such as temperature, density,

Slosh Dynamics. DOI: https://doi.org/10.1016/B978-0-443-44330-5.00006-2

viscosity, thermal conductivity, specific heat, and bulk modulus are introduced and interpreted in terms of molecular motion and intermolecular forces. These properties form the invariant material parameters that govern how fluids respond to deformation, pressure variation, and thermal gradients, irrespective of the specific flow configuration.

Building on this foundation, the chapter then distinguishes between intrinsic fluid properties and flow-dependent variables, such as pressure, velocity, shear stress, and surface tension, which arise from the instantaneous state of motion and boundary interactions. Special attention is given to interfacial phenomena, as surface tension and interfacial elasticity play a decisive role in free-surface stability, bubble dynamics, foam formation, and emulsification under dynamic excitation. The vapour phase above liquids and cavitation vapour within liquids are examined to clarify how phase change, nucleation, and compressibility effects emerge from thermodynamic and kinetic principles.

To provide a unifying mathematical framework, the chapter formulates the fundamental conservation laws of fluid dynamics – mass, momentum, and energy – in both integral and differential form. These governing equations constitute the backbone of analytical modelling and numerical simulation and are essential for understanding sloshing-induced pressure fields, force transmission, and thermal coupling. Their role is not merely formal, but deeply physical, linking observable macroscopic behaviour to universal conservation principles.

Finally, the chapter extends beyond classical single-phase descriptions to address foam and emulsion systems as inevitable outcomes of strong fluid excitation in realistic liquids. Foam formation, structural evolution, ageing, and interfacial mechanics are discussed as dynamic processes that alter sloshing response and damping characteristics. Emulsions are introduced as metastable, high interfacial-area systems whose formation, stability, and inversion are governed by interfacial thermodynamics and hydrodynamic forcing. By integrating these topics within a single physical narrative, the chapter provides a coherent basis for analysing sloshing not as a purely hydrodynamic phenomenon, but as a multiphase, thermodynamically driven, and structurally evolving system.

2.2 Physical nature of fluids

A fluid is a phase of matter defined by its inability to resist shear stress without continuously changing shape. This means that even a very small sideways force will cause it to flow, unlike a solid, which can hold its shape under small forces. The reason for this difference lies in the way molecules are arranged. In solids, atoms or molecules are fixed in an orderly pattern. In fluids, the molecules are not fixed in place; they move around each other freely, so the material can change shape and spread out to fill any container. Fluids include both liquids and gases. Both can flow, but they differ in density, compressibility, and the strength of forces between their molecules. These differences explain why liquids can form stable surfaces and why gases expand to fill all available space.

At the microscopic level, the behaviour of a fluid is controlled by the motion of its molecules and the forces acting between them (Kowalewski et al., 2016; Squires & Quake, 2005). In liquids, the molecules are close together, so they interact strongly. These interactions depend on the chemical nature of the fluid and can include attractions like van der Waals forces or hydrogen bonds (Fuller & Vermant, 2012; Müller et al., 2005; Oettel & Dietrich, 2008). Because the molecules are close, liquids are difficult to compress, and their volume does not change much with pressure (Durst, 2008; Hutter & Wang, 2016). Still, the molecules have enough thermal energy to move past one another, allowing the liquid to flow (Jou et al., 2011). In gases, the molecules are much farther apart, and the forces between them are very weak compared to their motion. Gas molecules move in straight lines most of the time, only changing direction when they collide with other molecules or with a surface. Because the spacing between molecules is large, gases are compressible and can change volume when pressure changes.

The speed and motion of molecules in a gas are not the same for every molecule but vary widely. This range of speeds follows patterns that can be predicted statistically. The average energy of motion is linked to temperature – when temperature rises, molecules move faster. In gases, pressure comes from the collisions of molecules with the walls of their container. In liquids, molecular motion is still linked to temperature, but the close spacing means that collisions happen so often that the liquid's volume changes very.

From a broader physics perspective, both liquids and gases can be described as continuous substances rather than as individual particles (Barker & Henderson, 1976; Tabor, 2012). This view allows scientists and engineers to use ideas like flow speed, density, and pressure to describe what is happening in the fluid with continuity phenomena, such as why water flows smoothly through a pipe, why air swirls around a moving car, and why clouds form in the atmosphere. The movement of fluids depends on their internal friction, known as viscosity, which measures how strongly the molecules resist sliding past each other. Liquids like honey have high viscosity, while air has very low viscosity. The difference in viscosity between fluids changes how they move and how they transfer heat and momentum.

From a fluid flow perspective, they also respond differently depending on how fast they are moving and how they interact with their surroundings. When the flow is gentle and smooth, it is called laminar. When it becomes chaotic and full of swirling motions, it is turbulent. Turbulence greatly increases mixing and affects how heat and particles move through the fluid. The change from laminar to turbulent flow depends on properties such as fluid speed, density, viscosity, and the size of the space the fluid moves through. In everyday life, both types of flow can be seen – for example, the smooth trickle of water from a tap at low speed versus the chaotic splashing at higher speed.

Liquids and gases can also behave very differently when their speed approaches the speed of sound in that medium. For gases, moving faster than sound produces sudden changes in pressure and density, such as shock waves. In liquids, changes in speed can create pressure variations that cause phenomena like cavitation, where vapour bubbles form and collapse.

2.2.1 Fundamental fluid properties

An important aspect of the development of sloshing research lies in the systematic study of numerical methods and their suitability for representing complex fluid–structure phenomena. Early investigations focused primarily on understanding the underlying physics of each problem and on constructing simplified analytical models that could capture the essential behaviour of the system. In this context, pioneering contributions such as those of Housner, grounded in a deep command of classical mechanics, introduced and formalised fundamental concepts including continuity, stress–strain relations, and dynamic equilibrium, which remain central to sloshing analysis. As the field progressed, these theoretical foundations enabled the effective use of numerical and seminumerical approximation techniques, notably finite difference and finite element methods, as well as more general computational engineering approaches such as the finite volume method that is widely employed in modern computational fluid dynamics. Within this framework, the study of sloshing naturally encompasses both the fluid phase and the confining structure, typically represented by rigid or flexible solid walls, and extends to their coupled interaction as a unified dynamical system. Accordingly, attention is directed towards the numerical strategies used to model this interaction and towards the practical considerations that arise when simulating sloshing phenomena within the context of real engineering applications.

Certain intrinsic characteristics of a continuous fluid remain invariant regardless of its motion or flow state. These fundamental attributes, termed the basic properties of the fluid, serve as essential parameters for describing and analysing fluid behaviour under various physical conditions. To facilitate a consistent treatment of these properties, it is necessary to establish a coherent framework that standardises the measurement and quantification of physical quantities involved in fluid mechanics. Such a framework ensures that equations describing fluid behaviour are dimensionally consistent and universally applicable. These fundamental properties typically include parameters such as density, viscosity, pressure, temperature, and specific volume, among others. Importantly, these properties are inherent to the fluid itself and do not depend on the particular flow regime or external forces acting upon the fluid. Liquids possess several fundamental properties that govern their behaviour and interaction with their surroundings. Density defines the mass per unit volume of a liquid, influencing buoyancy and hydrostatic pressure. Viscosity measures the internal resistance to flow, reflecting the liquid's molecular cohesion and affecting flow regimes and energy dissipation. Surface tension arises from molecular forces at the liquid–air interface, dictating phenomena such as droplet formation and capillary action. Compressibility describes the extent to which a liquid's volume changes under pressure; although typically low for liquids, it becomes significant in high-pressure applications. Vapour pressure characterises the equilibrium pressure exerted by a liquid's vapour phase at a given temperature, determining evaporation rates and boiling points. Temperature strongly influences all these properties, modulating density, viscosity, and vapour pressure, and thereby affecting fluid dynamics and heat transfer. Together, these intrinsic properties form the basis for understanding and predicting liquid behaviour in engineering systems, natural processes, and industrial applications. In the following subsections, we examine these aspects in detail.

2.2.1.1 Temperature

Temperature is one of the most fundamental physical quantities in both classical and modern physics, not merely as a measure of 'hotness' but as a conceptual bridge between energy, matter, and information (Tayurskii & Le Mehaute, 2011; Zhang et al., 2019). In thermodynamics, temperature defines the direction of spontaneous heat transfer: when two systems at different temperatures are brought into thermal contact and isolated from all other influences, energy in the form of heat will *always* flow from the hotter system (higher temperature) to the colder system (lower temperature) until both reach the same temperature. At this state of thermal equilibrium, no net heat transfer occurs, and the systems are said to share a common thermodynamic state variable – temperature.

This directional flow of heat is not an arbitrary empirical rule but a direct consequence of the Second Law of Thermodynamics. The Second Law asserts that in any natural process, the total entropy of an isolated system will either increase or remain constant – never decrease. Entropy, in this context, can be interpreted as a measure of energy dispersion or, in more modern formulations, as a measure of missing information about the microscopic state of the system. When heat flows from a hotter to a colder region, the overall entropy increases because the thermal energy becomes more evenly distributed among the available microscopic states. The reverse process – spontaneous heat flow from cold to hot without external work – would imply a decrease in entropy and is therefore forbidden by the Second Law.

While the Second Law dictates that heat flows from hot to cold, the *mechanism* of that transfer depends on the phase and structure of the medium. In liquids and gases, there are three primary heat transfer modes: conduction, convection, and radiation.

Conduction arises from direct microscopic interactions between neighbouring particles. In gases, these are primarily elastic collisions between molecules, transferring kinetic energy from faster-moving (hotter) molecules to slower-moving (colder) ones. In liquids, where molecular spacing is smaller and intermolecular forces are stronger, conduction occurs through both molecular collisions and vibrational coupling, often resulting in higher thermal conductivity than gases of comparable temperature.

Convection is the bulk transport of heat due to the macroscopic motion of the fluid. In gases, convection often dominates when density variations induced by temperature gradients lead to buoyancy-driven flow (natural convection), or when external forces, such as fans or pumps, impose fluid motion (forced convection). In liquids, convection can be even more efficient due to higher density and heat capacity, enabling large-scale transport of thermal energy over relatively short time intervals.

Radiation – the emission of electromagnetic waves – transfers energy without requiring direct contact or even a material medium. While all bodies above absolute zero emit thermal radiation, the relative contribution of radiation to heat transfer is generally small for moderate temperatures in liquids and gases compared to conduction and convection, but it becomes dominant in high-temperature systems, such as combustion gases or stellar atmospheres.

The difference between liquids and gases in heat transfer efficiency is substantial: liquids typically have higher thermal conductivities and heat capacities than gases, meaning they can absorb and transport more energy per unit temperature change. Gases, however, can exhibit faster large-scale convection due to lower density and viscosity, enabling rapid mixing over large distances.

It is important to mention that in liquid media, heat transfer occurs through molecular interactions and short-range energy exchange rather than long-range lattice vibrations or free electron transport. As temperature increases, changes in molecular mobility, intermolecular spacing, and microstructural organisation modify the efficiency of energy transport, causing thermal conductivity to vary with temperature. For most liquids, this variation is moderate but nonnegligible, particularly when large temperature gradients are present. Assuming constant thermal conductivity may therefore lead to inaccuracies in predicting heat flux and temperature fields. To capture the physical behaviour of liquid heat transfer more accurately, the temperature dependence of thermal conductivity can be empirically approximated as:

$$\kappa = \kappa_0 + \alpha (T - T_0) \tag{2.1}$$

where κ_0 is the conductivity at a reference temperature T_0, and α is the material scaling coefficient. Thermal conductivity directly impacts heat transfer rates in convective and conductive systems involving liquids, affecting cooling performance in engineering applications such as heat exchangers, chemical reactors, and energy storage devices.

2.2.1.2 *Thermal conductivity*

Thermal conductivity κ is a fundamental transport property of matter that quantifies its ability to conduct heat through molecular or lattice-scale interactions. It appears in Fourier's law of heat conduction, which is the phenomenological statement that the *heat flux* (the rate of heat transfer per unit area) is proportional to the negative of the temperature gradient in the material:

$$q = -\kappa \frac{\partial T}{\partial n} \tag{2.2}$$

Here, q is the component of heat flux normal to the surface $\frac{\partial T}{\partial n}$ is the spatial rate of temperature change in the normal direction, and κ is the thermal conductivity. The negative sign enforces the empirical observation – codified in the Second Law of Thermodynamics that heat flows spontaneously from regions of higher temperature to regions of lower temperature.

The mathematical form of Fourier's law closely parallels that of Newton's law of viscosity for momentum transport and Fick's law for mass diffusion. In all three cases, the transport process can be understood as the irreversible spread of a physical quantity (heat, momentum, or mass) down its gradient, with a proportionality constant – κ for heat, μ for viscosity, and D for mass diffusivity – that encapsulates the material's microscopic structure and dynamics. This parallelism is not accidental:

in the kinetic theory of gases, these transport coefficients emerge from the same molecular-collision picture, differing only in the specific property being exchanged between molecules.

Thermal conduction at the microscopic level is the result of energy transfer through particle interactions without macroscopic mass transport. The exact mechanism depends on the phase and structure of the material:

- In gases, conduction occurs primarily through *collisions between molecules.* A molecule with higher kinetic energy collides with one of lower kinetic energy, transferring part of its energy and thereby reducing local temperature differences. The mean free path, collision frequency, and molecular velocity distribution – all strongly temperature dependent – determine the rate of this process.
- In liquids, molecules are closer together and strongly influenced by intermolecular forces. Energy transfer occurs both through direct collisions and through the vibrational and rotational coupling of molecular clusters. Because liquids are generally incompressible and have high densities, their thermal conductivities are typically higher than those of gases but lower than those of metals.

2.2.1.3 Specific heat

Specific heat is a property that characterises a substance's ability to store thermal energy per unit mass per unit temperature change. In other words, it is defined as the amount of heat required to raise the temperature of one unit mass of a medium by one degree, and its value depends intrinsically on the thermodynamic process by which the temperature change occurs. For compressible fluids, the two most relevant definitions are the specific heat at constant volume C_v and the specific heat at constant pressure C_p. C_v quantifies the change in internal energy per unit mass per degree of temperature rise when volume is held fixed, while C_p measures the change in enthalpy per unit mass per degree under constant pressure, reflecting both the internal energy change and the work of volume expansion against the surroundings. Mathematically, for a unit mass of fluid,

$$\begin{cases} C_v = \left(\frac{\partial U}{\partial T}\right)_v \\ C_p = \left(\frac{\partial U}{\partial T}\right)_p \end{cases} \tag{2.3}$$

where U is specific internal energy and $H = U + pv$ is specific enthalpy.

The ratio of specific heats, $\gamma = \frac{C_p}{C_v}$, is a dimensionless quantity of central importance in compressible flow theory. From kinetic theory, γ is directly related to the molecular degrees of freedom (n) via

$$\gamma = \frac{n+2}{n} \tag{2.4}$$

For a monatomic ideal gas, $n = 3$, yielding $\gamma \approx 1.67$. For diatomic gases (e.g. nitrogen or oxygen) at moderate temperatures, rotational modes contribute, giving $n \approx 5$ and $\gamma \approx 1.4$. At higher temperatures, vibrational degrees of freedom may also be excited, decreasing γ. This molecular interpretation reveals that specific heat encapsulates microscopic energy storage mechanisms—translation, rotation, and vibration – linking macroscopic thermodynamic behaviour to quantum-mechanical energy quantisation.

The gas constant R bridges C_p and C_v through the relation

$$R = C_p - C_v \tag{2.5}$$

For a perfect gas, R can be expressed as $R = \frac{\overline{R}}{\mathfrak{M}}$, where $\overline{R}$ is the universal gas constant and $\mathfrak{M}$ is the molar mass. This makes R a process-independent property that directly connects the thermodynamic state variables in the ideal gas law, $pv = RT$.

In practical applications, specific heat is temperature dependent, particularly for gases at high temperatures where vibrational modes become active, or for liquids and solids near phase transitions. In fluids, specific heat governs how thermal energy is partitioned between internal energy and work, influencing compressibility effects, shock wave strength, acoustic propagation speed, and heat transfer rates. In convective flows, the product ρC_p determines the thermal capacity per unit volume, which, combined with thermal conductivity κ, defines the thermal diffusivity as:

$$\alpha = \frac{\kappa}{\rho C_p} \tag{2.6}$$

2.2.1.4 Bulk modulus of elasticity in liquids

The *bulk modulus of elasticity* (E) is a fundamental property that quantifies a fluid's resistance to uniform compression (Gholizadeh et al., 2012). It is defined as the ratio of a net pressure increase Δp applied uniformly to an element of fluid to the *fractional* (or *volumetric*) strain $\Delta V/V$ produced by that pressure change. In differential form, this is expressed as:

$$E = -V\frac{dp}{dV} \tag{2.7}$$

The negative sign indicates that an increase in pressure results in a decrease in volume. An alternative expression can be obtained by relating volume change to density change, noting that $\rho = m/V$. Differentiating, we find:

$$E = \rho\frac{dp}{d\rho} \tag{2.8}$$

Since $\frac{d\rho}{d\rho}$ represents the *relative* change in density, the bulk modulus is the inverse of isothermal compressibility β:

$$\beta = \frac{1}{E} \tag{2.9}$$

A high bulk modulus corresponds to a low compressibility, indicating the fluid strongly resists volume changes (Gholizadeh et al., 2011).

For example, at room temperature (≈293 Kelvin), water has a bulk modulus of approximately $E_{\mathrm{water}} \approx 2.2 \times 10^9$ Pascal, while air at the same conditions has a much smaller bulk modulus of $E_{\mathrm{air}} = 1 \times 10^5$ Pascal. This means that, under equal conditions, air is roughly 22,000 times more compressible than water. Physically, this difference arises from the molecular arrangement and intermolecular forces. In liquids such as water, molecules are closely packed, and any reduction in volume requires significant work against strong intermolecular cohesive forces (primarily hydrogen bonding in water). In gases such as air, molecules are far apart and interact weakly, allowing for large volume changes under modest pressure variations.

The bulk modulus is also connected to the *speed of sound* c in a fluid via:

$$c = \sqrt{\frac{E}{\rho}} \tag{2.10}$$

This relation shows that a higher bulk modulus results in faster acoustic wave propagation. For water at 20°, using $\rho \approx 998$ kg/m^3 and $E \approx 2.2 \times 10^9$ Pascal, the calculated sound speed is:

$$c \approx \sqrt{\frac{2.2 \times 10^9}{998}} \approx 1483\ m/s \tag{2.11}$$

For air at the same temperature, with $\rho \approx 1.204$ kg/m^3 and $E \approx 1 \times 10^5$ Pascal, the speed of sound is about 343 m/s. This large difference in sound speed illustrates how the bulk modulus influences the dynamic wave propagation in fluids.

Temperature and dissolved gases also affect the bulk modulus. In general, as temperature increases, liquid density decreases slightly, and cohesive intermolecular forces weaken, leading to a modest reduction in E. The presence of dissolved air or other gases can drastically lower the effective bulk modulus, because even small gas bubbles compress far more than the surrounding liquid. This is particularly important in hydraulic systems, where air entrainment reduces stiffness and degrades performance.

2.2.1.5 Density

The density ρ of a fluid quantifies the amount of mass contained within a unit volume, expressed in kilograms per cubic metre (kg/m^3) in the SI system. Mathematically, the density at a point can be defined as

$$\rho = \lim_{\Delta V \to 0} \frac{\Delta m}{\Delta V} \tag{2.12}$$

where Δm is the incremental mass of fluid within a small volume ΔV surrounding the point. In the continuum description of fluids, $\Delta V \to \delta V$ represents an infinitesimally small element over which the material properties can be assumed uniform.

Closely related to density is the specific weight ϖ, which represents the weight of fluid per unit volume. It is related to density by

$$\varpi = \rho g \tag{2.13}$$

where g is the local gravitational acceleration ($\approx$ 9.81 m/s^2). In SI units, ϖ is expressed in newtons per cubic metre (N/m^3). The use of $\approx$ for gravity is due to the fact that the gravitational acceleration diminishes slightly with increasing altitude and can be described by the relation

$$g = 9.80665 \times \left(\frac{R_e}{R_e + h}\right)^2 \tag{2.14}$$

where g is the gravitational acceleration at altitude h above sea level, R_e is the mean radius of the Earth. For altitudes small relative to R_e, the reduction in g is minimal – for example, at 10 km, a typical cruising altitude for aircraft, g is only about 0.3% less than at sea level.

2.2.2 Flow properties (dependent variables)

Dependent flow variables such as pressure, velocity, force, and shear stress characterise the fluid's instantaneous dynamic state within a flow field. Unlike fundamental fluid properties, these variables depend on the fluid's motion, boundary conditions, and external forces. Pressure represents the normal compressive force per unit area, velocity describes the fluid's local speed and direction, shear stress quantifies tangential forces due to viscosity, and force arises from the combined action of these stresses on surfaces. Together, they provide a detailed description of fluid behaviour essential for analysing and predicting flow phenomena.

2.2.2.1 Surface tension

Surface tension ξ is a property of liquids that governs the behaviour of their free surfaces or interfaces with other immiscible liquids. At its core, surface tension is due to molecular cohesion – the attractive forces between molecules of the same substance. Molecules deep within a liquid are surrounded by neighbours on all sides, so the attractive forces acting on them are balanced. However, molecules at the surface are exposed to the liquid on one side and to a different phase on the other. This creates an unbalanced inward pull, effectively making the surface behave like a stretched elastic membrane.

From a physical perspective, surface tension can be described in two equivalent ways: as the tangential force per unit length acting along a liquid's surface boundary, or as the surface free energy per unit area $\mathfrak{M}_s$. The latter description highlights that creating new surface requires work – molecules must be moved from the bulk liquid to the surface against the net inward molecular attraction. For example, stretching a soap film in a frame requires doing work to bring new molecules to the surface, increasing its total area. This is why soap films naturally contract to the smallest possible area, forming minimal surfaces such as spheres.

One of the most visible effects is capillarity – the rise or fall of a liquid in a narrow tube due to surface tension and wetting behaviour. When a glass capillary is placed in water, water molecules, which strongly adhere to glass, climb up the walls of the tube, forming a concave meniscus. The height h of this capillary rise can be expressed as

$$h \propto \frac{2\xi \cos\theta}{\rho g r} \tag{2.15}$$

where θ is the contact angle between the liquid surface and the solid wall, ρ the liquid density, g gravitational acceleration, and r the tube radius. For water, $\theta < 90°$, giving $\cos\theta > 0$ and producing an upward rise. In contrast, mercury does not wet glass ($\theta > 90°$), resulting in a convex meniscus and a depression of the liquid level in the tube.

Surface tension also controls droplet formation, bubble stability, and the breakup of liquid jets – processes critical in inkjet printing, fuel in engines, pulmonary function in biology, and microfluidics. At small scales, surface tension forces can dominate over gravity, while at large scales, they become negligible compared to inertial and gravitational forces. This balance is characterised by the Bond number Bo and capillary number Ca, dimensionless parameters that compare surface tension to gravitational and viscous effects, respectively. These parameters will be discussed alongside other dimensionless parameters in a dedicated section in this chapter.

Temperature strongly influences surface tension: as temperature rises, molecular kinetic energy increases, reducing cohesive forces, and thus σ decreases, vanishing at the liquid–gas critical point. Surfactants – molecules with both hydrophilic and hydrophobic parts – can further reduce surface tension by preferentially occupying the surface and disrupting cohesive interactions, a principle exploited in detergents, emulsions, and foams.

2.2.2.2 Pressure

In fluid mechanics, pressure is defined as the compressive normal force exerted per unit area by fluid particles on a surface. Mathematically, pressure p at a point within a liquid can be expressed as the limiting ratio of the infinitesimal normal force ΔF acting over an infinitesimal surface area ΔA, formally written as:

$$p = \lim_{\Delta A \to 0} \frac{\Delta F}{\Delta A} \tag{2.16}$$

As illustrated in Fig. 2.1, this definition assumes the fluid behaves as a continuum, where ΔA approaches an infinitesimal area over which fluid properties are well-defined and uniform. Pressure represents an isotropic compressive stress, meaning its magnitude is identical in all directions at a given point. This isotropy can be demonstrated by considering a fluid element shaped as a cube aligned with Cartesian axes, subjected only to hydrostatic pressure (i.e. shear stresses are zero). Let the pressures on the cube's faces along the x, y, and z directions be p_x, p_y and p_z, respectively. If a plane cuts the cube diagonally at an arbitrary angle, creating an inclined surface A_d, the pressure p_d acting normal to A_d must balance the components of p_x, p_y and p_z, to maintain equilibrium. By resolving forces along each coordinate axis and enforcing static equilibrium (sum of forces equals zero), it follows $p_x = p_y = p_z = p_d$, confirming pressure's directional uniformity. This fundamental property allows pressure to be treated as a scalar field within the fluid, crucial for analysing forces on submerged surfaces and deriving the hydrostatic pressure distribution, often expressed as:

$$p = p_0 + \rho g h \tag{2.17}$$

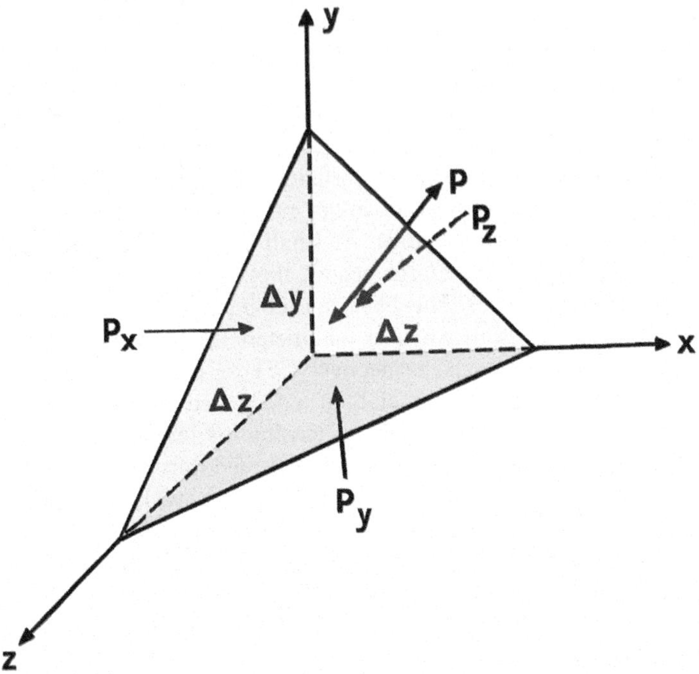

Figure 2.1 Pressure components acting on an infinitesimal fluid element, showing isotropic normal stresses resolved along Cartesian coordinate directions.

where p_0 is the reference pressure at the surface, ρ is the fluid density, g is the gravitational acceleration, and h is the depth below the surface. This relationship quantifies how pressure increases linearly with depth in a static fluid.

2.2.2.3 *Viscosity*

Viscosity is a property governing the internal resistance of a fluid to deformation and plays a pivotal role in the analysis of liquid behaviour and fluid motion, particularly near solid boundaries (Brush, 1962). It originates from intermolecular forces that arise when adjacent layers of fluid move at different velocities, resisting relative motion (Bhattad, 2023). For a Newtonian fluid, the shear stress τ_{yz} acting on a plane with normal in the y-direction and velocity in the x-direction is directly proportional to the transverse velocity gradient (as shown in Fig. 2.2):

$$\tau_{yz} = \mu \frac{\partial u}{\partial y} \tag{2.18}$$

where μ is the coefficient of dynamic viscosity. This relationship, known as Newton's law of viscosity, reflects the 'no-slip' condition (illustrated in Fig. 2.3) observed experimentally, whereby the velocity at a solid boundary is zero and increases with distance from the wall. The dynamic viscosity is primarily a function of temperature

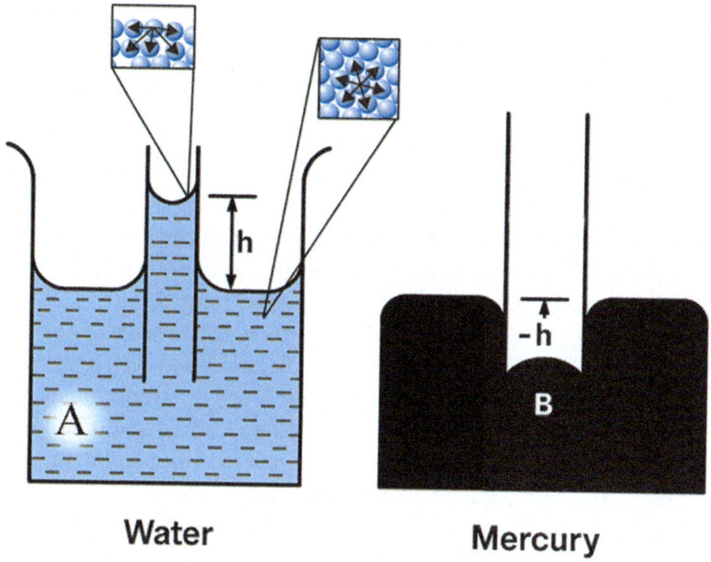

Figure 2.2 Capillary rise and depression in narrow tubes, illustrating the contrasting behaviour of water and mercury due to surface tension, wetting characteristics, and contact angle effects.

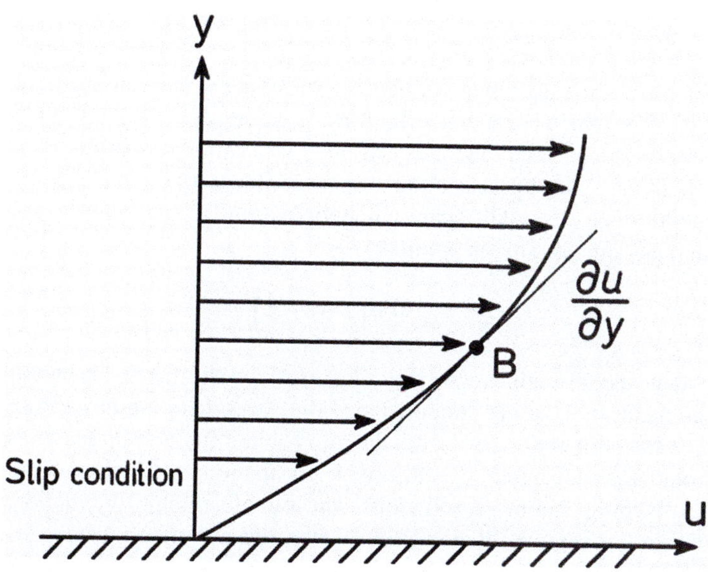

Figure 2.3 Velocity gradient development near a solid boundary, illustrating shear stress generation and the no-slip condition in a Newtonian fluid.

– decreasing with increasing temperature for liquids and increasing for gases – while pressure dependence is generally negligible under ordinary conditions. Near solid surfaces, viscosity effects dominate due to steep velocity gradients, whereas in regions far from boundaries, shear stresses in low-viscosity fluids may be negligible. A related quantity, the kinematic viscosity ν, defined as

$$\nu = \frac{\mu}{\rho} \tag{2.19}$$

where ρ is the fluid density, expresses viscous effects in terms of diffusivity of momentum, independent of applied forces. Kinematic viscosity, like dynamic viscosity, is strongly temperature dependent at constant pressure. Together, μ and ν form key parameters in characterising laminar–turbulent transitions, boundary-layer development, and energy dissipation in fluid flows (Song, 2018).

When a bottle partially filled with water is shaken, the hand experiences sharp and impulsive vibrations arising from the rapid acceleration and deceleration of a dense, coherent liquid column. The inertia of the liquid leads to strong impact pressures as the fluid repeatedly strikes the container walls. When a small amount of soap is added and the bottle is shaken again, air becomes dispersed throughout the liquid in the form of bubbles, and the mechanical response felt by the hand becomes markedly smoother. This change reflects a fundamental modification of the liquid's rheology: the dispersed gas phase disrupts large-scale coherent motion, redistributes stresses, and alters the effective viscosity of the mixture.

For a pure Newtonian liquid, the relationship between shear stress and deformation rate is given by:

$$\tau = \mu\dot{\gamma} \tag{2.20}$$

Here, $\dot{\gamma}$ is the shear rate. Once bubbles are introduced, the system becomes a two-phase suspension, and its resistance to flow is better described by an apparent or effective viscosity μ_{eff} that depends on the gas volume fraction ϕ and on the interfacial boundary conditions at the bubble surfaces (Krishna et al., 1999). For dilute concentrations of small, spherical bubbles with clean, shear-free interfaces, classical analysis due to Taylor predicts

$$\mu_{eff} = \mu\left(1 - \frac{5}{2}\phi\right) \tag{2.21}$$

indicating that the effective viscosity decreases with increasing bubble content because gas inclusions cannot support tangential stress and therefore reduce viscous dissipation (Llewellin et al., 2002). As the gas volume fraction increases and bubbles begin to deform and interact, additional dissipation mechanisms arise due to interfacial stretching, bubble oscillations, and bubble–bubble interactions (Orvalho et al., 2015). In this regime, higher order corrections are often introduced, leading to expressions of the form

$$\mu_{eff} = \mu\left(1 - \frac{5}{2}\phi + k\phi^2\right) \tag{2.22}$$

where the coefficient k accounts for nondilute effects and collective bubble dynamics. Together, these relations explain why a bubbly water–soap mixture responds more gently to shaking: coherent sloshing motions are broken into smaller, highly dissipative flows, impact pressures are reduced, and the forces transmitted to the hand become smoother and less impulsive.

It is important to mention that when the liquid composition and internal additives vary, sloshing can induce the formation of bubbles within the fluid. In systems containing surfactants or multiple immiscible liquids, this process may further lead to the development of foam and or emulsions. The presence of bubbles, foam, or emulsified phases alters the effective thermo-physical and rheological properties of the liquid, including density, viscosity, and energy dissipation characteristics. As a result, the sloshing dynamics and force transmission behaviour can deviate significantly from those of a single-phase liquid. These phenomena and their implications will be discussed in detail later in this chapter.

2.3 Thermal effects and energy scales

Temperature is a fundamental thermodynamic variable that governs the behaviour of matter across phases, profoundly influencing fluids at multiple scales. The impact of temperature manifests in molecular dynamics, phase stability, transport properties, and interfacial phenomena. To fully appreciate the complexity of thermal effects in fluids, it is instructive to consider distinct but interconnected aspects: the liquid phase, vapour both above and within liquids (cavitation vapour), the gaseous phase, and the interfacial region where liquid and gas coexist.

2.3.1 Thermal behavior of liquids

At the molecular-scale, liquids consist of densely packed molecules engaged in continuous, random thermal motion. As temperature T increases, the average kinetic energy per molecule scales approximately as

$$\frac{3}{2}k_B T \tag{2.23}$$

where k_B is Boltzmann's constant. This enhanced energy leads to more frequent and energetic molecular displacements, facilitating diffusional transport within the liquid matrix.

Diffusion in liquids is quantified by the self-diffusion coefficient

D, which measures the mean squared displacement of molecules over time. Experimentally and theoretically,

D exhibits a strong temperature dependence. Empirically, diffusion coefficients often obey an Arrhenius-type relationship:

$$D = D_0 e^{-\left(\frac{E_a}{RT}\right)} \tag{2.24}$$

where D_0 is a preexponential factor, E_a the activation energy associated with molecular rearrangements, and R the universal gas constant. This exponential dependence reflects the energy barrier molecules must overcome to relocate amid surrounding neighbours. Higher temperatures increase the probability of surmounting these barriers, resulting in exponentially enhanced diffusion rates.

Physically, this implies that at elevated temperatures, the increased molecular vibration amplitude and frequency reduce the effective cage formed by neighbouring molecules, allowing for easier exchange of molecular positions. This enhanced diffusion is critical in processes such as solute transport, chemical reaction rates, and homogenisation in liquid mixtures.

2.3.2 Volume expansion: thermal expansion coefficient

The volumetric expansion of liquids with temperature arises from the interplay between thermal motion and intermolecular forces. As molecules vibrate more vigorously, the average intermolecular separation increases, causing an expansion of the liquid volume V. The thermal expansion is characterised quantitatively by the volumetric thermal expansion coefficient β, defined as:

$$\beta = \frac{1}{V}\left(\frac{\partial V}{\partial T}\right)_P \tag{2.25}$$

where the partial derivative is taken at constant pressure P.

For most liquids, β is positive, meaning volume increases with temperature. Typical values range from 1 x 10^{-3} to 1 x 10^{-4}, varying with molecular structure and bonding. For example, water exhibits anomalous behaviour near 4°C, where β becomes negative, and density reaches a maximum, a phenomenon rooted in the hydrogen-bonded network structure that contracts upon initial heating before expanding at higher temperatures.

Thermal expansion has macroscopic consequences in fluid dynamics and engineering, influencing buoyancy-driven convection, thermal stresses in containment vessels, and precision fluidic devices. In addition, β varies with pressure, necessitating careful consideration in high-pressure systems.

2.3.3 Viscosity variation with temperature

Viscosity η is a measure of a fluid's resistance to deformation under shear stress, arising from momentum transport between adjacent fluid layers. Microscopically, viscosity in liquids stems from intermolecular friction: as molecules attempt to move past one another, cohesive forces hinder relative motion, dissipating mechanical energy as heat.

Counterintuitively, liquid viscosity decreases with increasing temperature, a trend opposite to that observed in gases. This behaviour is attributed to the weakening of intermolecular attractions as thermal agitation grows. Elevated temperature disrupts hydrogen bonding, van der Waals forces, and other cohesive interactions, reducing the effective 'stickiness' between molecules. Consequently, molecular layers slide more easily, lowering viscosity.

Mathematically, the temperature dependence of viscosity in many liquids can be approximated by empirical relations such as the Andrade equation:

$$\eta = Ae^{\left(\frac{B}{T}\right)} \tag{2.26}$$

where A and B are empirical constants specific to the fluid, or by the Vogel-Fulcher-Tammann relation for complex or glass-forming liquids:

$$\eta = \eta_0 e^{\left(\frac{B}{T-T_0}\right)} \tag{2.27}$$

with η_0, B, and T_0 fitting parameters. These expressions capture the exponential sensitivity of viscosity to temperature, emphasising the critical role of thermal energy in overcoming intermolecular potential wells that resist flow.

This reduction in viscosity with temperature has practical implications: it enhances fluid flow rates, affects lubrication regimes, and governs heat exchanger efficiency. In biological systems, for example, temperature dependent viscosity impacts blood flow and cellular transport.

2.4 Vapour phase above liquids: thermal and phase dynamics

Above the surface of any liquid, there exists a vapour phase composed of molecules that have escaped from the liquid into the surrounding space, forming a dilute gas whose density depends on temperature and the nature of the liquid (Smoluchowski, 1918; Thomson, 1872). In an open system, such as a container exposed to the atmosphere, the vapour mixes with the surrounding air, while in a closed system – such as a sealed tank – the vapour accumulates in the space above the liquid, eventually reaching a state of dynamic equilibrium with it. This equilibrium is characterised by the saturation vapour pressure, a thermodynamic property that depends sensitively on temperature and represents the pressure exerted by the vapour when the rates of molecular evaporation and condensation are equal. The origin of this behaviour lies at the molecular-level: within the liquid phase, molecules are bound together by intermolecular cohesive forces such as hydrogen bonding, dipole–dipole interactions, or van der Waals forces, depending on the fluid's composition. These forces create an energy barrier that must be overcome for a molecule to transition from the liquid into the vapour phase. Thermal motion within the liquid means that molecules possess a range of kinetic energies described statistically by a Maxwell–Boltzmann distribution; only those with energies exceeding the binding forces at the interface can escape into the vapour. This escape process is evaporation, and its rate depends on temperature, surface properties, and the surrounding pressure and humidity. Conversely, in the vapour phase above the liquid, molecules move freely along ballistic trajectories until they collide with other vapour molecules or return to the liquid surface. If such molecules lose enough kinetic energy upon collision with the surface, they can become recaptured by the liquid through condensation. When the rates of these two processes balance over time, the vapour achieves its equilibrium density and corresponding vapour pressure for that temperature, and the system is said to be saturated.

In a closed container partially filled with liquid, evaporation proceeds until the vapour pressure in the headspace matches the equilibrium value, at which point the number of molecules leaving the surface per unit time equals the number returning. If the container is then heated, the kinetic energy of molecules in the liquid increases, a larger fraction of them gain enough energy to overcome cohesive forces, and the rate of evaporation rises. This shifts the balance toward higher vapour density until the condensation rate catches up, establishing a new equilibrium at a higher vapour pressure. This strong dependence on temperature is a consequence of the exponential relationship between molecular energy distribution and the probability of exceeding the escape energy threshold. In an open system, such as a liquid exposed to air, the vapour phase is not confined, and vapour molecules can be transported away from the surface by diffusion into the surrounding gas or by bulk motion in the form of convection. Under these conditions, equilibrium vapour pressure still describes the maximum concentration of vapour in the air at a given temperature, but continuous removal of vapour molecules from the vicinity of the surface allows evaporation to persist until the liquid is depleted or environmental conditions change.

The energy required for a molecule to escape from the liquid phase into the vapour phase is known as the latent heat of vaporisation, and it reflects the work needed to separate a molecule from its neighbours without changing the liquid's temperature. This energy input does not appear as an increase in temperature because it is invested in breaking intermolecular bonds and creating the new phase rather than in accelerating molecular motion. As a result, evaporation produces a cooling effect in the remaining liquid because higher-energy molecules preferentially escape, lowering the average kinetic energy of the molecules left behind. In the opposite process, condensation, molecules in the vapour phase lose kinetic energy and become bound into the liquid, releasing the latent heat back into the surroundings.

From a microscopic perspective, vapour molecules behave according to the kinetic theory of gases, with their speeds distributed in accordance with the Maxwell–Boltzmann distribution. The mean kinetic energy of the molecules is proportional to temperature, so at higher temperatures, not only are more molecules able to escape the liquid, but the average speed of vapour molecules also increases. This has direct implications for transport properties in the vapour phase. In contrast to the liquid, where molecules are closely packed and interactions are dominated by cohesive forces, vapour molecules are widely separated, and their dynamics are governed by collisions that transfer momentum and energy between them. The frequency of these collisions determines macroscopic properties such as viscosity and thermal conductivity. Interestingly, in gases – including vapour – the viscosity tends to increase with temperature because faster-moving molecules collide more frequently and exchange momentum more effectively. This is opposite to the trend in most liquids, where viscosity decreases with temperature due to reduced relative effects of intermolecular attraction.

2.5 Cavitation vapour within liquids: nucleation and dynamics

Cavitation is a striking and multifaceted phenomenon in fluid physics in which vapour-filled cavities or bubbles form spontaneously within the bulk of a liquid under conditions where the local pressure falls below the liquid's vapour pressure or when localised heating promotes superheating beyond equilibrium boiling conditions, all without the need for an initially visible free liquid–vapour interface. Cavitation is a phase change dynamic occurring in situ inside a seemingly continuous liquid medium, driven by extreme local variations in pressure, temperature, and flow conditions. In hydrodynamic contexts, cavitation most often emerges when fluid accelerates rapidly through constrictions, around sharp blades, baffles, plates, in highly accelerated fuel tanks, or within regions of intense turbulence, producing localised pressure drops, vortex cores, or high strain-rate flows. When the instantaneous local pressure falls below the equilibrium vapour pressure of the liquid, molecules within the liquid are no longer held together solely by cohesive forces and instead transition to the vapour phase, forming microscopic nuclei. These nuclei may originate from preexisting submicron gas pockets trapped in surface roughness, dissolved gases coming out of solution, or thermal fluctuations creating momentary low-density clusters in the liquid. Once nucleation is initiated, the pressure imbalance allows the vapour region to expand, creating a bubble that may persist, grow, or collapse depending on surrounding conditions. Cavitation can also arise thermally when local heat deposition, such as from a high-energy laser pulse or ultrasonic absorption, causes rapid temperature spikes in small liquid volumes, lowering the vapour pressure threshold to the point that vapour formation is favoured even without significant bulk pressure reduction.

The initial formation of a stable vapour cavity requires overcoming an energetic nucleation barrier, which is linked to surface tension and the size of the vapour nucleus. Surface tension tends to suppress bubble formation by exerting a pressure penalty that scales inversely with bubble radius; as a result, very small cavities experience high internal pressures and tend to collapse. Only when a nucleus grows beyond a critical radius does the net pressure differential favour expansion, allowing the bubble to grow in response to ambient hydrodynamic forces. This interplay between surface tension, vapour pressure, and external liquid pressure sets the stage for the delicate balance governing cavitation inception.

The energy scales involved in bubble collapse are remarkable given the small spatial dimensions. Inertial collapse can concentrate energy so strongly that local pressures reach hundreds of megapascals, and temperatures can spike to several thousand kelvins for microseconds in the bubble interior. These extreme conditions have been observed to produce light emission in a phenomenon known as sonoluminescence, where an imploding bubble emits a flash of light whose duration is shorter than a billionth of a second. In more practical engineering contexts, the same concentration of energy at collapse leads to the generation of powerful pressure waves that propagate through the liquid, contributing to vibration, noise, and potential structural fatigue in nearby components.

2.6 Gas phase: thermal properties and momentum transport

Transitioning from the visible oscillations of liquid masses, we must now consider the role of the gaseous phase. In many engineering contexts, gases are not merely passive volumes but active participants that influence heat transfer and structural loading. This section shifts the focus to gas behaviour, exploring how its compressibility and flow patterns complement the liquid phase phenomena previously discussed.

The gas phase, whether it exists as vapour in equilibrium above a liquid surface or as an independent gaseous fluid occupying its own domain, exhibits thermodynamic and transport properties that are fundamentally distinct from those of the liquid phase, owing to the much greater average intermolecular spacing and the different energy scales governing molecular interactions. In gases, molecules are separated by distances many times their molecular diameters, moving in predominantly straight trajectories interrupted only by occasional collisions, a regime well described by kinetic theory. Unlike in liquids, where molecular motion is strongly constrained by cohesive forces, gas molecules can translate freely between collisions, and their mean free path (the average distance travelled before interacting with another particle) can range from tens of nanometres at atmospheric pressure to centimetres or more in low-pressure environments. This relative freedom of motion means that when temperature increases, the average molecular velocity increases in proportion to the square root of the absolute temperature, enhancing the rate of momentum exchange between fluid layers and thereby increasing the viscosity of the gas. This trend is the reverse of that found in liquids, where heating generally decreases viscosity by weakening intermolecular cohesion. For gases, higher temperatures lead to faster-moving molecules that more effectively transport momentum, resulting in viscosity growth with temperature. The same molecular kinetics govern thermal conductivity in gases, where heat is carried not by vibrational energy along a molecular, but by translational motion of molecules carrying internal energy between regions of different temperature. In this case, molecular collisions act as the mechanism for energy redistribution, and thermal conductivity, like viscosity, tends to increase with temperature because faster molecular speeds and longer energy jumps between collisions enhance the net rate of energy transport, although this increase is typically less steep than that of viscosity due to differences in how momentum and energy relax in gases.

From a thermodynamic standpoint, the state of a gas under most engineering and environmental conditions can be described with considerable accuracy by the ideal gas law, which relates pressure P, density ρ, and temperature T via $P = \rho RTP$ where R is the specific gas constant determined by the molar mass of the species. This relationship arises directly from kinetic theory by equating the macroscopic pressure to the microscopic momentum flux of molecules colliding with the walls of a container. The ideal gas equation captures the high compressibility of gases, a property that becomes crucial in high-speed flows where density changes cannot be neglected. In such compressible regimes, pressure, temperature, and density variations are intimately coupled, leading to phenomena such as shock waves, expansion fans, and choked flows.

In the context of vapour above a liquid, the gaseous phase is linked to the liquid phase through the processes of evaporation and condensation, which continuously exchange molecules between the two states. In a closed system at fixed temperature, these opposing processes eventually balance, establishing a dynamic equilibrium characterised by the saturation vapour pressure – a pressure that depends only on temperature for a given substance. At the molecular-level, evaporation occurs when molecules near the liquid surface acquire sufficient kinetic energy from thermal fluctuations to overcome cohesive forces and escape into the gas phase, while condensation occurs when gas phase molecules lose enough energy upon colliding with the surface to be captured by intermolecular forces and incorporated into the liquid. The equilibrium vapour pressure reflects the fraction of molecules energetic enough to escape, and thus rises steeply with temperature as more molecules meet this criterion. The density of the vapour phase increases accordingly, affecting transport properties such as thermal conductivity, viscosity, and diffusivity. Above the liquid surface, gas phase transport occurs via two main mechanisms: molecular diffusion, driven by concentration gradients, and convection, driven by bulk fluid motion and density differences, such as in natural convection above a warm liquid surface. These transport processes mediate not only the mass exchange between phases but also the heat exchange, since evaporation absorbs latent heat of vaporisation from the liquid, cooling it, while condensation releases this heat to the surroundings.

The latent heat associated with phase change in gases is a direct manifestation of the energy required to overcome intermolecular forces in the liquid without raising the temperature. This energy, the latent heat of vapourisation, is typically orders of magnitude larger than the sensible heat required to raise the liquid temperature by a few degrees, which is why evaporation is such an efficient cooling mechanism in both engineered systems, like evaporative coolers and distillation columns, and in natural processes, such as perspiration and oceanic heat transport. Conversely, when a gas condenses to a liquid – whether in the atmosphere forming clouds, in refrigeration cycles, or inside heat exchangers – it releases the same amount of latent heat, raising the temperature of the surrounding medium. In industrial systems, condensation of vapour in piping can cause operational problems if uncontrolled, but it can also be exploited for energy recovery and desalination. The ability of a gas to transition back to liquid depends not only on temperature but also on pressure; sufficiently high pressures can force a gas into the liquid state even at temperatures above its standard boiling point. This is the principle behind liquefaction of gases such as natural gas, oxygen, and nitrogen for storage and transport.

When considering gases not in equilibrium with a liquid, such as atmospheric air, combustion products, or industrial process gases, their behaviour is still governed by the same kinetic and thermodynamic principles, but without the continual phase exchange at a nearby interface. In these contexts, viscosity, thermal conductivity, and specific heats remain functions of temperature and composition, while density and compressibility effects depend strongly on pressure. Real gases deviate from ideal gas behaviour at high pressures or low temperatures, where intermolecular forces and finite molecular size become significant, requiring more sophisticated equations of state such as the van der Waals, Redlich–Kwong, or Peng–Robinson models. These corrections capture the onset of condensation and other nonideal behaviours, including supercritical fluid states where the distinction between gas and liquid vanishes. In high-speed gas flows, such as those in jet engines or supersonic wind tunnels, the interplay of compressibility, heat capacity ratio, and shock-induced temperature changes determines performance and stability. In low-pressure environments, such as outer space or high-altitude flight, the long mean free paths alter transport properties, and continuum models must give way to molecular-level descriptions such as the Boltzmann equation or direct simulation Monte Carlo methods.

Overall, the gas phase occupies a unique position in fluid physics, combining the freedom of molecular motion with sensitivity to temperature and pressure variations, making it both highly dynamic and responsive in ways that liquids are not. Whether as a vapour layer in delicate balance with its parent liquid or as a stand-alone gaseous medium flowing freely through engineered or natural systems, its physical behaviour is dictated by the statistical mechanics of molecular motion, the thermodynamic relationships between macroscopic state variables, and the transport processes that redistribute mass, momentum, and energy. Understanding these mechanisms enables prediction and control of a vast range of processes, from the formation of clouds to the operation of high-performance gas turbines, and underpins applications in energy, climate science, propulsion, and chemical processing.

2.7 Fundamental conservation systems in fluid dynamics

Fluid mechanics is founded on a small set of fundamental conservation principles that govern the motion and evolution of fluids. These principles express the conservation of mass, momentum, and energy and provide a universal framework for describing fluid behaviour across a wide range of physical conditions.

Within the continuum mechanics formulation, fluids are treated as continuous media, allowing these conservation laws to be expressed in differential and integral forms and applied to diverse flow configurations and boundary conditions. Together, they constitute the governing equations of fluid dynamics and form the basis for both analytical and numerical modelling. Because these principles control the essential mechanisms underlying fluid motion, they are of central importance to the present study. In particular, for sloshing dynamics, where fluid motion is strongly influenced by inertia, confinement, and energy exchange, a clear formulation of the conservation equations is indispensable. Accordingly, the following sections introduce and discuss these governing equations in detail.

2.7.1 Continuity equation

Fluid mechanics is fundamentally governed by several core principles: the conservation of mass, Newton's laws of motion, and the laws of thermodynamics. Each principle follows a consistent conceptual framework, stating that any physical property – be it mass, momentum, or energy – within a defined fluid volume changes due to two primary processes: the net flux of that property across the volume's boundaries and internal production or consumption within the volume itself. These governing principles are most effectively formulated in terms of rates. Since they involve both boundary fluxes and volumetric variations, their mathematical representation naturally includes surface and volume integrals. Although integral (global) forms are sometimes sufficient, local (differential) forms are often required to resolve detailed spatial behaviour within the fluid domain.

As illustrated in Fig. 2.4, the continuity equation is the mathematical embodiment of the law of mass conservation. Consider a fixed rectangular control volume aligned with the coordinate axes, of dimensions Δx, Δy, and Δz, through which fluid flows primarily in the x-direction in a quasi-one-dimensional manner. Let $m(t)$ be the mass enclosed at time t. At the inlet face (x), the fluid has velocity u_0 and density ρ_0; at the outlet face $x + \Delta x$, these quantities are $ux + \Delta x$ and $\rho_0 + \Delta\rho$, respectively. Here, velocity and density are continuous field variables that depend on spatial representatives x, y, z, and temporal representative t. Over a short time interval Δt, the balance between the mass entering and leaving the control volume causes $m(t)$ to change to $m(t + \Delta t)$. In the limit as Δx, Δy, Δz, and Δt approach zero, this balance yields the differential form of the continuity equation, expressing mass conservation at every point in the flow field.

Building upon the previously established framework, we now rigorously express the principle of mass conservation for a fixed control volume over a finite time interval Δt. To illustrate, consider the mass entering the control volume specifically

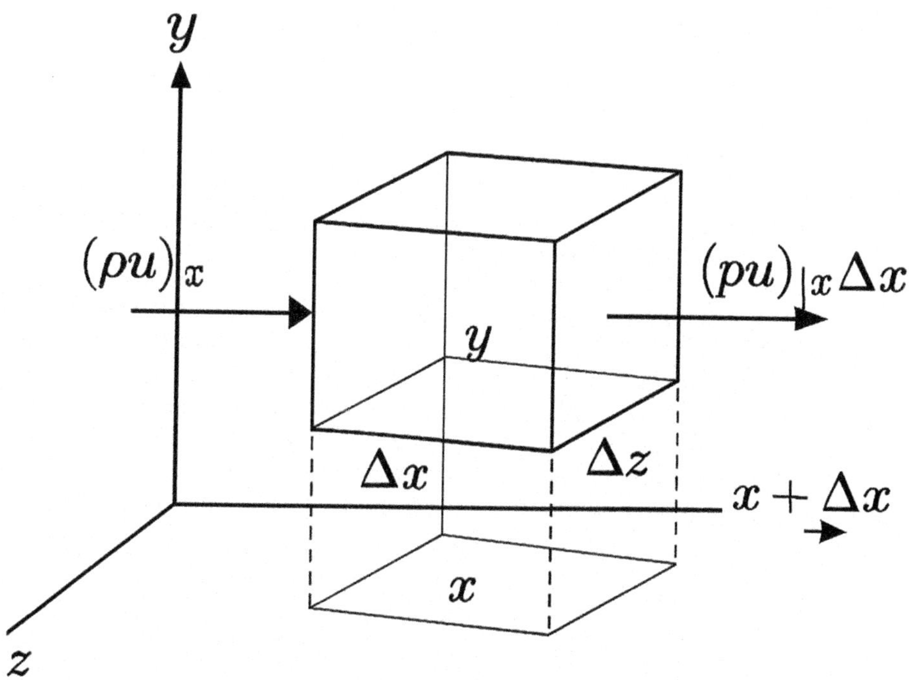

Figure 2.4 Infinitesimal control volume used in the derivation of the continuity equation, showing mass flux entering and leaving through differential faces.

through the face oriented normal to the z-direction. The mass influx during the interval Δt can be quantified as the product of several key factors: the local fluid density ρ, the velocity component u_z normal to that face, the area of the face A, and the duration of the time interval Δt. Physically, this corresponds to the volume of fluid swept through the boundary, which is approximated by the distance the fluid travels during Δt, given by $u_z \Delta t$, multiplied by the cross-sectional area A. This calculation assumes that both the velocity u_z and the density ρ remain essentially constant over this infinitesimally small spatiotemporal region. As such, the entered mass flow to the system can be represented as:

$$\text{Inflow rate} = (\rho u)|_x \Delta y \Delta z \Delta t \tag{2.28}$$

The mass that flows out through the surface at $x + \Delta x$ during the time interval Δt can be evaluated by treating $u|_{x+\Delta x} \Delta t\ t$ as the distance travelled per unit time. At the same time, the exited mass can be described as:

$$\text{Outflow rate} = (\rho u)|_{x+\Delta x} \Delta y \Delta z \Delta t \tag{2.29}$$

As such the Accumulated Quantity of fluid within the system is the difference of Inflow Rate and Outflow Rate.

$$\text{Mass flow} = (\rho|_{t+\Delta t} - \rho|_t) \Delta x \Delta y \Delta z \tag{2.30}$$

Based on the above, this analysis considers the flow rate in relation to the examined system. The continuity equation may take the form

$$\frac{\partial(\rho u)}{\partial x} = -\frac{\partial \rho}{\partial t} \tag{2.31}$$

By extending the analysis to three-dimensional flow, we define v and ww as the velocity components in the y- and z-directions, respectively, complementing the u-component along the x-axis. The conservation of mass within a differential control volume over a finite time interval Δt requires accounting for the mass fluxes through all six faces oriented normal to the x, y, and z coordinate axes. The instantaneous mass flow rate through each face is quantitatively represented as the product of the local fluid density ρ, the velocity component normal to the face, and the face area. Integrating these fluxes over the time interval Δt yields the total mass inflow and outflow across each boundary (as illustrated in equations x and y, respectively). Summing these contributions from all orthogonal directions facilitates the derivation of the three-dimensional continuity equation, which mathematically enforces the principle of local mass conservation within the fluid domain.

$$\text{Inflow rate} = (\rho u)|_x \Delta y \Delta z \Delta t + (\rho u)|_y \Delta z \Delta x \Delta t + (\rho u)|_z \Delta x \Delta y \Delta t \tag{2.32}$$

$$\begin{aligned}\text{Outflow rate} = {} & (\rho u)|_{x+\Delta x} \Delta y \Delta z \Delta t + (\rho u)|_{y+\Delta y} \Delta z \Delta x \Delta t \\ & + (\rho u)|_{z+\Delta z} \Delta x \Delta y \Delta t\end{aligned} \tag{2.33}$$

Similarly, assuming infinitesimal increments (Δx, Δy, Δz, Δt)$\rightarrow$0, and applying the divergence theorem, we obtain the differential form of the mass conservation equation.

The mass conservation equation governing the fluid flow can be expressed in its most general differential form as:

$$\frac{\partial(\rho u_x)}{\partial x} + \frac{\partial(\rho u_y)}{\partial y} + \frac{\partial(\rho u_z)}{\partial z} = -\frac{\partial \rho}{\partial t} \tag{2.34}$$

Eq. (2.33) expresses the continuity equation in its most general form for three-dimensional fluid flow. Here, the vector denotes the mass flux of the fluid, encompassing both magnitude and direction of mass transport. Up to this stage, the analysis has addressed the fully unsteady and compressible flow regime, where flow variables, including density ρ, can vary as functions of both spatial coordinates and time.

We now turn to the special case of steady flow, characterised by flow properties that remain constant at any fixed spatial location over time. Under these conditions, all time-dependent partial derivatives vanish, causing the temporal term in Eq. (2.33) to reduce to zero. Consequently, the continuity equation simplifies, reflecting the absence of transient changes in the flow field.

For a steady state case, Eq. (2.33) can be reduced to

$$\frac{\partial(\rho u_x)}{\partial x} + \frac{\partial(\rho u_y)}{\partial y} + \frac{\partial(\rho u_z)}{\partial z} = 0 \tag{2.35}$$

And for constant density flow Eq. (2.34) can be reduced to

$$\frac{\partial u_x}{\partial x} + \frac{\partial u_y}{\partial y} + \frac{\partial u_z}{\partial z} = 0 \tag{2.36}$$

2.7.2 Momentum conservation equation

The conservation of momentum in fluid mechanics is a direct consequence of Newton's second law of motion, which states that the rate of change of momentum of a system is equal to the net force acting on it. When applied to a fluid, this principle governs how velocity fields evolve under the influence of pressure gradients, viscous stresses, and body forces such as gravity.

Consider a fixed control volume V bounded by a closed surface S. The total momentum of the fluid inside the control volume is $\int_V \rho u dV$, where ρ is the fluid density and u is the velocity vector. The rate of change of momentum within the control volume is balanced by the net momentum flux across the boundaries and the sum of external forces acting on the fluid.

The integral form of the momentum conservation equation is written as:

$$\frac{d}{dt}\int_V \rho u dV + \int_S \rho u(u \cdot n) dS = \int_S \sigma \cdot n dS + \int_V \rho b dV \tag{2.37}$$

where σ is the stress tensor, n is the outward normal to the surface, and b represents body forces per unit mass.

The stress tensor is decomposed into an isotropic pressure component and a deviatoric viscous component:

$$\sigma = -PI + \tau \tag{2.38}$$

where τ is the viscous stress tensor.

Applying the divergence theorem and assuming smooth fields, the local differential form of momentum conservation becomes:

$$\frac{\partial(\rho u)}{\partial t} + \nabla\cdot(\rho u \otimes u) = -\nabla P + \nabla\cdot\tau + \rho b \tag{2.39}$$

For a Newtonian fluid, the viscous stress tensor is linearly related to the rate of deformation:

$$\tau = \mu(\nabla u + (\nabla u)^T) - \frac{2}{3}\mu(\nabla\cdot u)I \tag{2.40}$$

Substituting this relation yields the Navier–Stokes equations, which represent the most general form of momentum conservation for viscous fluids.

For incompressible flow with constant viscosity and density, the equation simplifies to:

$$\rho\left(\frac{\partial u}{\partial t} + u\cdot\nabla u\right) = -\nabla P + \mu\nabla^2 u + \rho b \tag{2.41}$$

2.7.3 Energy conservation equation

The conservation of energy in fluid mechanics is governed by the first law of thermodynamics, which states that energy can neither be created nor destroyed, only transferred or converted between different forms. In a flowing fluid, energy exists primarily as internal energy, kinetic energy, and potential energy.

Consider a fixed control volume. The total energy per unit mass is defined as:

$$E = e + \frac{1}{2}\,|u|^2 + gz \tag{2.42}$$

where e is internal energy, $\frac{1}{2}\,|u|^2$ is kinetic energy, and gz is gravitational potential energy.

The integral form of the energy conservation equation is:

$$\frac{d}{dt}\int_V \rho E dV + \int_S \rho E\,(u\cdot n)dS = \int_S q\cdot ndS + \int_S (\sigma\cdot u)\cdot ndS + \int_V \rho b\cdot udV \tag{2.43}$$

where q is the heat flux vector.

$$q = -\kappa\nabla T \tag{2.44}$$

As such, the differential form of the energy equation becomes:

$$\frac{\partial(\rho E)}{\partial t} + \nabla\cdot(\rho u E) = -\nabla\cdot q + \nabla\cdot(\tau\cdot u) - \nabla\cdot(Pu) + \rho b\cdot u \tag{2.45}$$

In many practical applications, this equation is rewritten in terms of temperature rather than total energy. For incompressible flow with constant properties, the temperature equation reduces to:

$$\rho C_p\left(\frac{\partial T}{\partial t} + u\cdot\nabla T\right) = \kappa\nabla^2 T + \phi \tag{2.46}$$

where ϕ is the viscous dissipation function.

2.8 Foam formation and characteristics

Foams are encountered far more frequently in fluid systems than is often recognised, particularly in situations involving agitation, vibration, or rapid acceleration. In the context of sloshing dynamics, foam formation is not a marginal phenomenon but a realistic and, in many cases, unavoidable outcome. Whenever a liquid containing surface-active species is subjected to oscillatory motion, impact, or sudden changes in acceleration, the generation and entrainment of gas bubbles becomes highly probable. Once formed, these bubbles can accumulate, interact, and evolve into foam structures that significantly alter the macroscopic behaviour of the fluid (Al-Darweesh et al., 2024; Nazari et al., 2025; Stemmet et al., 2008).

From a physical perspective, a foam is a dispersed colloidal system in which a gas phase is distributed within a continuous liquid phase. In sloshing problems, this typically arises when air is entrained into an aqueous or fuel-based liquid during motion. The presence of even modest concentrations of surfactants, additives, or impurities can dramatically increase the persistence of the resulting bubbles, allowing them to survive long enough to interact with the bulk flow. As a result, the fluid can no longer be treated as a single-phase continuum with fixed properties, and the sloshing response may deviate substantially from that predicted by classical models (Stevenson, 2012).

A simple experimental observation illustrates this effect clearly and may be readily reproduced by the reader. Consider a rigid bottle partially filled with pure water. When the bottle is shaken or subjected to oscillatory motion, the resulting sloshing produces sharp, impulsive forces that are readily transmitted to the hand. Now repeat the experiment after adding a small amount of liquid soap to the water. Upon shaking the bottle again, air is rapidly entrained, bubbles form throughout the liquid, and a transient foam develops. The perceived reaction force becomes noticeably smoother, less impulsive, and more damped. This striking difference arises not from a change in liquid density but from the formation of a dispersed gas phase that alters the effective rheology, compressibility, and energy dissipation mechanisms of the system.

In sloshing dynamics, foam formation affects the flow in several interconnected ways. Firstly, the presence of bubbles increases the effective compressibility of the fluid. Gas-filled structures deform under pressure fluctuations, absorbing energy that would otherwise contribute to violent impact loads on container walls. Secondly, foam introduces additional dissipation mechanisms. Bubble deformation, film stretching, and interfacial motion convert kinetic energy into surface energy and viscous losses. Thirdly, the redistribution of mass within the container is altered. Foam tends to accumulate near free surfaces and regions of intense agitation, modifying the effective free-surface shape and, consequently, the natural frequencies of the sloshing system.

These effects are particularly relevant in engineering applications where sloshing loads are critical. In liquid fuel tanks, for example, agitation during vehicle motion, launch manoeuvres, or attitude control can entrain gas into the fuel. The formation of foam alters the pressure response at tank walls, affects fuel delivery, and may influence sensor readings. In extreme cases, excessive foaming can lead to cavitation-like behaviour, pump starvation, or inaccurate mass flow measurements. Similar concerns arise in marine tanks, chemical reactors, and transport containers subjected to vibration.

It is important to emphasise that foams differ fundamentally from emulsions, despite superficial similarities. In a foam, the dispersed phase is compressible, and the continuous liquid phase occupies only a fraction of the total volume. Moreover, foams are intrinsically time-dependent. Once formed, they evolve through liquid drainage, bubble growth, and structural rearrangement, eventually collapsing as films rupture. These processes occur on time scales that may be comparable to, or even shorter than, the characteristic sloshing period. Consequently, the fluid properties relevant to sloshing are not constant but evolve dynamically during motion.

The formation of foam during sloshing is typically initiated by hydrodynamic entrainment. Oscillatory motion generates waves, jets, and turbulent eddies that trap air pockets beneath the liquid surface. These pockets are stretched and broken into bubbles, which may coalesce or accumulate depending on the local flow conditions. In pure liquids, such bubbles tend to rise and burst rapidly. However, in realistic fluids containing surfactants, additives, or contaminants, the gas–liquid interface becomes stabilised, allowing bubbles to persist and interact.

Once bubbles are present, the system undergoes a transition from a simple liquid with a free-surface to a complex two-phase structure. At relatively low gas fractions, the bubbles remain largely spherical and are dispersed within the liquid. As agitation continues, bubbles accumulate near the free-surface, forming a foam layer. With further gas entrainment and drainage of liquid from the bubble network, the foam becomes drier and more structured, with bubbles deforming into polyhedral shapes separated by thin liquid films.

From the standpoint of sloshing dynamics, this evolution is critical. Wet foams behave as highly viscous, energy-absorbing layers that damp surface motion. Dry foams, while mechanically fragile, can persist long enough to modify pressure distributions and alter impact dynamics. In both cases, the presence of foam changes the effective boundary conditions at the liquid–gas interface, invalidating assumptions of a sharp, clean free-surface.

Another important aspect is the coupling between foam evolution and sloshing-induced stresses. Sloshing motion continuously deforms bubble films, stretching and compressing them. These deformations can either stabilise the foam through interfacial elasticity or accelerate its breakdown through rupture, depending on the chemical composition of the liquid. As a result, foam formation and decay are not passive processes but are actively coupled to the imposed motion.

In practical terms, this coupling implies that foam may appear intermittently during sloshing, forming under intense motion and collapsing during quiescent periods. Such transient behaviour complicates both experimental measurement and numerical modelling. Standard single-phase sloshing models cannot capture these effects, while fully resolved multiphase simulations remain computationally demanding.

For these reasons, foam formation deserves careful consideration in any realistic analysis of sloshing phenomena. Its impact extends beyond aesthetic or secondary effects and can fundamentally alter force transmission, damping characteristics, and system response. In the sections that follow, we therefore examine foam formation, evolution, and stabilisation in greater detail, with the aim of providing a physical framework that can be integrated into the broader understanding of sloshing dynamics.

2.8.1 Definition and classification of foams

In its most general sense, a foam is defined as a dispersed system in which a gas phase is distributed within a continuous medium, most commonly a liquid. While this definition encompasses a wide range of physical systems, the term *foam* is typically used to describe a highly aerated structure composed of numerous gas bubbles that accumulate, interact, and persist, often near a liquid surface. What distinguishes a foam from a simple bubbly liquid is not merely the presence of gas inclusions, but the high gas content and close packing of bubbles, which gives rise to a connected, mechanically responsive structure. In practical terms, a foam behaves neither as a simple fluid nor as a static solid, but as a dynamic soft material whose properties depend on its internal architecture.

Unlike emulsions, for which well-established classification frameworks exist, foams lack a universally accepted taxonomy. One of the most widely used and physically intuitive approaches is classification based on bubble size. At one extreme, nano-bubbles and submicron bubbles may remain suspended for extended periods and often behave more like gas emulsions than classical foams. At the other extreme, macro-bubbles rise rapidly under buoyancy and burst upon reaching the free-surface. From an engineering and industrial perspective, the most relevant foams are composed of bubbles with characteristic diameters in the range

$$10\,\mu m \le d_b \le 1\,mm \tag{2.47}$$

as bubbles in this size range are sufficiently buoyant to accumulate yet sufficiently small to form dense, interacting assemblies. It is within this regime that stable foam layers develop and exert a significant influence on fluid behaviour.

A complementary and often more informative classification is based on the gas volume fraction, φ_g, which quantifies the fraction of the total foam volume occupied by gas. At relatively low gas fractions, typically

$$0.65 \le \varphi_g \le 0.72 \tag{2.48}$$

the system is referred to as a *wet foam*. In this regime, bubbles are largely spherical and are separated by comparatively thick liquid films, with the liquid phase forming a continuous network throughout the structure. As drainage proceeds and the gas fraction increases, bubbles are forced into closer contact and begin to deform. When the gas fraction exceeds approximately

$$\varphi_g > 0.9 \tag{2.49}$$

the foam enters the *dry foam* regime, in which bubbles adopt polyhedral shapes separated by thin, nearly planar liquid films known as lamellae. These films meet at Plateau borders and nodes, forming a characteristic cellular architecture dictated by surface energy minimisation. This transition from wet to dry foam marks a profound change in mechanical response, stability, and transport properties, and provides a useful framework for analysing foam behaviour across a wide range of applications.

2.8.2 Mechanisms of foam formation

Foams may be generated intentionally or inadvertently through several physical pathways. Despite their diversity, all formation mechanisms involve the creation of new gas–liquid interfacial area, a process that requires energy. Foam formation may arise through a variety of physical pathways, either deliberately engineered or occurring unintentionally during fluid motion and processing. Despite their apparent diversity, all foam-generation mechanisms share a common physical requirement: the creation of new gas–liquid interfacial area, an energetically unfavourable process that must be driven by external work or thermodynamic instability. One of the most fundamental and widely encountered pathways is gas precipitation, which occurs when a gas dissolved in a liquid becomes supersaturated and separates from the solution in the form of bubbles.

The thermodynamic basis of gas precipitation is governed by Henry's law, which relates the equilibrium concentration of a dissolved gas to its partial pressure above the liquid,

$$D_g = k_H P \tag{2.50}$$

where D_g is the dissolved gas concentration, k_H is Henry's constant (which depends on both the gas and the solvent), and P is the gas partial pressure. When a liquid initially in equilibrium is subjected to a rapid pressure reduction, the equilibrium solubility decreases instantaneously, while the actual dissolved gas concentration remains temporarily unchanged. This mismatch produces a supersaturated state, which is thermodynamically unstable and drives the nucleation of gas bubbles.

Bubble nucleation may occur homogeneously within the bulk liquid or, more commonly, heterogeneously at solid surfaces, impurities, or surface defects, where the energetic barrier to nucleation is reduced. The driving force for bubble growth is the difference between the gas pressure inside the bubble and the surrounding liquid pressure. For a spherical bubble of radius R_b the internal pressure is elevated by capillary effects according to the Young–Laplace relation,

$$P_{\text{bubble}} = P_{\text{liquid}} + \frac{2\gamma_b}{R_b} \tag{2.51}$$

where γ_b is the gas–liquid surface tension. This elevated pressure implies that very small bubbles are inherently unstable and tend to dissolve unless stabilised by interfacial effects.

In pure liquids, the absence of surface-active species leads to rapid bubble collapse or bursting once bubbles reach the free-surface. The thin liquid films separating adjacent bubbles drain quickly, and no restoring interfacial forces exist to counter local thinning or rupture. Consequently, gas precipitation in pure liquids typically produces only transient foams, whose lifetimes are measured in seconds or less.

In contrast, liquids containing amphiphilic molecules, such as surfactants or certain polymers, can support persistent foam formation following gas precipitation. These species adsorb at the newly created gas–liquid interfaces, reducing the surface tension and thereby lowering the work required to generate additional interfacial area,

$$\Delta W = \gamma_b \Delta A \tag{2.52}$$

More importantly, surfactant adsorption imparts mechanical stability to the interfacial films by generating surface elasticity and resistance to deformation. As bubbles nucleate and grow, neighbouring bubbles can remain separated by stabilised liquid films, allowing them to accumulate and interact rather than immediately coalesce or rupture.

As gas precipitation continues, bubbles rise under buoyancy and collect near the liquid surface, where their close packing leads to the formation of a foam layer. The persistence of this foam depends not only on the rate of gas release but also on the balance between interfacial stabilisation, liquid drainage, and bubble growth. Thus gas precipitation provides a clear illustration of how a purely thermodynamic instability, when coupled with interfacial chemistry, can give rise to a complex, evolving foam structure.

Foam formation may also arise from mechanical agitation, a mechanism that is particularly relevant in dynamically excited liquids and industrial flows. When a liquid is subjected to stirring, shaking, vibration, or turbulent motion, the free-surface becomes unstable and is repeatedly deformed, folded, and fragmented. Under such conditions, pockets of air are readily entrained into the liquid bulk. These entrained gas volumes are initially large and irregular, but the action of shear stresses and turbulent eddies rapidly stretches and fragments them into smaller bubbles. The characteristic bubble size is governed by a balance between disruptive hydrodynamic forces and restorative capillary forces, often expressed through a critical Weber number,

$$W_e = \frac{2\rho u^2 R_b}{\xi} \tag{2.53}$$

When inertial stresses exceed surface tension forces, air pockets are torn apart, generating a population of bubbles that can persist within the liquid.

Once formed, these bubbles rise under buoyancy and accumulate near the free-surface, where their concentration increases and bubble–bubble interactions become frequent. In the absence of stabilising agents, the bubbles coalesce or burst rapidly, and only a transient froth is observed. However, in liquids containing surface-active components, agitation-driven entrainment becomes an efficient route to stable foam formation. A classic example is whipped cream, where mechanical agitation serves a dual role. Firstly, it entrains air into the liquid, creating bubbles through repeated deformation and breakup of air pockets. Secondly, the same shear forces disrupt fat globules present in the cream, releasing partially crystalline fat molecules that migrate to the gas–liquid interface. These molecules adsorb strongly, forming a viscoelastic interfacial layer that reduces surface tension and imparts mechanical rigidity to the bubble films.

The stabilisation mechanism in such systems is not purely chemical but structural and mechanical. Adsorbed fat particles and amphiphilic components create interfaces with high surface viscosity and elasticity, enabling the thin liquid films between adjacent bubbles to resist drainage, stretching, and rupture. As agitation continues, bubbles are continuously generated faster than they collapse, leading to the build-up of a persistent foam. From a physical standpoint, agitation-induced foam formation highlights the intimate coupling between hydrodynamics and interfacial science: turbulence creates interfacial area, while interfacial chemistry determines whether that area can survive.

2.8.3 *Thermodynamics of interfacial area creation*

The formation of foam is fundamentally a thermodynamic process governed by the creation of new gas–liquid interfacial area. Any increase in this area is necessarily accompanied by an increase in free energy, reflecting the energetic penalty associated with bringing gas and liquid molecules into contact. As established earlier in Eq. (2.51), the minimum reversible work required to generate additional interfacial area is directly proportional to the surface tension of the interface. This relationship defines the primary energetic barrier to foam formation: liquids with high gas–liquid surface tension strongly resist the generation of bubbles, whereas systems in which the surface tension is reduced can form interfacial area more readily under mechanical agitation or thermodynamic instability (illustrated in Fig. 2.5).

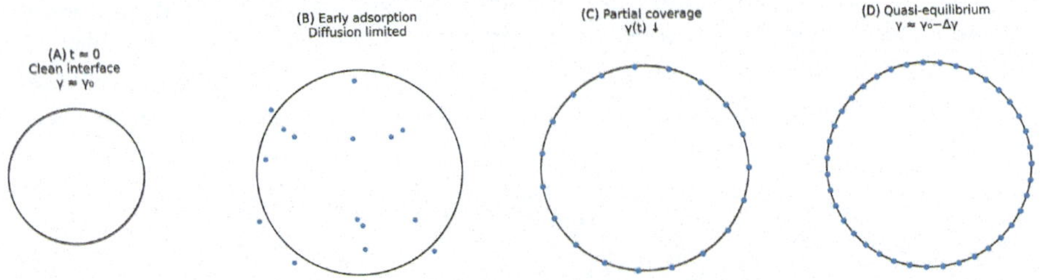

Figure 2.5 Schematic representation of dynamic surfactant adsorption at a newly created gas–liquid interface and the time-dependent evolution of surface tension.

In real foam formation, however, the interfacial area is created extremely rapidly. Processes such as turbulence, jet impingement, and surface wave breaking generate new surfaces on time scales of milliseconds or less. Under these conditions, the relevant interfacial property is not the equilibrium surface tension, but the dynamic surface tension, $\gamma_b(t)$, which reflects the transient state of surfactant adsorption at the newly formed interface. Immediately after interface creation, the surface is largely devoid of surfactant, and the surface tension is close to that of the pure solvent. As time progresses, surfactant molecules diffuse from the bulk and adsorb at the interface, progressively lowering γ_b.

The time dependence of surface tension may be described schematically as

$$\gamma_b(t) = \gamma_b^0 - \Delta\gamma_b f(t) \tag{2.54}$$

where γ_b^0 is the surface tension of the clean interface, $\Delta\gamma_b$ is the maximum surface tension reduction achievable by surfactant adsorption, and $f(t)$ is a kinetic function that depends on surfactant diffusivity, concentration, and adsorption dynamics. For rapidly created interfaces, $f(t)$ may remain small during the critical moments of bubble formation, meaning that the energetic cost of interface creation is higher than suggested by equilibrium measurements. This observation explains why surfactants with fast adsorption kinetics are particularly effective foam formers, even if their equilibrium surface tensions are not exceptionally low.

Surfactants facilitate foam formation in two distinct but complementary ways. Firstly, by reducing the surface tension, they lower the work required to create interfacial area, thereby promoting the initial generation of bubbles. Secondly, by adsorbing to the interface, they modify the mechanical and transport properties of the thin liquid films that separate adjacent bubbles. However, it is important to emphasise that surface tension reduction alone is not sufficient to ensure foam persistence. A liquid with low surface tension may still produce only transient foam if the interfacial films lack mechanical stability.

Once bubbles are formed, the newly created interfacial area must be maintained against processes that tend to reduce it, such as film drainage, rupture, and bubble coalescence. These processes are driven by the system's tendency to minimise free energy by eliminating interfacial area. The persistence of foam therefore depends on stabilisation mechanisms that oppose this tendency. One key stabilising factor is the development of interfacial elasticity, which arises when local variations in surfactant concentration generate restoring stresses. If a film is locally stretched, the surface concentration of surfactant decreases in that region, increasing the local surface tension and creating a gradient that drives liquid back into the stretched area. This mechanism converts chemical potential differences into mechanical resistance.

From a thermodynamic standpoint, the competition between interfacial energy minimisation and stabilising forces can be expressed through a balance of pressures acting across thin liquid films. The capillary pressure tending to thin the film is proportional to (\gamma), while repulsive interactions between the opposing film surfaces generate a disjoining pressure, $\Pi(h)$, that increases as the film thickness (h) decreases. A quasistable film thickness is achieved when

$$\Pi(h) = P_{\text{capillary}} \tag{2.55}$$

Surfactants and polymers influence both sides of this balance by modifying surface tension, interfacial structure, and intermolecular forces.

The thermodynamics of interfacial area creation thus extends beyond the simple energetic cost expressed by Eq. (2.51). In foam formation, the rate at which surface tension evolves, the ability of surfactants to redistribute under deformation, and the resulting mechanical response of interfacial films are all equally important. Foam persistence emerges not from a single thermodynamic parameter, but from the dynamic interplay between energy input, interfacial kinetics, and stabilising forces acting at nanometre length scales.

2.8.4 Foam evolution and drainage

Once a foam is formed, it does not represent a static material but rather a dynamically evolving structure whose morphology, composition, and mechanical response change continuously in time. This evolution is governed primarily by the redistribution of liquid within the foam and by the rearrangement of bubbles as the system seeks to reduce its overall free energy. The dominant drainage mechanisms depend strongly on the foam's gas volume fraction (i.e. φ_g), which controls both the geometry of the bubble network and the pathways available for liquid transport. As φ_g increases, the foam undergoes a transition from a relatively wet, fluid-like system to a dry, highly structured cellular material, accompanied by a fundamental change in the physical processes that govern liquid removal.

In the early stages of foam evolution, when the gas fraction remains moderate, the foam is classified as a wet foam. In this regime, bubbles are approximately spherical and are separated by relatively thick liquid films, with the liquid phase forming a continuous network throughout the structure. Liquid drainage in wet foams is dominated by gravity-driven flow, which occurs through the interconnected system of films, channels, and interstitial regions between bubbles. From a hydrodynamic perspective, the foam can be treated as a deformable porous medium, where the bubbles act as obstacles to flow and the liquid percolates through the available void space under the influence of gravity.

A useful semiquantitative description of this process is obtained by analogy with flow through packed beds, leading to a Carman–Kozeny-type relation for the volumetric drainage rate (Q),

$$Q \propto \frac{(1-\varphi_g)^3}{\mu \varphi_g^2} \tag{2.56}$$

This expression captures several essential features of foam drainage. Firstly, the strong dependence on $1 - \varphi_g$ reflects the fact that liquid content decreases rapidly as the gas fraction increases, reducing the available cross-sectional area for flow. Secondly, the inverse dependence on μ highlights the role of viscous resistance: more viscous liquids drain more slowly, prolonging the lifetime of the wet foam. Thirdly, although not explicit in this simplified form, the permeability of the foam is implicitly linked to the characteristic bubble size, with smaller bubbles producing narrower flow channels and therefore lower permeability.

As drainage proceeds, liquid is progressively removed from the foam structure, and the gas fraction increases. This increase in φ_g has a pronounced feedback effect on drainage: as the foam becomes drier, the drainage rate decreases sharply. Physically, this occurs because the liquid pathways become thinner and more tortuous, while viscous dissipation increases. The system approaches a critical gas fraction, typically around ($\varphi_g \approx 0.72$), at which bubbles are forced into contact and begin to deform. Beyond this point, the assumption of spherical bubbles breaks down, and the foam undergoes a structural transition towards a dry foam morphology.

In dry foams, gravity-driven drainage becomes increasingly ineffective. The remaining liquid is confined primarily to thin films separating adjacent bubbles and to the channels where these films meet, known as Plateau borders. At this stage, the dominant mechanism for liquid removal is capillary drainage, driven by pressure gradients arising from differences in interfacial curvature within the foam. These curvature-induced pressure differences are described by the Young–Laplace equation,

$$\Delta P = \frac{\gamma_b R_{b1} R_{b2}}{R_{b1} + R_{b2}} \tag{2.57}$$

where R_{b1} and R_{b2} are the principal radii of curvature of the interface.

In the context of dry foams, this relation explains why liquid is drawn out of thin films and into Plateau borders. The liquid films separating bubbles are nearly planar and therefore exhibit very large radii of curvature, resulting in negligible capillary pressure. In contrast, the Plateau borders have highly curved liquid–gas interfaces, with radii of curvature comparable to the local channel dimensions. As a result, the pressure within the liquid at the Plateau borders is lower than that within the films, establishing a pressure gradient that drives liquid flow from the films toward the borders. Once collected in the Plateau borders, the liquid drains downward under gravity to lower regions of the foam.

The rate of capillary-driven drainage depends on the balance between capillary pressure gradients and viscous resistance within the narrow liquid channels. A simplified scaling estimate for the capillary drainage velocity (u_c) may be written as

$$u_c \sim \frac{\gamma_b}{\mu} \frac{R_c}{L} \tag{2.58}$$

where R_c is a characteristic radius of curvature of the Plateau border and L is a characteristic length over which the pressure gradient acts. This expression highlights that capillary drainage is enhanced by higher surface tension and lower viscosity, and that geometric factors associated with bubble size and packing play a critical role.

As liquid continues to drain, the thickness of the lamellae separating bubbles decreases, often reaching values on the order of tens of nanometres. At these length scales, classical hydrodynamics alone is insufficient to describe the system, and intermolecular forces become dominant. If this balance cannot be maintained, the film continues to thin until it reaches a critical thickness, typically of order (5)–(10)

nm, at which thermal fluctuations or local stresses trigger rupture. Film rupture leads to bubble coalescence, releasing stored interfacial energy and causing a local rearrangement of the foam structure.

Throughout both gravity-driven and capillary drainage stages, the evolution of the foam is strongly coupled to bubble coarsening processes, such as gas diffusion between bubbles of different sizes. As bubbles grow larger, the curvature of Plateau borders increases, enhancing capillary suction and accelerating drainage. This coupling creates a feedback loop in which drainage promotes coarsening, and coarsening, in turn, accelerates drainage. The resulting structural rearrangements impose local stresses on thinning films, increasing the likelihood of rupture even before the critical thickness is reached.

The transition from gravity-dominated to capillarity-dominated drainage thus represents a fundamental shift in the physical mechanisms governing foam evolution. In wet foams, drainage is relatively rapid and controlled primarily by bulk hydrodynamics and viscosity. In dry foams, drainage becomes slower, more spatially heterogeneous, and increasingly sensitive to interfacial curvature and molecular-scale interactions. This multistage evolution underscores the inherently hierarchical nature of foam dynamics, in which macroscopic flow, mesoscale geometry, and nano-scale forces act in concert to determine the lifetime and mechanical behaviour of the foam.

2.8.5 Structural evolution, interfacial mechanics, and ageing of foams

As a foam evolves following its initial formation, progressive loss of liquid from the structure forces bubbles into ever closer contact, triggering a sequence of structural transitions that fundamentally alter the foam's geometry, mechanics, and stability. In the early stages, when sufficient liquid remains, bubbles are largely spherical and separated by relatively thick liquid films. Continued drainage reduces the liquid fraction and increases the gas volume fraction, such that neighbouring bubbles come into mechanical contact and must deform in order to accommodate one another. This deformation is driven by the system's tendency to minimise total interfacial free energy, and it results in a transition from spherical bubbles to polyhedral gas cells whose faces are nearly planar. In this configuration, three films meet along edges at angles close to (120 degrees), a geometric consequence of surface tension forces balancing at junctions and a hallmark of energy minimisation in foams. The resulting dry foam structure consists of a highly ordered three-dimensional network composed of thin liquid films, known as lamellae, connected by Plateau borders and nodes that act as conduits for the remaining liquid. Although this cellular architecture appears mechanically rigid, it is intrinsically metastable. The thin films continue to drain and thin over time, rendering the structure increasingly susceptible to rupture and rearrangement.

Superimposed on these geometric transitions are bubble coarsening processes, which play a central role in foam ageing by increasing the average bubble size and reducing the number of bubbles in the system. One such mechanism is disproportionation, closely analogous to Ostwald ripening in emulsions, which arises from differences in capillary pressure between bubbles of unequal size. The internal pressure of a spherical bubble of radius exceeds the ambient pressure by an amount determined by the Young–Laplace relation

$$P = P_0 + \frac{2\gamma_b}{R_b} \tag{2.59}$$

Smaller bubbles therefore possess higher internal pressure than larger ones, creating a thermodynamic driving force for gas to diffuse from smaller bubbles towards larger neighbours through the liquid phase or across thin films. Disproportionation progressively eliminates small bubbles, coarsens the foam texture, and alters the geometry of Plateau borders, thereby enhancing drainage and accelerating structural weakening.

A second coarsening pathway is coalescence, which occurs when the liquid film separating two bubbles drains to a critical thickness and ruptures. Film rupture leads to the rapid formation of a capillary bridge that merges the two gas volumes into a single, larger bubble. This event releases stored surface energy and often induces transient oscillations and local rearrangements in the surrounding foam. Both disproportionation and coalescence act to increase bubble size, reduce film area, and intensify liquid drainage, creating a feedback loop in which coarsening promotes further destabilisation of the foam structure.

The persistence of foams against these ageing mechanisms depends critically on the presence and behaviour of surfactants, which influence both interfacial thermodynamics and film mechanics. Surfactants perform a dual function: they lower surface tension to facilitate foam formation and impart elasticity to the thin films that separate bubbles. When a liquid film is locally stretched by deformation or flow, the surface concentration of surfactant decreases in that region, leading to an increase in local surface tension. This generates a restoring stress that resists further deformation and draws liquid back towards the stretched area. This elastic behaviour underpins the Gibbs–Marangoni mechanism, in which surface tension gradients drive interfacial flows that heal incipient thinning and suppress rupture. Pure liquids lack this mechanism entirely, which explains why they are unable to sustain stable foams. The magnitude of surface elasticity depends strongly on surfactant concentration and adsorption kinetics and typically reaches a maximum near the critical micelle concentration, beyond which additional surfactant resides primarily in micelles rather than at the interface, contributing little to film elasticity.

As foam films thin to nanometre scales during drainage, colloidal interactions between the opposing interfaces become dominant and play a decisive role in determining film stability. Electrostatic repulsion between charged interfaces, steric hindrance arising from adsorbed polymers or surfactant layers, and short-range structural forces collectively contribute to the disjoining pressure, which acts to oppose further thinning of the film. If the repulsive interactions are sufficiently strong, film thinning slows dramatically, delaying rupture and extending foam lifetime. Conversely, weak disjoining pressure allows continued thinning until thermal fluctuations or mechanical disturbances trigger failure.

The combined effects of structural transitions, coarsening mechanisms, interfacial elasticity, and colloidal forces ultimately govern the observable performance of foams, which is commonly assessed through standardised experimental methods. One widely used technique is the Ross–Miles test, in which a liquid jet plunges into a solution to generate foam and the initial foam height is recorded as a measure of foamability. Foam stability is characterised by the remaining foam height after a fixed time interval. These metrics reflect different aspects of foam physics: foamability is controlled primarily by dynamic surface tension and the kinetics of surfactant adsorption during rapid interface creation, whereas stability depends on film elasticity, surface viscosity, drainage resistance, and the magnitude of disjoining pressure. Together, these intertwined physical processes define the complex, time-dependent behaviour of foams as structured, metastable colloidal systems.

2.9 Emulsions

Emulsions are colloidal dispersions in which one liquid is present as droplets within a second, immiscible liquid. Their defining feature is not simply the coexistence of two liquids, but the presence of an interface that is actively engineered to resist separation. Because creating droplets multiplies interfacial area by orders of magnitude, an emulsion is intrinsically a high interfacial energy material. It can therefore exist for practical times only when interfacial structures such as adsorbed surfactants, proteins, polymers, or particles generate barriers that slow the natural relaxation back to two bulk phases.

Although emulsions are commonly introduced through familiar examples such as milk or mayonnaise, the underlying physics becomes especially transparent when viewed through sloshing-driven emulsification. The butter-like material produced in the Chapter 1 demonstration is not a culinary curiosity; it is a direct manifestation of interfacial physics under strong mechanical forcing. When a partially filled container is shaken, the resulting sloshing flow generates intense, time-dependent shear and extensional deformation, repeatedly stretching and fragmenting the oil-water interface. This process continuously creates new surface area, and the system responds by partitioning interfacial active molecules to the newly formed interface. In the butter case, naturally occurring milk proteins and phospholipids adsorb rapidly and provide interfacial stabilisation, while temperature-dependent fat crystallisation can further reinforce the droplet structure by introducing a mechanical skeleton within the dispersed phase. Under certain conditions, the balance of interfacial curvature preferences and stabiliser solubility can even promote phase inversion, transforming the identity of the continuous phase. Thus the sloshing experiment compactly illustrates how hydrodynamics, interfacial chemistry, and phase behaviour cooperate to generate a stable or semistable dispersed structure.

From a materials science standpoint, an emulsion is best described as a structured nonequilibrium system whose properties emerge from the competition among interfacial energy, hydrodynamic forcing, diffusion, and interparticle interactions. The canonical pair is oil and water: their mutual immiscibility produces a sharp interface with substantial interfacial tension, and the formation of droplets is energetically costly. Yet mechanical energy input can overcome this cost by converting macroscopic work into interfacial area, while stabilisers reduce the interfacial tension and build protective interfacial layers. The resulting droplet population is characterised by a size distribution that controls optical appearance, rheology, and stability. For this reason, emulsions are often classified by droplet size into macroemulsions, nanoemulsions, and microemulsions, each associated with distinct optical behaviour and distinct dominant destabilisation mechanisms.

A central theme in emulsion science is kinetic stability. Many technologically important emulsions are not thermodynamically stable, meaning that complete phase separation is the equilibrium outcome. Nevertheless, they can persist for weeks, months, or years because separation requires specific pathways whose rates can be dramatically slowed. The principal destabilisation routes include creaming or sedimentation driven by density differences, droplet aggregation and flocculation driven by interparticle attractions, coalescence driven by interfacial film rupture, and Ostwald ripening driven by curvature-dependent solubility and molecular transport. Modern formulation strategies are therefore aimed less at changing the final equilibrium state and more at increasing kinetic barriers that retard these processes.

The practical relevance of these concepts extends far beyond food. Many everyday necessities, including cosmetics, sunscreens, detergents, inks, coatings, agrochemical sprays, and pharmaceutical formulations, are emulsions or emulsion-derived materials. Their performance is controlled by droplet size distribution, interfacial composition, bulk rheology, and stability under storage and use. Accordingly, this article reframes the lecture content into a research-style overview focused on biphasic liquid–liquid emulsions, emphasising droplet size-based classification, formation routes from high-energy methods such as sloshing to low-energy inversion strategies, rational stabiliser selection, and mechanistic understanding of destabilisation pathways.

2.9.1 What exactly is an emulsion

An emulsion is a colloidal system in which one liquid is dispersed as a multitude of small droplets within another liquid with which it is intrinsically immiscible. In both everyday products and scientific formulations, the most common and technologically relevant case involves oil and water. Although these two liquids do not mix at the molecular-level, mechanical energy and interfacial chemistry can be used to force one phase to fragment into droplets that are temporarily stabilised within the other. An emulsion is therefore not a simple mixture, but a deliberately constructed, nonequilibrium material whose properties are governed by interfacial physics (Saleh, 2022; Sun et al., 2025).

A typical emulsion consists of three essential components. The first is the dispersed phase, which is the liquid broken up into droplets. In many practical systems, this is oil, although in some cases it is water. The second component is the continuous phase, the surrounding liquid matrix that fills the space between droplets and largely determines the macroscopic behaviour of the emulsion, including its viscosity, electrical conductivity, optical appearance, and flow characteristics. The third and indispensable component is the emulsifier system, which comprises one or more additives that enable the formation of droplets and suppress their tendency to merge back together. Without an emulsifier, any droplets created during mixing would rapidly coalesce, and the system would revert to two separate liquid layers.

When oil droplets are dispersed within water, the system is referred to as an oil-in-water emulsion. In this case, water forms the continuous phase, and oil exists as discrete droplets suspended throughout it. Conversely, when water droplets are dispersed within oil, the system is known as a water-in-oil emulsion, with oil acting as the continuous phase. These two configurations are not simply mirror images of one another. For a given oil-water pair and a given emulsifier formulation, one arrangement is usually more stable than the other and therefore forms preferentially during emulsification. This preference arises from how the emulsifier partitions between oil and water and how it organises at the interface, which, in turn, determines the curvature and mechanical strength of the droplet surface.

The distinction between oil-in-water and water-in-oil emulsions has profound physical consequences. Because the continuous phase forms a connected pathway through the material, it controls transport properties. For example, oil in water emulsions typically exhibit electrical conductivities similar to water, while water in oil emulsions is far less conductive. Likewise, the ability to dilute an emulsion with additional liquid depends on the identity of the continuous phase: an oil-in-water emulsion can be readily diluted with water but not with oil, whereas the opposite is true for a water-in-oil system. Even tactile perception reflects this difference, with oil-in-water emulsions often described as creamy and water-in-oil emulsions as greasy, a direct consequence of which phase is in contact with surfaces during flow and spreading.

Under carefully controlled changes in system conditions, emulsions can undergo inversion, a process in which the roles of the dispersed and continuous phases are exchanged. Inversion may be induced by altering temperature, changing the relative proportions of oil and water, modifying salt concentration, or adjusting the composition of the emulsifier system. From a physical perspective, inversion corresponds to a change in the preferred curvature of the oil-water interface. When the interfacial layer favours curvature towards the oil side, oil droplets form in water; when curvature towards the water side is favoured, water droplets form in oil. Inversion is therefore a powerful demonstration of how subtle changes in interfacial energetics can lead to dramatic changes in macroscopic structure.

A defining and often misunderstood characteristic of emulsions is their kinetic stability. From a thermodynamic standpoint, emulsions are generally unstable. Creating droplets dramatically increases the interfacial area between oil and water, and this increase carries a significant energetic penalty because the oil-water interface has a positive interfacial tension. The true equilibrium state of an oil-water system is complete phase separation into two bulk layers with minimal interfacial area. Emulsions persist not because this thermodynamic driving force has been removed, but because the pathways leading back to equilibrium are slow.

The persistence of an emulsion is therefore controlled by energy barriers that impede separation. Emulsifiers play a central role in generating these barriers. By adsorbing at the oil-water interface, emulsifier molecules reduce interfacial tension, making droplet formation easier during mixing. More importantly for stability, they create interfacial films that resist deformation, drainage, and rupture when droplets collide. As a result, processes such as droplet coalescence, gravitational separation, and molecular transport between droplets occur over long time-scales rather than instantaneously.

Several distinct physical mechanisms contribute to emulsion breakdown, and the emulsifier system is designed to slow each of them. Droplets may migrate under gravity, leading to creaming or sedimentation depending on their density relative to the continuous phase. They may form clusters through attractive interactions, increasing the likelihood of collision. Upon close approach, the thin liquid film between droplets must drain and rupture for coalescence to occur, a process strongly influenced by the rheological and structural properties of the interfacial layer. Even in the absence of collisions, droplets can change size through molecular diffusion driven by curvature-dependent pressure differences, a process known as Ostwald ripening. An emulsion remains usable so long as these mechanisms are sufficiently suppressed.

In this sense, an emulsion is best viewed as a carefully delayed return to equilibrium. Its usefulness lies precisely in the fact that separation is slow on the timescale of storage, processing, and application. This perspective explains why emulsions are so central to modern materials and products. By controlling droplet size, interfacial composition, and continuous phase properties, scientists and engineers can design oil-water systems that appear stable and homogeneous, even though they are fundamentally metastable. Understanding what an emulsion truly is therefore requires not only a compositional description but a physical appreciation of interfacial energy, hydrodynamic forces, and kinetic barriers that together sustain this ubiquitous class of soft matter systems.

2.9.2 Classification by droplet size and optical appearance

Emulsions are commonly classified according to droplet size, because droplet dimensions exert a decisive influence on optical appearance, mechanical behaviour, and the dominant mechanisms governing stability and breakdown. While all emulsions share the basic feature of dispersed liquid droplets suspended within a continuous immiscible liquid, differences in droplet size introduce qualitatively different physical regimes. These regimes are traditionally described as macroemulsions, miniemulsions or nanoemulsions, and microemulsions, each representing a distinct balance between interfacial energy, stabilisation mechanisms, and transport processes.

2.9.2.1 Macroemulsions (ordinary emulsions)

Macroemulsions, often referred to as *ordinary emulsions*, are the most familiar and widely encountered class. They typically contain droplets with characteristic diameters ranging from approximately 100 nanometres up to several micrometres, and in some cases even larger. Because these droplet sizes are comparable to or larger than the wavelength of visible light, macroemulsions scatter light very efficiently. As a result, they exhibit the characteristic white, opaque, milk-like appearance that is readily recognised in products such as milk, cream, mayonnaise, butter, margarine, lotions, and many cosmetic creams.

From a physical perspective, macroemulsions are kinetically stable colloidal systems. They are not thermodynamically stable, since the large interfacial area between oil and water carries a substantial energetic penalty. However, emulsifiers adsorbed at the droplet interface create barriers that slow the processes leading to separation. The broad droplet size distributions commonly found in macroemulsions arise naturally from high-energy emulsification methods such as stirring, homogenisation, or sloshing, where droplet breakup and coalescence occur simultaneously. This polydispersity has important consequences for stability, as it enhances processes such as creaming and Ostwald ripening, but these processes can often be sufficiently retarded to yield products with long shelf lives.

In most engineering and industrial applications, macroemulsions represent the practical workhorse. Their relatively large droplets simplify formulation, reduce the need for excessive emulsifier concentrations, and allow for straightforward processing at large scales. For this reason, much of classical emulsion science and industrial practice has focused on understanding and controlling macroemulsion behaviour.

2.9.2.2 *Miniemulsions and nanoemulsions*

Between macroemulsions and molecularly structured systems lies an intermediate class often described as miniemulsions or nanoemulsions. These emulsions typically possess droplet diameters in the range of roughly ten to one hundred nanometres. At these dimensions, light scattering is significantly reduced, and the optical appearance of the emulsion changes markedly. Depending on the degree of size uniformity and the refractive index contrast between oil and water, nanoemulsions may appear translucent or even optically clear, a property that is attractive for applications such as beverages, pharmaceuticals, and personal care products.

Like macroemulsions, nanoemulsions are generally kinetically stable rather than thermodynamically stable. Their smaller droplet size confers certain advantages, including improved resistance to gravitational separation. Because sedimentation and creaming velocities decrease strongly with droplet size, nanoemulsions can remain spatially homogeneous for long periods even when density differences between oil and water are significant. However, the small droplet radius introduces new challenges. High-interfacial curvature leads to elevated Laplace pressure within droplets, which increases the driving force for molecular transport through the continuous phase. As a result, nanoemulsions are often more susceptible to Ostwald ripening, a process in which smaller droplets shrink while larger ones grow.

The generation of nanoemulsions typically requires either intense mechanical energy input or carefully designed low-energy routes such as phase inversion strategies. Emulsifier selection becomes particularly critical, as rapid adsorption to newly formed interfaces is essential to prevent immediate coalescence during droplet formation. Although nanoemulsions occupy a narrower and more controlled size range than macroemulsions, they remain metastable systems whose long-term behaviour is dictated by kinetic barriers rather than equilibrium thermodynamics.

2.9.2.3 Microemulsions

Microemulsions form a fundamentally different category despite sometimes overlapping in droplet size with nanoemulsions. They often involve characteristic structural dimensions on the order of a few tens of nanometres, yet their defining feature is not size alone but thermodynamic stability. Under appropriate conditions of surfactant and cosurfactant composition, microemulsions can form spontaneously upon simple mixing of oil and water, without the need for significant mechanical energy input.

Physically, microemulsions are best regarded not as collections of discrete droplets in the classical sense, but as highly organised interfacial structures, often described as swollen micelles or bicontinuous phases. The oil-water interfacial tension in these systems is extremely low, sometimes approaching zero, allowing large interfacial areas to exist without an energetic penalty. As a result, microemulsions are equilibrium phases rather than kinetically trapped states.

The scientific description of microemulsions is correspondingly more complex. Their behaviour is governed by surfactant packing constraints, interfacial curvature elasticity, and detailed phase behaviour that depends sensitively on temperature, composition, and ionic strength. Because they require relatively high concentrations of surfactants and cosurfactants, microemulsions are less common in bulk consumer products, but they are of considerable interest in specialised applications such as drug delivery, enhanced oil recovery, and nanostructured materials synthesis.

2.9.2.4 Practical perspective

Although macroemulsions, nanoemulsions, and microemulsions can all be described under the broad umbrella of emulsified oil-water systems, they represent distinct physical regimes with different stabilisation strategies and performance characteristics. In most food, engineering, and consumer product applications, macroemulsions dominate because they balance ease of production, moderate emulsifier demand, and sufficient kinetic stability for practical use. Nanoemulsions offer advantages where optical clarity and enhanced physical stability are required, while microemulsions occupy a more specialised niche defined by equilibrium thermodynamics and complex interfacial organisation. Understanding this classification by droplet size and optical appearance provides a crucial framework for rational emulsion design and interpretation.

2.9.3 Identifying oil in water versus water in oil emulsions in practice

Correctly identifying whether an emulsion is composed of oil dispersed in water or water dispersed in oil is a fundamental practical task in emulsion science and formulation. This distinction is far more than a matter of nomenclature. The identity of the continuous phase governs a wide range of physical properties, including electrical conductivity, dilution behaviour, rheology, surface interactions, and sensory perception. Consequently, determining the continuous phase provides immediate insight into how an emulsion will behave during processing, storage, and application.

In practice, a number of simple yet physically meaningful tests are routinely employed to distinguish oil-in-water emulsions from water-in-oil emulsions. These tests exploit the fact that the continuous phase forms a connected pathway throughout the material and therefore dominates transport and interfacial interactions.

One of the most straightforward methods is the dilution test. An emulsion can be readily diluted by adding more of its continuous phase, because the added liquid simply expands the surrounding medium without disrupting droplet integrity. By contrast, adding the dispersed phase typically leads to poor mixing or even phase separation. Thus an emulsion that dilutes easily with water but not with oil can be identified as an oil-in-water system, while the opposite behaviour indicates a water-in-oil emulsion. Although deceptively simple, this test directly reflects the connectivity of the continuous phase and is widely used in both laboratory and industrial settings.

Electrical conductivity measurements provide another powerful diagnostic. Water, particularly when it contains dissolved ions, is a good electrical conductor, whereas oils are generally insulating. As a result, oil-in-water emulsions often exhibit conductivities close to that of the aqueous phase, since the water forms a continuous conductive network. In contrast, water in oil emulsions typically displays very low conductivities, because the water is confined to isolated droplets that do not provide continuous pathways for charge transport. Even relatively small differences in conductivity can therefore be used to identify the emulsion type with confidence.

A closely related approach involves the use of selective dyes. Water-soluble dyes readily dissolve and spread throughout the continuous aqueous phase of an oil-in-water emulsion, producing strong and uniform colouration. In a water-in-oil emulsion, the same dye remains confined to the dispersed water droplets, leading to weak or highly localised colouration. Conversely, oil-soluble dyes preferentially colour water in oil emulsions, while having little effect on oil in water systems. Dye tests are particularly useful because they provide immediate visual confirmation and can be combined with microscopy for more detailed analysis.

Optical microscopy, especially when combined with refractive index contrast, offers further insight. When an emulsion is observed under transmitted or reflected light, droplets may appear brighter or darker than the surrounding medium, depending on their refractive index relative to the continuous phase. By adjusting illumination conditions or using phase contrast techniques, it is often possible to identify which phase is continuous and which is dispersed. While this method requires more specialised equipment and expertise, it is valuable when dealing with complex or highly viscous systems where simple dilution or dye tests may be ambiguous.

Surface interaction tests based on wetting behaviour and contact angle measurements provide another physically grounded means of differentiation. When a small droplet of emulsion is placed on a low-energy surface, such as a hydrophobic polymer, its spreading behaviour depends strongly on the identity of the continuous phase. Oil in water emulsions typically form droplets with relatively high contact angles on low-energy surfaces, whereas water in oil emulsions tend to spread more readily or even completely wet the surface. On high energy surfaces such as clean glass, the opposite trend is often observed. These differences arise because the continuous phase is the phase indirect contact with the surface and therefore controls interfacial energetics.

Furthermore, sensory and rheological observations provide useful qualitative indicators. Oil-in-water emulsions are commonly perceived as creamy, light, or refreshing, reflecting the lubricating properties of water at interfaces and during flow. Water in oil emulsions, by contrast, often feels greasy or heavy, as oil dominates the contact with surfaces and skin. While sensory perception is subjective and should not be relied upon alone, it frequently corroborates conclusions drawn from more quantitative tests.

Together, these practical signatures translate conceptual lecture level descriptions into robust formulation diagnostics. They are routinely employed in research laboratories, quality control environments, and manufacturing lines to verify emulsion type, detect unintended inversion, and guide formulation adjustments. Importantly, each test is rooted in fundamental physical principles associated with phase connectivity, transport, and interfacial behaviour, reinforcing the central role of the continuous phase in determining the properties and performance of emulsified systems.

2.9.4 Core thermodynamic driver and why emulsifiers are necessary

The fundamental physical challenge in forming an emulsion arises from the creation of a large oil–water interfacial area. When two immiscible liquids are present as separate bulk phases, the total interfacial area between them is minimal. Emulsification transforms one of these phases into a multitude of small droplets dispersed throughout the other, increasing the interfacial area by several orders of magnitude (McClements & Jafari, 2018). This increase carries a substantial energetic cost, which can be expressed to first order as

$$\Delta G \approx \gamma_{em} \Delta A \tag{2.60}$$

where γ_{em} is the interfacial tension between oil and water, and ΔG is the increase in interfacial area. Because ΔG becomes enormous when bulk liquids are converted into droplets, the associated increase in free energy is large and positive. From a thermodynamic standpoint, this means that emulsification is intrinsically unfavourable, and the equilibrium state of an oil–water system is complete phase separation.

Emulsification therefore requires the input of mechanical work, supplied through processes such as stirring, homogenisation, or sloshing, to overcome this energetic penalty. However, mechanical energy alone is not sufficient to create a usable emulsion. In the absence of additional stabilising agents, any droplets formed during mixing would rapidly merge, reducing interfacial area and returning the system towards equilibrium. The key to both droplet formation and persistence lies in reducing the interfacial tension and constructing barriers that slow interfacial relaxation. This is the essential role of emulsifiers.

2.9.4.1 Surfactants and interfacial tension reduction

Surfactants are the most important class of emulsifying agents because of their ability to adsorb rapidly at the oil–water interface and reduce interfacial tension. A surfactant molecule typically contains a hydrophilic segment with affinity for water and a hydrophobic segment with affinity for oil. When present in an oil–water system, surfactants spontaneously migrate to the interface, orienting themselves so as to minimise the free energy of the system. This adsorption can reduce the interfacial tension from values of several tens of millinewtons per metre to much smaller values, depending on the chemical nature and concentration of the surfactant.

Lowering the interfacial tension has two profound and closely related consequences. Firstly, droplet formation becomes easier. Because the minimum work required to generate new interfacial area is proportional to γ_{em}, a reduction in interfacial tension directly lowers the mechanical energy needed to break one liquid into droplets within the other. For a given mixing intensity, systems with lower interfacial tension can therefore achieve finer dispersions. Secondly, the thermodynamic driving force for droplet coalescence is reduced. Coalescence eliminates interfacial area, and the energetic benefit of this process scales with γ_{em}. When γ_{em} is small, the free energy gain associated with droplet merging is diminished, making coalescence less favourable and easier to suppress kinetically.

A simple scaling argument illustrates why surfactants lead to smaller droplets. Consider two emulsification processes in which the same amount of mechanical work is supplied, one carried out without surfactant and the other with a surfactant present. In the first case, the high interfacial tension severely limits the amount of interfacial area that can be generated for a given energy input. In the second case, where the interfacial tension is reduced, the same mechanical work can produce a much larger interfacial area. Since total interfacial area scales inversely with average droplet size for a fixed dispersed phase volume, a larger area corresponds directly to smaller droplets. This reasoning explains why the introduction of surfactants often reduces droplet sizes by one or even two orders of magnitude under otherwise similar processing conditions.

2.9.4.2 Emulsifiers as multicomponent systems

While surfactants are central to emulsification, practical emulsifier systems are rarely composed of a single molecular species. In real formulations, the term ‘emulsifier’ usually refers to a multicomponent mixture designed to perform several complementary functions at the interface and in the surrounding phases.

Primary surfactants, whether ionic or nonionic, are responsible for rapid adsorption to the newly created interface and for the initial reduction of interfacial tension. Their relatively small molecular size allows them to diffuse quickly, which is essential during droplet breakup when a fresh interface is generated on very short time-scales.

However, surfactants alone often provide only limited resistance to droplet coalescence, particularly under shear or during long-term storage.

To enhance stability, polymers are frequently incorporated into emulsifier systems. Polymers adsorbed at the interface can extend into one or both liquid phases, creating steric barriers that resist close droplet approach. In addition, polymer layers impart viscoelastic character to the interface, increasing its resistance to deformation and rupture. Because polymers desorb much more slowly than small surfactant molecules, they can significantly improve long-term kinetic stability.

In some systems, solid particles are used as stabilising agents, leading to so-called Pickering emulsions. When particles of appropriate size and wettability adsorb at the oil–water interface, they form a rigid or semirigid armour around droplets. The energy required to detach such particles from the interface is often extremely large compared with thermal energy, making particle-stabilised droplets remarkably resistant to coalescence.

Cosurfactants are also commonly employed to fine-tune interfacial properties. These species can adjust interfacial packing density, modify preferred curvature, and improve compatibility between different stabilising components. Through careful formulation, cosurfactants help optimise both droplet formation and stability across a range of temperatures and compositions.

In practical terms, a stable macroemulsion typically requires on the order of one to three per cent emulsifier relative to the dispersed phase, although the exact amount depends on the desired droplet size, the nature of the oil and water phases, and the intensity of the emulsification process. Smaller droplets demand greater interfacial area and therefore higher stabiliser coverage, while harsher processing conditions require emulsifier systems capable of rapid adsorption and strong interfacial reinforcement. Together, these considerations highlight that emulsifiers are not optional additives but essential components that make emulsions physically viable by moderating interfacial thermodynamics and imposing kinetic barriers to separation.

2.9.5 Sloshing-driven emulsification: from macroscopic flow to droplet microstructure

Sloshing provides a particularly clear and physically rich example of how macroscopic fluid motion can be converted into microscopic interfacial structure. In a partially filled container, external agitation induces large-amplitude motion of the free-surface, accompanied by rapidly varying flow fields, transient jets, and repeated impacts of the liquid against the container walls. These unsteady and highly inhomogeneous flow conditions generate intense local shear and extensional deformation, which are especially effective at disrupting an oil–water interface and promoting droplet formation.

At the microscale, several complementary mechanisms contribute to emulsification under sloshing. Interfacial stretching produces thin liquid ligaments and sheets that are inherently unstable and readily fragment into droplets. Simultaneously, strong shear and extensional flows act on existing droplets, deforming them until capillary forces can no longer restore their spherical shape, at which point breakup occurs. In addition, the violent motion of the free-surface can entrain air and generate cavitation-like events, both of which enhance local mixing and repeatedly renew the oil–water interface. This continual creation of a fresh interface is crucial, as it allows surface-active species to adsorb and stabilise newly formed droplets before they can merge.

The efficiency of droplet breakup during sloshing is governed by the competition between hydrodynamic stresses and capillary pressure. This balance is conveniently expressed through the Weber number,

$$W_e = \frac{\rho U^2 L}{\gamma_{\mathrm{em}}} \tag{2.61}$$

where U is a characteristic velocity associated with the flow, L is a characteristic length scale such as droplet diameter or ligament thickness, and γ_{em} is the oil–water interfacial tension. When the Weber number is small, capillary forces dominate, and deformed interfaces relax back to minimise surface area. When the Weber number is large, hydrodynamic stresses overwhelm capillary restoration, leading to sustained deformation and eventual breakup. Sloshing tends to increase U through rapid acceleration and deceleration of the fluid, thereby driving the system towards higher Weber numbers and more efficient droplet fragmentation.

The presence of surfactants or other emulsifying agents further amplifies this effect by reducing the interfacial tension, which simultaneously increases the Weber number and lowers the energetic cost of creating new interfacial area. As a result, the combined action of vigorous sloshing and interfacial tension reduction pushes the system towards smaller droplet sizes and a substantially larger total interfacial area. Importantly, because sloshing repeatedly renews the interface, it allows emulsifiers to adsorb dynamically during droplet formation, preventing immediate coalescence and locking in the dispersed microstructure.

From this perspective, the butter-forming demonstration can be interpreted as a coupled, multi-stage process. Sloshing first generates a population of oil droplets by converting macroscopic kinetic energy into interfacial area. Natural emulsifiers present in the system rapidly adsorb to these interfaces, stabilising the droplets against immediate coalescence. Subsequent droplet–droplet interactions, partial coalescence, and temperature dependent solidification of the dispersed phase then transform the initially fluid emulsion into a macroscopic, structured material. Sloshing therefore provides a compelling physical illustration of how hydrodynamics, interfacial chemistry, and phase behaviour interact to control emulsion formation and evolution.

2.9.6 High-energy and low-energy emulsification routes

The formation of an emulsion requires the creation of new oil–water interfacial area, an energetically unfavourable process that must be driven either by mechanical work or by exploiting changes in interfacial thermodynamics. From a practical and conceptual standpoint, emulsification strategies are therefore divided into high-energy and low-energy routes. Although both approaches ultimately aim to generate and stabilise small droplets, they rely on fundamentally different physical principles and impose different constraints on formulation and processing.

2.9.6.1 High energy methods

High-energy emulsification methods are the most widely used techniques in both industrial production and laboratory practice. These approaches rely on the direct input of mechanical energy to overcome interfacial tension and forcibly fragment one liquid phase into droplets within the other. Common examples include mechanical stirring and rotor–stator mixers, high-pressure homogenisation, sonication, membrane emulsification, microfluidic devices, and a variety of fluid-shear and turbulent mixing systems.

In all high-energy methods, droplet formation is achieved through hydrodynamic stresses generated by shear, turbulence, and extensional flow. These stresses deform droplets or interfacial structures until capillary forces are no longer able to restore the interface, leading to breakup. The resulting droplet size distribution is governed by a balance between the intensity of the applied stresses and the stabilising action of interfacial tension and emulsifiers. As a general rule, producing smaller droplets requires higher-energy densities because the resistance to further breakup increases as droplets become smaller and capillary pressure rises.

An essential feature of high-energy emulsification is the role of surfactant transport. During droplet breakup, a fresh oil–water interface is created on very short time-scales. For the newly formed droplets to remain stable, emulsifier molecules must rapidly adsorb to this interface. If adsorption is slow relative to droplet formation, newly created droplets will coalesce almost immediately, limiting the effectiveness of the process. Consequently, high-energy emulsification places strong demands on emulsifier chemistry, favouring surface-active species with fast diffusion and adsorption kinetics.

High-pressure homogenisation and sonication are particularly effective at generating fine emulsions because they create extremely high local energy dissipation rates. However, these methods can be energetically costly and may lead to unwanted side effects such as heating, degradation of sensitive ingredients, or incorporation of air. Despite these limitations, high-energy routes remain the dominant choice for producing macroemulsions and many nanoemulsions, especially when robust, scalable processing is required.

2.9.6.2 Low energy methods

In contrast, low-energy emulsification methods exploit changes in thermodynamic or phase behaviour to generate fine dispersions with minimal mechanical input. Rather than relying on strong hydrodynamic stresses to break droplets apart, these approaches manipulate the properties of the oil–water–surfactant system so that interfacial tension is transiently reduced or interfacial curvature preferences are altered. Once a finely dispersed structure is formed, it is then ‘locked in’ by restoring conditions that favour stability.

One of the most prominent low-energy strategies is the phase inversion temperature method, which is commonly used with nonionic surfactants containing ethoxylated chains. The water affinity of these surfactants decreases with increasing temperature, due to changes in hydrogen bonding between the ethylene oxide groups and water. At low temperatures, the surfactant is more hydrophilic and favours oil dispersed in water. As the temperature increases, the surfactant becomes progressively more oil-soluble, eventually favouring water dispersed in oil. Near a critical temperature range, the oil–water interfacial tension reaches a minimum, and the interface becomes highly flexible. If the system is emulsified near this condition and then cooled, a fine oil-in-water emulsion can be obtained with relatively little mechanical energy.

A related class of techniques involves the emulsion inversion point, where inversion is triggered not by temperature but by changes in composition, such as altering the oil-to-water ratio, salt concentration, or surfactant content. These compositional changes modify surfactant solubility and interfacial curvature, again allowing transient access to conditions of low interfacial tension and high interfacial mobility. Under these circumstances, small droplets can form spontaneously or with gentle mixing.

Low-energy methods offer several attractive advantages. Because they rely less on intense mechanical forcing, they can substantially reduce energy consumption and minimise thermal or mechanical damage to sensitive components. In addition, they often yield narrower droplet size distributions than high-energy methods, which can be beneficial for kinetic stability by reducing the driving force for processes such as Ostwald ripening. However, low-energy routes are typically more sensitive to formulation details and operating conditions, and their applicability may be limited to specific classes of surfactants and oils.

2.9.7 Which phase becomes continuous: Bancroft logic and its limits

One of the earliest and most widely cited principles in emulsion science concerns the question of which phase becomes continuous when two immiscible liquids are emulsified. A commonly used heuristic, often attributed to Bancroft, states that the phase in which the emulsifier is more soluble tends to become the continuous phase. In practical terms, if the emulsifying agent is predominantly water-soluble, the system is more likely to form an oil-in-water emulsion, whereas if the emulsifier is more oil-soluble, a water-in-oil emulsion is favoured. This guideline has proved extremely useful in formulation practice, as it links macroscopic emulsion structure to molecular-level solubility preferences of surface-active species.

From a physical standpoint, this behaviour can be interpreted in terms of interfacial curvature and interfacial tension gradients. When an emulsifier preferentially partitions into one phase, it reduces the interfacial tension more effectively on that side of the interface. The resulting asymmetry in interfacial tension promotes curvature of the interface towards the phase with the higher residual interfacial tension. As a consequence, droplets of that phase are formed and dispersed within the other, which becomes continuous. In this sense, the preferred solubility of the emulsifier determines not only where the emulsifier resides but also the sign of the interfacial curvature and, ultimately, the identity of the continuous phase.

Despite its conceptual appeal and practical value, this rule must be applied with caution. Real emulsions are rarely ideal systems composed of a single surfactant and two simple liquids. Industrial and laboratory formulations often contain mixtures of surfactants with different affinities, polymers that adsorb slowly but strongly at the interface, cosurfactants that modify packing density, dissolved salts that alter electrostatic interactions, and oils with complex or poorly defined compositions. Each of these factors can influence interfacial structure in ways that are not captured by simple solubility arguments.

In addition, external conditions play a critical role. The solubility of many emulsifiers, particularly nonionic surfactants, is strongly temperature dependent, meaning that the preferred emulsion type can change with temperature even at fixed composition. Interfacial packing constraints may limit how surfactant molecules organise at the interface, altering curvature preferences independently of bulk solubility. Furthermore, the processing history of the system, including the order of component addition, mixing intensity, and rate of emulsification, can kinetically trap a system in a morphology that is not predicted by equilibrium considerations.

For these reasons, Bancroft logic is best regarded as a useful starting hypothesis rather than a universal law. It provides valuable intuition for initial formulation design and for anticipating likely emulsion types, but it does not replace experimental verification. In both research and manufacturing contexts, reliable identification of the continuous phase ultimately requires direct observation and diagnostic testing. Recognising the strengths and limitations of this heuristic allows it to be used effectively, while avoiding overreliance on a rule that cannot fully account for the complexity of real emulsified systems.

2.9.8 Destabilisation pathways and stabilisation strategies in emulsions: from failure mechanisms to rational design

Understanding how emulsions fail is as important as understanding how they are formed. The vast majority of emulsions encountered in food, pharmaceutical, cosmetic, and engineering applications are metastable systems. Although they may appear uniform and stable over practical time-scales, they are not in thermodynamic equilibrium and will ultimately evolve towards phase separation if given sufficient time. Emulsion science is therefore fundamentally concerned with identifying the physical pathways by which destabilisation occurs and with developing strategies to slow these processes to an acceptable degree.

2.9.8.1 Creaming and sedimentation

One of the most readily observed modes of emulsion destabilisation is gravitational separation, which manifests as creaming or sedimentation depending on the relative densities of the dispersed droplets and the continuous phase. When droplets are less dense than the surrounding medium, they rise under buoyancy, leading to creaming;

when they are denser, they settle, resulting in sedimentation. Importantly, these processes do not initially involve droplet coalescence and are therefore reversible at early stages.

The velocity of droplet migration is strongly dependent on droplet size and the viscosity of the continuous phase. Larger droplets experience greater buoyant forces and migrate more rapidly, while higher continuous phase viscosity increases hydrodynamic resistance and slows motion. Consequently, reducing droplet size and increasing the viscosity of the continuous phase are two of the most effective ways to suppress gravitational separation. In practical formulations such as creams and sauces, thickeners are routinely employed not to alter thermodynamic stability, but to reduce droplet mobility and thereby delay visible separation.

2.9.8.2 Flocculation and aggregation

Beyond simple gravitational effects, droplets can interact with one another through a variety of attractive forces, leading to flocculation or aggregation. Flocculation refers to the formation of loosely bound clusters of droplets that remain distinct entities separated by thin films of continuous phase. These structures are often reversible and may be disrupted by gentle agitation. Aggregation increases the effective size of dispersed entities, which, in turn, accelerates creaming or sedimentation by enhancing buoyant motion.

From a kinetic perspective, flocculation increases collision frequency and promotes spatial heterogeneity within the emulsion. Even if droplets do not immediately merge, the formation of clusters increases the likelihood of close droplet–droplet contact, creating conditions favourable for more severe destabilisation pathways such as coalescence. Attractive interactions may arise from van der Waals forces, depletion interactions induced by dissolved polymers, or insufficient electrostatic or steric repulsion. Controlling these interactions is therefore central to emulsion stability.

2.9.8.3 Coalescence

Coalescence represents a more advanced and often irreversible stage of emulsion breakdown. It involves the merging of two or more droplets into a single larger droplet, thereby reducing the total interfacial area. Coalescence requires drainage and rupture of the thin liquid film that separates approaching droplets. Once this film fails, the interface disappears locally, and the droplets fuse.

Coalescence leads to an increase in the mean droplet size, which not only alters optical and rheological properties but also accelerates gravitational separation. The dominant factor controlling coalescence is the mechanical strength and viscoelastic character of the interfacial layer, which is determined primarily by emulsifier composition and interfacial structure. Interfaces that behave as weak, mobile films drain rapidly and rupture easily, whereas interfaces with high surface viscosity or elasticity resist deformation and suppress film rupture. For this reason, the design of interfacial layers is a central focus in stabilising practical emulsions.

2.9.8.4 Ostwald ripening

Even in the absence of droplet collisions, emulsions can evolve through Ostwald ripening, a molecular-scale transport process driven by interfacial curvature. Smaller droplets experience higher Laplace pressure and therefore possess a higher chemical potential than larger droplets. As a result, molecules of the dispersed phase tend to dissolve preferentially from smaller droplets into the continuous phase and subsequently redeposit onto larger droplets. This process broadens the droplet size distribution over time and can ultimately lead to macroscopic separation.

A widely used kinetic description of this phenomenon is associated with the Lifshitz–Slyozov–Wagner framework, which predicts that ripening rates increase with higher solubility and diffusivity of the dispersed phase in the continuous phase and decrease with lower interfacial tension. An important and sometimes counter-intuitive consequence is that very small droplets can ripen more rapidly than larger ones, making ultrafine emulsions particularly vulnerable unless carefully designed. Ostwald ripening therefore imposes fundamental limits on the stability of nano-scale emulsions.

2.9.9 Stabilisation strategies: slowing each failure mode

Because emulsions fail through multiple, often coupled pathways, effective stabilisation strategies must address each mechanism explicitly rather than relying on a single approach.

Electrostatic stabilisation is achieved using ionic surfactants or charged polymers, which generate repulsive interactions between droplets by forming electrical double layers at the interface. These repulsions increase the effective interaction distance between droplets and suppress flocculation and coalescence. However, electrostatic stabilisation is sensitive to environmental conditions: increasing salt concentration compresses the electrical double layer, while pH changes can alter surface charge. As a result, this strategy must be carefully matched to the intended operating conditions.

Steric stabilisation relies on nonionic surfactants, polymers, or grafted chains that extend into the continuous phase. When two coated droplets approach, the overlap of these chains produces an entropic repulsion that resists close contact. Steric stabilisation is particularly attractive in practical formulations because it remains effective across a wide range of ionic strengths and compositions, making it robust in complex environments.

Increasing continuous phase viscosity is a widely used, though indirect, stabilisation strategy. By thickening the surrounding medium, droplet diffusion, collision frequency, and gravitational migration are all reduced. This approach is ubiquitous in lotions, sauces, and creams, but must be balanced against sensory, processing, and application requirements.

Interfacial reinforcement targets coalescence directly by strengthening the droplet interface. Mixed surfactant systems can pack more efficiently and create mechanically robust interfacial layers. Interfacial complexes formed between oil-soluble and water-soluble species can further enhance film strength. Polymers adsorbed at the interface resist desorption and deformation, while at higher surfactant loadings, liquid crystalline structures may form that dramatically increase interfacial elasticity and viscosity.

Pickering stabilisation uses solid particles of appropriate size and wettability that adsorb strongly at the oil–water interface. Once attached, these particles form a quasirigid armour around droplets, making coalescence energetically prohibitive. The effectiveness of this mechanism depends on particle size relative to droplet size, particle stability, and contact angle. Many clay minerals and silica derivatives are exploited industrially to achieve robust stabilisation without excessive surfactant use.

Suppressing Ostwald ripening requires strategies that reduce the chemical potential gradient driving molecular transport. These include selecting oils with very low solubility in the continuous phase, reducing interfacial tension while maintaining strong interfacial films, narrowing the droplet size distribution, and adding highly hydrophobic components to the dispersed phase to generate osmotic resistance to dissolution.

2.9.10 Practical surfactant selection and formulation mapping

The selection of an appropriate surfactant system is one of the most critical steps in the practical formulation of emulsions. While the underlying physics of interfacial tension, curvature, and kinetic barriers is complex, formulation scientists have long relied on semiempirical frameworks to translate this complexity into actionable design rules. Among these, the hydrophilic–lipophilic balance (HLB) concept remains one of the most widely used and practically useful tools, particularly for emulsions stabilised by nonionic surfactants.

The HLB number provides a single scalar measure of the relative affinity of a surfactant for water versus oil. Low HLB values correspond to more oil-soluble, lipophilic surfactants, whereas high HLB values indicate more water-soluble, hydrophilic species. Although this description is a simplification of molecular interactions at the oil–water interface, it offers an intuitive starting point for surfactant selection. Importantly, many oils can be associated with a characteristic required HLB, defined as the HLB value of a surfactant or surfactant mixture that yields a stable oil-in-water emulsion under specified conditions. This required HLB is not a universal material constant but an operational parameter that depends on oil chemistry, temperature, emulsifier concentration, and processing history.

In practice, formulation mapping using HLB proceeds through a systematic experimental approach. Rather than relying on a single surfactant, formulators typically prepare blends of surfactants whose combined HLB values span a broad range. Because HLB values are approximately additive on a mass-fraction basis, blending two surfactants with known HLB numbers allows precise tuning of the effective HLB of the emulsifier system. A series of emulsions is then prepared using identical oil-to-water ratios and emulsification conditions, while varying only the HLB of the surfactant blend. These samples are subsequently evaluated for stability over fixed time-scales, such as immediately after preparation, after 24 hours, and after longer storage periods.

Stability assessment in this context is necessarily multifaceted. Visual observation of phase separation provides a first indication, but more informative diagnostics include droplet size analysis, microscopy, creaming or sedimentation measurements, and rheological characterisation. In many cases, an optimal HLB window emerges in which the emulsion exhibits minimal droplet growth, limited gravitational separation, and acceptable texture. This window is interpreted as corresponding to a favourable balance between surfactant solubility, interfacial packing, and curvature preference.

Amongst the most versatile surfactant families used in HLB-based formulation studies are sorbitan esters and their ethoxylated derivatives. Sorbitan esters, often associated with lower HLB values, are relatively lipophilic and favour water-in-oil emulsions or interfacial structures with curvature towards the water phase. Ethoxylation of these molecules introduces poly(ethylene oxide) chains, increasing hydrophilicity and shifting the HLB to higher values. By controlling the degree of ethoxylation and the nature of the fatty acid moiety, a wide and nearly continuous range of HLB values can be accessed within a chemically related surfactant family. This chemical consistency is advantageous, as it allows systematic variation of hydrophilic–lipophilic balance without introducing drastically different molecular architectures.

Despite its widespread use, the HLB concept has well-recognised limitations. It is fundamentally semiempirical and does not explicitly account for many factors known to influence emulsion stability, such as temperature dependent surfactant solubility, interfacial rheology, mixed adsorption behaviour, or the presence of polymers, salts, and cosurfactants. Complex oils, particularly those containing polar components or broad molecular weight distributions, often deviate from simple required HLB predictions. In addition, HLB values are typically determined under specific experimental conditions, meaning that values reported in the literature may not translate directly to new systems.

For these reasons, HLB-based formulation should be viewed not as a definitive solution, but as an initial mapping tool that narrows the search space for more detailed investigation. When combined with microscopic observation, droplet size measurements, and targeted stability testing, HLB mapping provides valuable insight into how surfactant chemistry influences interfacial structure and emulsion performance. It also serves as a practical bridge between molecular-scale surfactant properties and macroscopic emulsion behaviour, enabling rational iteration rather than trial-and-error formulation.

2.9.11 Beyond simple emulsions: inversion and multiple emulsions

Emulsion behaviour extends well beyond the classical picture of one liquid dispersed as droplets within another. By deliberately manipulating formulation variables and processing conditions, emulsions can be driven into more complex structural regimes that offer enhanced or entirely new functionalities. Among these, emulsion inversion and multiple emulsions represent two particularly important extensions of conventional emulsion science, with significant implications for both fundamental understanding and technological application.

Emulsion inversion refers to a transformation in which the identities of the dispersed and continuous phases are exchanged. An oil-in-water emulsion may invert into a water-in-oil emulsion, or vice versa, when key system parameters are altered. Such inversion can be induced by changes in temperature, composition, or ionic strength, each of which modifies interfacial properties and surfactant behaviour. For example, in systems stabilised by nonionic surfactants, increasing temperature often reduces the affinity of the surfactant for water, shifting its effective solubility towards the oil phase and promoting inversion. Similarly, altering the ratio of oil-to-water or adding salt can change interfacial curvature preferences and drive a switch in emulsion type.

From a practical standpoint, emulsion inversion is a powerful tool because it enables controlled modulation of macroscopic properties. Switching the continuous phase can dramatically alter texture, electrical conductivity, lubrication behaviour, and mass transport characteristics. In formulation and processing, inversion is sometimes exploited deliberately to generate fine emulsions using low-energy routes, after which the system is 'locked' into the desired configuration. However, inversion can also occur unintentionally, leading to product failure. Understanding the conditions under which inversion occurs is therefore essential for both innovation and quality control.

An even greater level of structural complexity is achieved in multiple emulsions, often described as emulsions within emulsions. Common examples include water-in-oil-in-water and oil-in-water-in-oil systems, in which droplets of one liquid are encapsulated within droplets of another, which are themselves dispersed in a continuous phase. These hierarchical structures create discrete internal compartments separated by multiple interfaces, enabling advanced functions such as encapsulation, controlled release, and protection of sensitive ingredients. In food systems, multiple emulsions have been explored for fat reduction and flavour delivery, while in pharmaceuticals and cosmetics, they offer routes to staged or sustained release of active compounds.

Despite their promise, multiple emulsions are inherently challenging to design and stabilise. Each interface requires stabilisation, often with surfactants of opposing affinities, and interactions between the different emulsifier layers can lead to complex and sometimes unpredictable behaviour. In addition, the presence of multiple interfaces increases the number of possible destabilisation pathways, including coalescence at either the inner or outer interface, leakage of encapsulated material, and accelerated Ostwald ripening. As a result, multiple emulsions are typically far more sensitive to processing conditions, formulation details, and environmental changes than simple emulsions, making them a continuing focus of active research.

Taken together, the study of inversion, hierarchical structures, destabilisation pathways, and stabilisation strategies defines the practical limits and opportunities of emulsion science. While true thermodynamic stability is rarely achievable in oil–water systems, a deep understanding of the physical mechanisms governing emulsion formation and failure allows formulators to design systems that remain functional over the time-scales required for real-world use. In this sense, successful emulsions are not those that eliminate instability entirely, but those in which instability is controlled, delayed, and harnessed to deliver desired performance.

2.10 Dimensionless numbers

Dimensionless numbers occupy a central position in fluid mechanics because they condense complex physical interactions into compact, scale-independent parameters. At their core, these numbers express ratios of competing physical effects, such as inertia to viscosity, gravity to surface tension, or viscous stresses to interfacial forces. By removing physical units through appropriate normalisation, dimensionless numbers allow fluid behaviour to be compared across different fluids, geometries, and length scales. This universality is the primary reason they are constructed and used so extensively in theoretical analysis, experimentation, and numerical modelling.

From a governing-equation standpoint, dimensionless numbers emerge naturally when the Navier–Stokes equations and associated transport equations are nondimensionalised. Velocity, length, time, pressure, and temperature are rescaled by characteristic reference values that reflect the dominant physics of the problem. The resulting equations reveal a limited set of nondimensional groups that determine the qualitative and quantitative nature of the flow. Rather than depending on absolute values of viscosity, density, or surface tension, the system behaviour is governed by how strongly one physical mechanism competes with another. This insight is fundamental: two flows with identical dimensionless numbers are dynamically similar, even if their actual sizes, velocities, or fluids differ greatly.

In general fluid flow, one of the most widely used dimensionless numbers is the Reynolds number Re, which compares inertial forces to viscous forces.

$$Re = \frac{\rho u L}{\mu} \tag{2.62}$$

It determines whether a flow remains laminar, transitions to turbulence, or becomes fully turbulent (illustrated in Fig. 2.6). Low Reynolds number flows are dominated by viscosity, leading to smooth, reversible motion, as seen in microfluidics or creeping flows around small particles. High Reynolds number flows are dominated by inertia, resulting in flow separation, vortices, and turbulence. In sloshing systems, the Reynolds number controls the internal flow structure beneath the free-surface, influencing boundary-layer development along container walls,

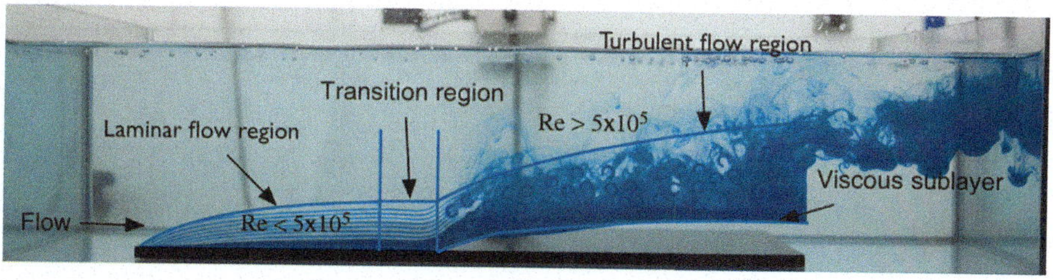

Figure 2.6 Flow over a flat plate illustrating laminar, transitional, and turbulent boundary-layer regimes as a function of Reynolds number.

energy dissipation, and the transition from smooth wave motion to violent breaking and impact.

The Mach number (M) is a dimensionless parameter that defines the ratio of the velocity of an object (or a fluid stream) to the local speed of sound within that medium. It is an essential metric in aerodynamics for determining the compressibility of a fluid.

$$M = \frac{u}{a} \tag{2.63}$$

The Prandt number (Gr) quantifies the relationship between the buoyancy forces and the viscous forces which act to resist motion.

$$Gr = \frac{g\beta\Delta TL^3}{\nu^2} \tag{2.64}$$

Grashof number performs a role analogous to that of the Reynolds number (Re) in forced convection; it dictates the transition from stable, orderly flow to more complex, turbulent regimes (illustrated in Fig. 2.7). This effect is elegantly demonstrated when observing the thermal patterns within the space (annulus) between a heated circular cylinder and its cooler, square outer casing.

Another fundamental group is the Froude number Fr, which compares inertial forces to gravitational forces (illustrated in Fig. 2.8).

$$Fr = \frac{u}{\sqrt{gL}} \tag{2.65}$$

Here, L is the characteristic length. This number is especially important for free-surface flows, where gravity acts as the restoring force for surface waves. In sloshing mechanics, the Froude number governs wave propagation speed, resonance conditions, and the relationship between excitation frequency and natural sloshing modes. Low Froude numbers correspond to gravity-dominated motion, where waves remain relatively gentle and hydrostatic effects prevail. High Froude numbers indicate strong inertial forcing, leading to steep waves, run-up along container walls, and potentially hydraulic jumps or wave breaking. Because sloshing is inherently a free-surface phenomenon, the Froude number is often more relevant than the Reynolds number for predicting global surface motion.

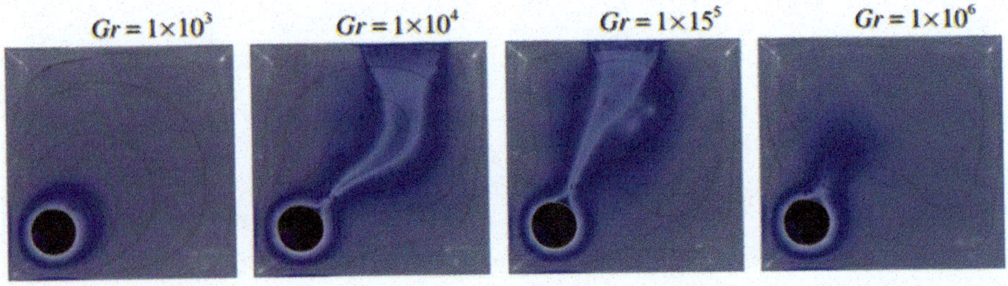

Figure 2.7 Isotherm contours at increasing Grashof numbers for natural convection in the annulus between an eccentric heated circular cylinder and a cooled square enclosure.

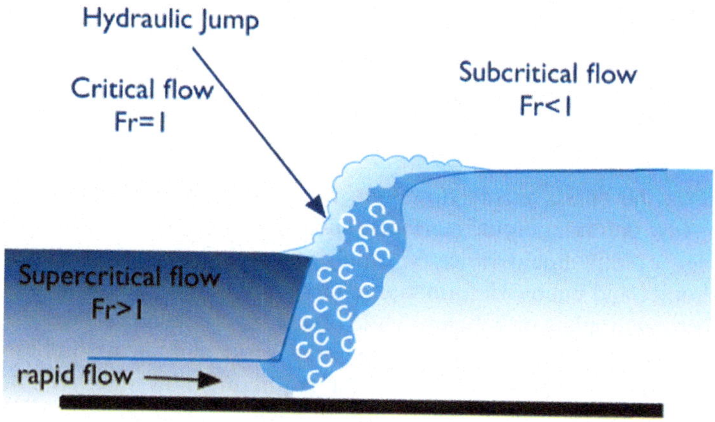

Figure 2.8 Free-surface flow regimes showing subcritical, critical, and supercritical states characterised by the Froude number.

When interfacial phenomena become important, as in sloshing with air entrainment, foaming, or emulsification, surface tension introduces additional force balances. The Bond number Bo compares gravitational forces to surface tension forces and therefore quantifies whether interface curvature is controlled primarily by weight or by molecular cohesion.

$$Bo = \frac{\Delta\rho g L^2}{\xi} \tag{2.66}$$

Here, $\Delta\rho$ is the density difference between the two fluid phases.

In small containers or under microgravity conditions, the Bond number is low, and surface tension dominates, causing interfaces to adopt shapes that minimise surface area. In large tanks under terrestrial gravity, the Bond number is high, and gravity flattens interfaces. Sloshing problems often span intermediate Bond numbers, where gravity shapes the overall wave motion while surface tension controls local curvature, wave breaking, and droplet formation.

Closely related is the Weber number *We*, which compares inertial forces to surface tension forces.

$$We = \frac{\rho u^2 L}{\xi} \tag{2.67}$$

The Weber number is particularly important when interfaces deform rapidly, such as during violent sloshing, jet breakup, or droplet ejection from a free-surface. High Weber numbers indicate that inertia overcomes surface tension, leading to fragmentation, splashing, and atomisation. In contrast, low Weber numbers imply that surface tension suppresses interface deformation. In sloshing tanks, the Weber number helps determine whether surface waves remain coherent or disintegrate into droplets and sprays during impact events.

Viscous effects at interfaces are captured by the Capillary number *Ca*, which compares viscous stresses to surface tension.

$$Ca = \frac{\mu \overline{u}}{\xi} \tag{2.68}$$

Here, $\overline{u}$ is the characteristic flow velocity. This number is central in problems involving slow deformation of interfaces, such as coating flows, bubble motion, and the drainage of thin liquid films. In emulsions and foams, the Capillary number governs droplet and bubble deformation under shear. Low Capillary numbers imply nearly spherical droplets or bubbles, while higher values lead to elongation, breakup, or coalescence. In sloshing systems containing surfactants or multiphase mixtures, the Capillary number becomes critical for predicting how internal flows modify interfacial area and stability.

Thermal effects introduce additional dimensionless groups. The Prandtl number *Pr* compares momentum diffusivity to thermal diffusivity and controls the relative thickness of velocity and thermal boundary layers.

$$Pr = \frac{\mu C_p}{k} \tag{2.69}$$

Although sloshing is often treated as isothermal, temperature gradients can arise in cryogenic tanks, nuclear systems, or energy storage applications. In

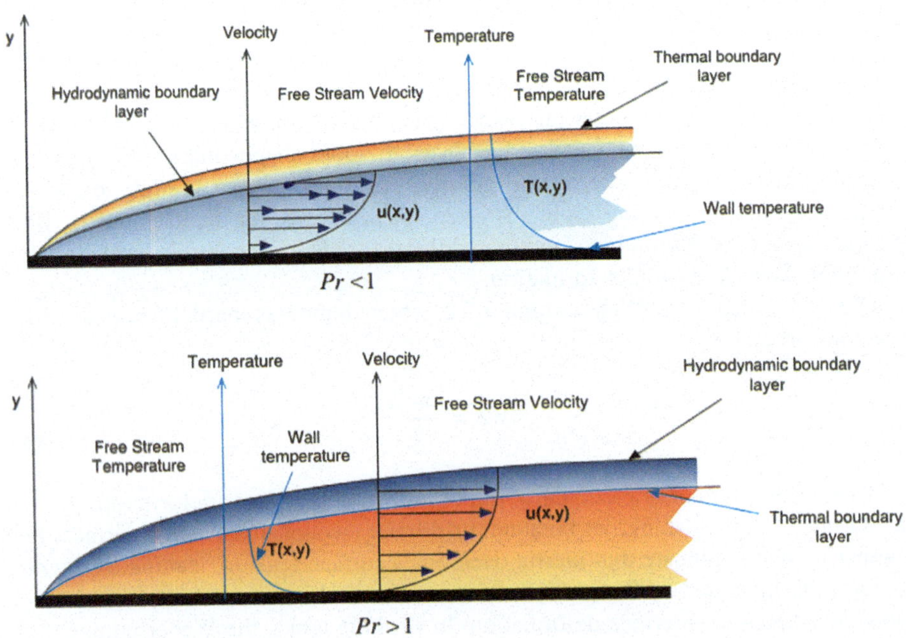

Figure 2.9 Comparison of thermal and hydrodynamic boundary-layer thicknesses for systems with Prandtl numbers less than one and greater than one.

such cases, the Prandtl number influences heat transfer rates at walls and interfaces, indirectly affecting fluid properties such as viscosity and surface tension (illustrated in Fig. 2.9).

Coupled with the Nusselt number Nu, which quantifies convective heat transfer relative to conduction, these parameters determine thermal response under oscillatory motion.

$$\mathrm{Nu} = \frac{\text{Convective Heat Transfer}}{\text{Conductive Heat Transfer}} = \frac{\mathrm{hL}}{k} \tag{2.70}$$

where h is the convective heat transfer coefficient.

In multiphase systems such as emulsions and foams, dimensionless numbers acquire additional significance because multiple interfaces and phases coexist. The Ohnesorge number Oh, which combines viscosity, surface tension, and density, characterises the relative importance of viscous dissipation during droplet or bubble deformation.

$$Oh = \frac{\mu}{\sqrt{\rho \xi L}} \tag{2.71}$$

It is particularly useful for predicting whether breakup leads to clean fragmentation or is damped by viscosity. In foams, where thin liquid films separate gas bubbles, dimensionless film thickness and drainage numbers determine stability and ageing behaviour. These parameters dictate how quickly liquid drains from Plateau borders, how films thin and rupture, and how foam structure evolves over time.

Surfactants add further complexity by altering interfacial properties dynamically. Dimensionless groups incorporating surface elasticity and adsorption kinetics describe how surface tension varies with deformation rate. In sloshing systems containing surfactants, these effects can significantly enhance damping by suppressing interfacial motion and increasing energy dissipation. Dimensionless numbers thus provide a framework for incorporating chemical and physicochemical effects into hydrodynamic analysis without resorting to case-specific empirical descriptions.

Beyond interpretation, dimensionless numbers are indispensable for scaling laws and model testing. Laboratory experiments on sloshing are often conducted at a reduced-scale for practical reasons. By matching key dimensionless numbers, particularly the Froude and Weber numbers, researchers ensure dynamic similarity between model and full-scale systems. This approach allows reliable extrapolation of experimental results to real tanks, vehicles, or industrial equipment. In numerical simulations, dimensionless formulation improves generality and numerical stability, enabling results to be interpreted independently of specific unit systems.

In emulsions and foams, dimensionless numbers also guide formulation and processing. Droplet size distributions, stability against coalescence, and rheological response under flow can often be correlated with Capillary, Weber, and Reynolds numbers. This provides a rational basis for process design in mixing, homogenisation, spraying, and foaming operations. Rather than relying solely on trial and error, engineers can predict regime transitions and optimise operating conditions by targeting specific nondimensional ranges.

2.10.1 Physical interpretation of dimensionless numbers

Dimensionless numbers arise naturally when governing equations are nondimensionalised using characteristic scales of length, velocity, time, and force. Each such number represents a ratio of competing physical effects, rather than an abstract mathematical construct.

For example, the Reynolds number expresses the ratio of inertial forces to viscous forces. Physically, it measures the ability of fluid inertia to overcome molecular momentum diffusion. At low Reynolds numbers, viscous stresses rapidly damp velocity gradients, leading to smooth and orderly motion. At high Reynolds numbers, inertia dominates, allowing disturbances to grow and giving rise to turbulence. The Reynolds number therefore acts as a control parameter for flow regime rather than merely a scaling constant.

Similarly, the Froude number compares inertial forces to gravitational forces. In free-surface flows, this ratio governs the interaction between fluid motion and gravity-driven wave propagation. When inertia is weak compared to gravity, surface disturbances remain small and adjust rapidly. When inertia becomes dominant, surface deformation increases, and large-amplitude waves or breaking may occur. In this sense, the Froude number encapsulates the competition between kinetic energy and gravitational potential energy.

The Weber number compares inertial forces to surface tension forces. Its physical meaning lies in the ability of the flow to deform interfaces. Low Weber numbers indicate that surface tension resists deformation, preserving smooth interfaces. High Weber numbers imply that inertia overwhelms surface tension, leading to breakup, splashing, or atomisation. In sloshing problems involving small containers or low-gravity environments, surface tension may become as important as gravity, making the Weber number a key descriptor.

Other dimensionless numbers, such as the Mach number or Strouhal number, similarly express ratios between characteristic velocities, wave speeds, and time scales. What unifies all dimensionless numbers is that they identify the dominant mechanisms controlling the flow, allowing complex behaviour to be understood in terms of a small number of governing parameters.

2.11 Similarity principles in fluid mechanics

Similarity principles formalise the idea that two flows can be considered dynamically equivalent if the relevant dimensionless parameters are the same. These principles underpin model testing, experimental scaling, and the generalisation of numerical results.

Similarity is traditionally divided into three categories: geometric, kinematic, and dynamic similarity. Each represents a progressively stronger requirement.

2.11.1 Geometric similarity

Geometric similarity requires that all corresponding dimensions of two systems are proportional by a constant scale factor. Shapes, angles, and relative positions must be preserved. For a model tank and a full-scale tank to be geometrically similar, their length ratios, aspect ratios, and free-surface shapes must be identical when scaled.

Geometric similarity alone is insufficient to guarantee similar flow behaviour, but it is a necessary prerequisite. Without geometric similarity, the spatial distribution of forces and boundary conditions differs, invalidating any direct comparison. In sloshing experiments, geometric similarity ensures that the container shape, fill depth ratio, and free-surface curvature are scaled consistently.

2.11.2 Kinematic similarity

Kinematic similarity requires that the velocity field in the model is a scaled version of that in the prototype. This implies similarity of streamlines, pathlines, and time-dependent motion patterns. Velocities at corresponding points must maintain a constant ratio, and time scales must scale consistently.

Kinematic similarity ensures that the flow evolves in a similar manner, but it does not guarantee that the forces responsible for that motion are balanced in the same way. As a result, kinematic similarity is rarely sufficient on its own for predictive modelling in fluid mechanics.

2.11.3 Dynamic similarity

Dynamic similarity is the strongest and most physically meaningful requirement. It demands that the ratios of all relevant forces acting on corresponding fluid elements are identical in the model and prototype. This condition is achieved when all governing dimensionless numbers are matched.

For viscous internal flows, dynamic similarity requires matching the Reynolds number. For free-surface flows, matching the Froude number is often essential. When surface tension plays a role, the Weber number must also be considered. In compressible flows, the Mach number becomes critical.

In practice, it is often impossible to match all dimensionless numbers simultaneously, especially when multiple physical mechanisms are active. This limitation leads to the need for prioritisation based on the dominant physics of the problem.

2.11.4 Scaling of model experiments

Model experiments are widely used to investigate fluid behaviour that is difficult or expensive to study at full scale. Dimensional analysis provides the framework for scaling such experiments meaningfully.

In a typical scaling exercise, characteristic length, velocity, and time scales are chosen, and the governing dimensionless numbers are identified. The experiment is then designed to match the most important of these numbers between the model and the prototype. Secondary effects may be distorted, but their influence is assessed and minimised.

For example, in laboratory sloshing experiments, gravity-driven wave motion is often the primary concern. In such cases, Froude similarity is prioritised, ensuring that the ratio of inertial to gravitational forces is preserved. Reynolds number similarity may not be fully achieved due to viscosity differences or practical constraints, but as long as the flow remains in a regime where viscous effects are secondary, the results remain valid.

Scaling also affects time. Under Froude similarity, time scales with the square root of length, meaning that oscillation frequencies in a scaled model are higher than in the full-scale system. This has important implications for interpreting experimental data and comparing it with numerical simulations.

2.11.5 Application to sloshing and free-surface flows

Sloshing dynamics represents a particularly rich application of dimensional analysis and similarity principles because multiple physical mechanisms interact simultaneously. The dominant nondimensional parameters depend on the specific regime of interest.

In gravity-dominated sloshing, the Froude number governs wave propagation, resonance frequencies, and free-surface deformation. Matching the Froude number ensures that the balance between inertia and gravity is preserved, leading to dynamically similar wave patterns and impact loads.

Viscous effects become important in damping sloshing motion and controlling energy dissipation. The Reynolds number influences boundary-layer development, flow separation near walls, and decay rates of oscillations. In many practical sloshing problems, Reynolds numbers are sufficiently high that viscous effects act mainly as a secondary damping mechanism.

Surface tension becomes relevant when characteristic length scales are small, such as in microgravity environments, small containers, or partially filled tanks with shallow depths. In such cases, the Weber number determines the stability of the free-surface and the formation of capillary waves. Neglecting surface tension under these conditions can lead to significant modelling errors.

For violent sloshing involving wave breaking or liquid impact on tank walls, additional parameters related to impact pressure and air entrainment may become important. While these phenomena are difficult to scale perfectly, dimensional analysis still provides guidance on which effects can be meaningfully reproduced in reduced-scale experiments.

2.11.6 Limitations of dimensional analysis

Despite its power, dimensional analysis is not a universal solution. It relies on several assumptions and has inherent limitations that must be acknowledged.

Firstly, dimensional analysis does not provide numerical values for coefficients. Empirical constants and closure relations must be determined experimentally or through detailed simulations. Dimensionless numbers identify relevant mechanisms but do not fully specify system behaviour.

Secondly, similarity may breakdown when multiple competing mechanisms are equally important. In sloshing problems involving gravity, viscosity, surface tension, and compressibility simultaneously, it is often impossible to match all relevant dimensionless numbers in a single experiment. This forces compromises and introduces scale distortions.

Thirdly, dimensional analysis assumes that the governing physics is known and complete. If an important mechanism is overlooked, the resulting dimensionless groups may be incomplete or misleading.

Finally, some phenomena are inherently scale-dependent. Turbulence, multiphase interactions, and air–liquid coupling can exhibit behaviour that does not scale cleanly with simple dimensionless parameters.

Here is a rewritten concluding remarks section, clearer and more direct about what follows, while keeping a natural, human, and academically modest tone.

2.12 Concluding remarks

This chapter has presented the fundamental physical principles of fluid dynamics that underpin sloshing phenomena, focusing on the behaviour of a single liquid phase within a partially filled container. The governing concepts of mass and momentum conservation, pressure distribution, free-surface evolution, and interfacial effects have been introduced to establish a clear physical understanding of fluid motion under external excitation. These principles form the first and essential layer in the description of sloshing dynamics.

The treatment in this chapter has intentionally been limited to the fluid phase in order to isolate the dominant mechanisms governing liquid motion and free-surface deformation. This provides a necessary reference framework before considering more complex configurations encountered in practical systems. In real engineering applications, sloshing often involves additional phases, including gas–liquid interfaces, multiple liquid phases, and interactions with deformable or flexible solid boundaries. The presence of these phases introduces new sources of coupling, energy exchange, and dynamic feedback that cannot be captured by fluid mechanics alone.

Building on the fluid phase foundations established here, Chapter 3 advances the discussion by introducing the solid-phase foundations relevant to sloshing dynamics. The next chapter focuses on the essential principles of solid mechanics required to describe structural response, load transmission, and fluid–structure interaction, thereby extending the analysis to systems where sloshing involves the coupled behaviour of liquid and solid phases.

References

Al-Darweesh, J., Saleh Aljawad, M., Shahzad Kamal, M., Mahmoud, M., & Al-Yousef, Z. (2024). The impact of chelating agent pH on the stability and viscosity of CO_2 foam under harsh reservoir conditions. *Journal of Molecular Liquids, 412*, 125847. https://doi.org/10.1016/j.molliq.2024.125847.

Barker, J. A., & Henderson, D. (1976). What is "liquid"? Understanding the states of matter. *Reviews of Modern Physics, 48*, 587. https://doi.org/10.1103/RevModPhys.48.587.

Bhattad, A. (2023). Review on viscosity measurement: Devices, methods and models. *Springer Journal of Thermal Analysis and Calorimetry, 148*(14), 6527–6543. https://doi.org/10.1007/s10973-023-12214-0.

Brush, S. G. (1962). Theories of liquid viscosity. *Chemical Reviews, 62*(6), 513–548. https://doi.org/10.1021/cr60220a002.

Durst, F. (2008). *Fluid mechanics: An introduction to the theory of fluid flows* (1st ed.). Berlin, Heidelberg: Springer, 1–723. 10.1007/978-3-540-71343-2.

Fuller, G. G., & Vermant, J. (2012). Complex fluid-fluid interfaces: Rheology and structure. *Annual Review of Chemical and Biomolecular Engineering, 3*, 519–543. https://doi.org/10.1146/annurev-chembioeng-061010-114202.

Gholizadeh, H., Burton, R., & Schoenau, G. (2011). Fluid bulk modulus: A literature survey. *International Journal of Fluid Power, 12*(3), 5–15. https://doi.org/10.1080/14399776.2011.10781033.

Gholizadeh, H., Burton, R., & Schoenau, G. (2012). Fluid bulk modulus: Comparison of low pressure models. *International Journal of Fluid Power, 13*(1), 7–16. https://doi.org/10.1080/14399776.2012.10781042.

Hutter, K., & Wang, Y. (2016). *Fluid and thermodynamics: Volume 1: Basic fluid mechanics*. Cham: Springer. https://doi.org/10.1007/978-3-319-33633-6.

Jou, D., Casas-Vázquez, J., & Criado-Sancho, M. (2010). *Thermodynamics of fluids under flow* (2nd ed.). Dordrecht: Springer. https://doi.org/10.1007/978-3-319-33633-6.

Kowalewski, T. A., Nakielski, P., Pierini, F., Zembrzycki, K., & Pawłowska, S. (2016). Micro and nano fluid mechanics. In M. Kleiber, T. Burczynski, K. Wilde, J. Gorski, K. Winkelmann, & L. Smakosz (Eds.). *Advanced mechanics: Theoretical, computational and interdisciplinary issues* (pp. 27–34). London: CRC Press. 10.1201/b20057-7.

Krishna, R., Urseanu, M. I., van Baten, J. M., & Ellenberger, J. (1999). Rise velocity of a swarm of large gas bubbles in liquids. *Chemical Engineering Science, 54*(2), 171–183. https://doi.org/10.1016/s0009-2509(98)00245-0.

Llewellin, E. W., Mader, H. M., & Wilson, S. D. R. (2002). The rheology of a bubbly liquid. *Proceedings of the Royal Society A, 458*, 987–1016. https://doi.org/10.1098/rspa.2001.0924.

McClements, D. J., & Jafari, S. M. (2018). Improving emulsion formation, stability and performance using mixed emulsifiers: A review. *Advances in Colloid and Interface Science, 251*, 55–79. https://doi.org/10.1016/j.cis.2017.12.001.

Müller, M., Solenthaler, B., Keiser, R., & Gross, M. (2005). Particle-based fluid-fluid interaction. In *SCA '05: Proceedings of the 2005 ACM SIGGRAPH/Eurographics symposium on Computer animation*. New York, NY: Association for Computing Machinery, 237–244. https://doi.org/10.1145/1073368.1073402.

Nazari, F., Potanin, A., & Mohammadigoushki, H. (2025). Flow of a two-dimensional liquid foam: Impact of surfactant type and boundary conditions. *Journal of Non-Newtonian Fluid Mechanics, 339-340*, 105417. https://doi.org/10.1016/j.jnnfm.2025.105417.

Oettel, M., & Dietrich, S. (2008). Colloidal interactions at fluid interfaces. *Langmuir: the ACS Journal of Surfaces and Colloids, 24*(4), 1425–1441. https://doi.org/10.1021/la702794d.

Orvalho, S., Ruzicka, M. C., Olivieri, G., & Marzocchella, A. (2015). Bubble coalescence: Effect of bubble approach velocity and liquid viscosity. *Chemical Engineering Science, 134*, 205–216. https://doi.org/10.1016/j.ces.2015.04.053.

Saleh, T. A. (2022). Overview of surface and interface science. *Interface Science and Technology, 34*, 1–38. https://doi.org/10.1016/B978-0-12-849876-7.00001-4.

Smoluchowski, M. V. (1918). Versuch einer mathematischen Theorie der Koagulationskinetik kolloider Lösungen. *Zeitschrift für Physikalische Chemie, 92U*(1), 129–168. https://doi.org/10.1515/zpch-1918-9209.

Song, H. (2018). *Engineering fluid mechanics*. Singapore: Springer, 1–256. https://doi.org/10.1007/978-981-13-0173-5.

Squires, T. M., & Quake, S. R. (2005). Microfluidics: Fluid physics at the nanoliter scale. *Reviews of Modern Physics, 77*(3), 977–1026. https://doi.org/10.1103/RevModPhys.77.977.

Stemmet, C. P., Bartelds, F., van der Schaaf, J., Kuster, B. F. M., & Schouten, J. C. (2008). Influence of liquid viscosity and surface tension on the gas–liquid mass transfer coefficient for solid foam packings in co-current two-phase flow. *Chemical Engineering Research and Design, 86*(10), 1094–1106. https://doi.org/10.1016/j.cherd.2008.04.007.

Stevenson, P. (2012). *Foam engineering: Fundamentals and applications*. Wiley. 10.1002/9781119954620.

Sun, J., Liu, S., Liao, G., Wang, S., Lu, Y., Wu, C., He, Y., Sun, N., & Li, W. (2025). Influencing factors and predictive models of oil–water emulsions: A comprehensive review and future outlook. *Annals of the New York Academy of Sciences, 1552*(1), 94–116. https://doi.org/10.1111/nyas.70071.

Tabor, D. (2012). *Gases, Liquids and Solids and Other States of Matter* (3rd ed.). Cambridge University Press. https://doi.org/10.1017/CBO9781139170253.

Tayurskii, D., & Le Mehaute, A. (2011). The concept of temperature in the modern physics. In J. C. Moreno Pirajn (Ed.). *Thermodynamics - Systems in equilibrium and non-equilibrium* (pp. 21–34). InTech. https://doi.org/10.5772/22082.

Thomson, W. (1872). On the equilibrium of vapour at a curved surface of liquid. *Proceedings of the Royal Society of Edinburgh, 7*, 63–68. https://doi.org/10.1017/s0370164600041729.

Zhang, D., Qi, H., Xu, J., Yang, Y., & Fu, B. (2019). Modeling and analysis of aeroelasticity and sloshing for liquid rocket. *IEEE Access, 7*, 4453–4465. https://doi.org/10.1109/ACCESS.2018.2884544.

Fundamentals of solid mechanics

3

3.1 Introduction

Sloshing is fundamentally a fluid-dynamic phenomenon, governed by the motion of a free surface under external excitation and constrained by container geometry. At its core, it is the fluid that stores, transports, and dissipates energy. Nevertheless, in all realistic engineering systems, the fluid does not exist in isolation. It is contained, guided, and constrained by a surrounding structure whose mechanical response may, in turn, influence the evolution of the fluid motion. This reciprocal interaction between fluid and structure forms an essential part of the overall physics of sloshing and motivates the inclusion of solid mechanics and structural vibration theory in the present work (Hung et al., 1992; Hung & Long, 1996; Babu & Bhattacharyya, 1996).

In many practical applications—such as fuel tanks, containment vessels, transportation systems, and marine structures—the structural components enclosing the fluid are designed to be sufficiently stiff that their deformation is small when compared with the characteristic dimensions of the container. For this reason, sloshing analyses are often initiated under the assumption of rigid walls, allowing the dominant fluid behaviour to be studied in isolation. This idealisation is both convenient and, in many cases, adequate for preliminary analysis. However, as excitation levels increase, container walls become thinner, or operational frequencies approach structural natural modes, the mechanical response of the structure can no longer be ignored. Even modest structural flexibility may alter boundary conditions perceived by the fluid, modify resonance characteristics, and introduce additional pathways for energy exchange and dissipation.

The purpose of this chapter is therefore not to elevate solid mechanics to the same conceptual level as fluid physics in sloshing analysis, but rather to provide a coherent and self-contained overview of the structural principles that underpin fluid–structure interaction in sloshing systems. The solid is treated as a secondary but indispensable component of the coupled problem. Its role is to respond to fluid-induced pressures, transmit forces to supports, and, under certain conditions, participate dynamically in the system response through vibration and deformation.

To achieve this objective, the chapter begins with a concise but rigorous formulation of stress, strain, and constitutive behaviour within the framework of linear elasticity. These concepts are introduced not as abstract mathematical constructs, but as physical measures of how structural components react locally to the pressures generated by moving fluids. The governing equations of equilibrium and compatibility are presented to establish the conditions under which stress and deformation fields are physically admissible, thereby providing the necessary background for subsequent dynamic formulations (Gantenbein & Buland, 1991; Hung & Pan, 1995b, 1996).

Slosh Dynamics. DOI: https://doi.org/10.1016/B978-0-443-44330-5.00001-3

The variational principles of solid mechanics are then introduced, including the principle of virtual work, the principle of minimum total potential energy, and Hamilton's principle. These formulations are particularly important in the context of sloshing, as they provide a unified and systematic framework for deriving equations of motion for coupled systems. Variational approaches are also the natural foundation for many modern numerical methods and reduced-order models used in fluid–structure interaction, and their inclusion here ensures conceptual continuity with later developments (Calayir & Dumanoğlu, 1993; Hung et al., 1990; Hung & Pan, 1994, 1995a; Schulkes, 1992).

A distinction is made between discrete and continuous representations of structural dynamics. From a modelling perspective, discrete systems offer simplicity, transparency, and computational efficiency. They allow complex structures to be idealised as assemblies of lumped masses, springs, and dampers, capturing the dominant dynamic behaviour with minimal mathematical overhead. In sloshing analysis, such models are frequently employed to represent the gross motion of tanks, supports, and mounts, and to introduce simplified representations of fluid–structure coupling. For this reason, discrete vibration theory is treated explicitly and illustrated in a manner consistent with typical engineering practice.

At the same time, it is recognised that discretisation is an approximation whose validity depends on the problem at hand. Certain structural components – most notably tank walls, baffles, and panels – are inherently continuous, and their deformation cannot always be represented adequately by a small number of lumped parameters. In these cases, the structural response must be described using continuum mechanics, leading naturally to plate and shell theories. This is particularly relevant in sloshing problems involving thin-walled containers, where structural vibration modes may interact with fluid free-surface modes, sometimes leading to coupled resonances or amplification phenomena (Calayir & Dumanoğlu, 1993; Hung et al., 1990; Hung & Pan, 1994, 1995a; Schulkes, 1992).

Accordingly, the chapter introduces the linear vibration theory of plates within the classical thin-plate framework. The governing equations are derived systematically from equilibrium and energy principles, culminating in the biharmonic plate equation and its dynamic extension. Although the mathematical development is necessarily formal, emphasis is placed on physical interpretation and modelling relevance, especially in relation to distributed hydrodynamic loading and modal behaviour. The treatment is intended to equip the reader with sufficient theoretical understanding to recognise when continuum effects are important and how they may be incorporated into sloshing analyses.

3.2 Stress strain

Consider an elastic body of arbitrary solid phase under external forces that are in static equilibrium. Select any material point within the interior of the body, and introduce a Cartesian coordinate system with axes x, y, and z. Around this point, imagine an infinitesimally small parallelepiped element (dx, dy, dz) whose faces are aligned with the coordinate planes. The internal forces transmitted across each face of this element are described by *stresses*, which represent the intensity of force per unit area at that location.

Each stress vector acting on a face can be decomposed into two components:

1. Normal stress—acting perpendicular to the face.
2. Shear stress—acting tangentially to the face.

For a given face, three stress components act simultaneously (as shown in Fig. 3.1). In the standard notation, the first subscript specifies the direction of the outward normal to the face under consideration, and the second subscript specifies the direction of the stress vector itself. This collection of stress components, defined at every point in the material, forms the stress tensor, which fully characterises the state of stress at that point.

For brevity, repeated subscripts will be omitted in subsequent discussions. Thus normal stresses will be denoted with a single subscript indicating their direction (e.g. σ_{xx} for normal stress in the x-direction). The following sign convention will be used: a positive normal stress corresponds to tension, while a negative value corresponds to compression.

For shear stresses, the sign depends on the relationship between the direction of the outer normal to the face and the direction of the shear component on that face. If both the outer normal and the shear stress vector point in either the positive or negative directions of the coordinate axes, the shear stress is positive. If one points in a positive direction and the other in a negative direction, the shear stress is negative. Under this convention, all stress components are positive.

On any face, the three stress components (one normal and two shear) combine to form a surface traction vector. The complete set of these components acting on the faces of the infinitesimal element defines the stress tensor, $\boldsymbol{T}_\sigma$:

$$\boldsymbol{T}_\sigma = \begin{bmatrix} \sigma_x & \tau_{xy} & \tau_{xz} \\ \tau_{yx} & \sigma_y & \tau_{yz} \\ \tau_{zx} & \tau_{zy} & \sigma_z \end{bmatrix} \tag{3.1}$$

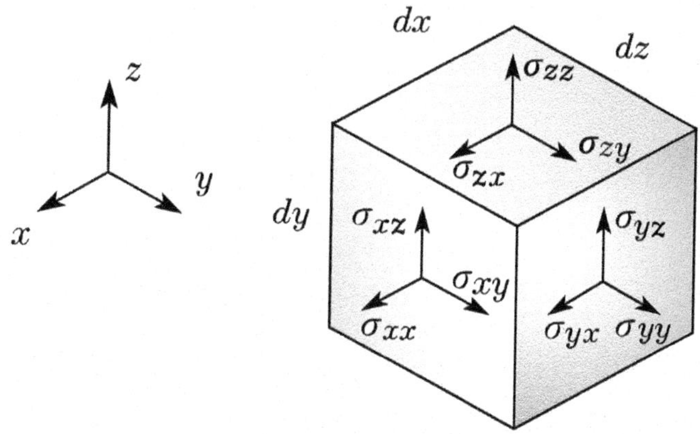

Figure 3.1 Stress components acting on an infinitesimal material element in a Cartesian coordinate system, illustrating normal and shear stress components on mutually perpendicular faces and the symmetry of the stress tensor.

The stress tensor is symmetric with respect to its principal diagonal due to the reciprocity of shear stresses:

$$\tau_{xy} = \tau_{yx},\ \tau_{xz} = \tau_{zx},\ \tau_{yz} = \tau_{zy} \tag{3.2}$$

This symmetry arises from the requirement that the net torque on the infinitesimal element must vanish in static equilibrium.

Consider the elastic body in Fig. 3.2, which is supported so that rigid body motions both translations and rotations – are completely prevented. When external forces –are applied, the body undergoes deformation, and each material point experiences a small elastic displacement. For example, let point M have initial coordinates $(x,\ y,\ z)$ in the undeformed configuration. After deformation, it moves to a new position M' with coordinates:

$$x' = x + u,\ y' = y + v,\ z' = z + w \tag{3.3}$$

Here, u, v, and w are the components of the displacement vector MM' along the x, y, and z axes, respectively. In the general case, these displacement components vary with position and are functions $u(x,\ y,\ z)$, $v(x,\ y,\ z)$, and $w(x,\ y,\ z)$

Consider an infinitesimal material element in the form of a rectangular parallelepiped centred at a generic point in the body. When the deformation of the body is sufficiently small, the change in shape of this element can be described as a superposition of six independent elementary deformation modes. These modes represent the most fundamental ways in which a small material element can deform and together provide a complete kinematic description of infinitesimal deformation.

The first three deformation modes correspond to extension or contraction of the element along the three Cartesian coordinate directions. In these modes, the edges of the element remain mutually orthogonal, but their lengths change. Such deformations are characterised by the normal, or linear, strain components. For an edge initially aligned with the x-direction, the normal strain is defined as the relative change in length of that edge. Denoting the original length by dx and the change in length by du_x, the normal strain in the x-direction is defined as

$$\varepsilon_{xx} = \frac{du_x}{dx} \tag{3.4}$$

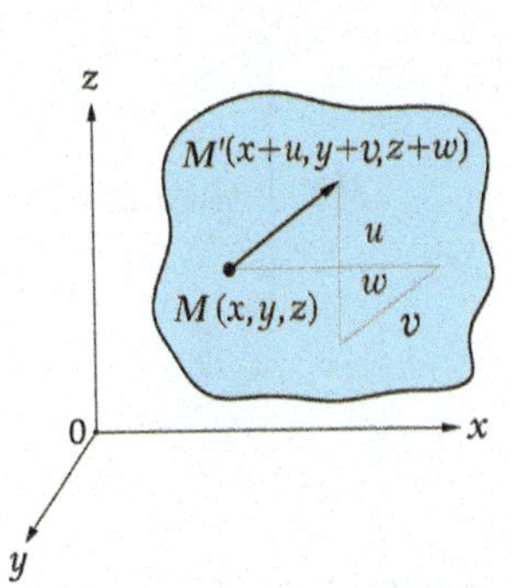

Figure 3.2 Displacement of a material point in an elastic body, showing the transition from the undeformed configuration to the deformed configuration under applied loading.

Analogous definitions apply to the normal strains in the y- and z-directions, denoted by ε_{yy} and ε_{zz}, respectively. These quantities measure the local stretching or compression of the material along each coordinate axis.

For small deformations, the displacement field may be assumed to vary smoothly in space, allowing the displacement components to be expanded in a first-order Taylor series about the centre of the material element. Under this approximation, the normal strain components can be expressed directly in terms of displacement gradients. Specifically, if u_x, u_y, and u_z denote the displacement components in the x, y, and z directions, the normal strains are given by

$$\begin{Bmatrix} \varepsilon_{xx} \\ \varepsilon_{yy} \\ \varepsilon_{zz} \end{Bmatrix} = \begin{Bmatrix} \dfrac{\partial u_x}{\partial x} \\ \dfrac{\partial u_y}{\partial y} \\ \dfrac{\partial u_z}{\partial z} \end{Bmatrix} \tag{3.5}$$

These expressions provide a clear physical interpretation: a normal strain represents the spatial rate of change of displacement along the same direction.

In addition to changes in length, a material element may also deform through changes in shape without a change in volume. These deformations are described by shear strains and constitute the remaining three elementary deformation modes. Shear strains measure the change in angle between two line elements that were initially perpendicular. Each shear strain is associated with a specific coordinate plane and is denoted by γ_{xy}, γ_{yz}, or γ_{zx}, where the subscripts indicate the plane in which the angular distortion occurs.

To illustrate the nature of shear deformation (illustrated in Fig. 3.3), consider two material line elements initially aligned with the x- and y-directions, respectively. In the undeformed configuration, these lines intersect at a right angle. After deformation, this angle generally changes due to displacements occurring in both coordinate directions. The shear strain in the xy-plane is defined as the total change in this right angle. This change arises from two contributions: one due to the variation of the x-displacement in the y-direction, and the other due to the variation of the y-displacement in the xxx-direction.

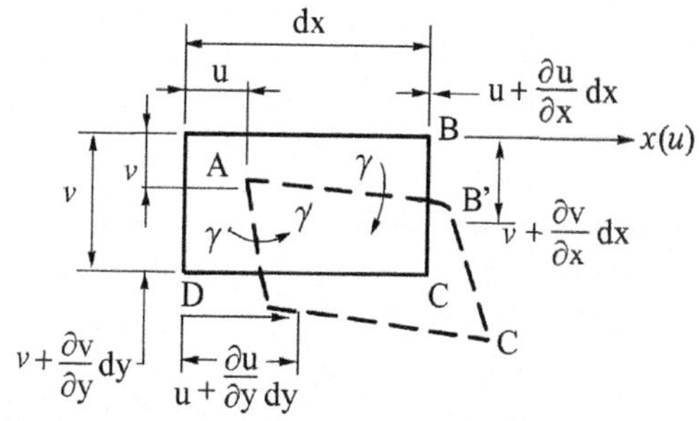

Figure 3.3 Kinematic description of an infinitesimal element under small deformation, indicating normal strains and shear strains arising from displacement gradients.

Let the initial right angle between the two line elements be denoted by $\pi/2$, and let the angle after deformation be reduced or increased by small angular changes α' and α'', associated with the rotations of the two line elements. The total change in angle is therefore $\alpha' + \alpha''$. For infinitesimal deformations, these angular changes are small, and the engineering shear strain in the xy-plane is defined as

$$\gamma_{xy} = \alpha' + \alpha'' \tag{3.6}$$

To express this in terms of the in-plane displacement components $u(x, y)$ and $v(x, y)$, we examine the change in orientation of the edges:

$$\gamma_{xy} = \frac{\frac{\partial v}{\partial x}}{1 + \frac{\partial u}{\partial x}} + \frac{\frac{\partial u}{\partial y}}{1 + \frac{\partial v}{\partial y}} \tag{3.7}$$

Because we are considering small deformations, the displacement gradients $\frac{\partial u}{\partial y}$ and $\frac{\partial v}{\partial y}$ are much smaller than unity, allowing us to neglect their contributions in the denominators and products. Under this assumption, the shear strain simplifies to the familiar symmetric form:

$$\gamma_{xy} = \frac{\partial v}{\partial x} + \frac{\partial u}{\partial y} \tag{3.8}$$

This expression shows that shear strain in the *xy*-plane is the sum of the angular distortions produced by displacement gradients in the two orthogonal directions.

In a similar manner, the shear strain in *xz* and *yz* planes as

$$\gamma_{xz} = \frac{\partial u}{\partial z} + \frac{\partial w}{\partial x} \tag{3.9}$$

$$\gamma_{yz} = \frac{\partial v}{\partial z} + \frac{\partial w}{\partial y} \tag{3.10}$$

analogous to the stress tensor, the complete state of strain at a point can be represented by the strain tensor:

$$T_\varepsilon = \begin{bmatrix} \varepsilon_x & \frac{1}{2}\gamma_{xy} & \frac{1}{2}\gamma_{xz} \\ \frac{1}{2}\gamma_{yx} & \varepsilon_y & \frac{1}{2}\gamma_{yz} \\ \frac{1}{2}\gamma_{zx} & \frac{1}{2}\gamma_{zy} & \varepsilon_z \end{bmatrix} \tag{3.11}$$

3.3 Constitutive equations

The constitutive equations form the essential link between kinematics and statics in solid mechanics, establishing how a material responds to applied loads by relating the stress components to the corresponding strains. While the equilibrium equations enforce force balance and the kinematic relations describe admissible deformations, the constitutive law captures the intrinsic mechanical behaviour of the material itself.

For a linear elastic, isotropic solid undergoing small deformations, this relationship is expressed through the generalised form of Hooke's law. Linearity implies that stress is proportional to strain, while isotropy indicates that the material response is identical in all directions, independent of the orientation of the coordinate system. Under these assumptions, the stress–strain relationship is fully characterised by two independent elastic constants. In three dimensions, the normal strain components are influenced not only by the normal stress acting in the same direction but also by the normal stresses acting in the orthogonal directions. This coupling reflects the Poisson effect, whereby stretching in one direction produces contraction in the transverse directions. The normal strain components are therefore related to the normal stresses by:

$$\begin{Bmatrix} \varepsilon_x \\ \varepsilon_y \\ \varepsilon_z \end{Bmatrix} = \begin{Bmatrix} \frac{1}{E}(\sigma_x - \upsilon(\sigma_y + \sigma_z)) \\ \frac{1}{E}(\sigma_y - \upsilon(\sigma_x + \sigma_z)) \\ \frac{1}{E}(\sigma_z - \upsilon(\sigma_y + \sigma_x)) \end{Bmatrix} \tag{3.12}$$

where E is Young's modulus and υ is Poisson's ratio. Young's modulus quantifies the stiffness of the material in uniaxial loading, while Poisson's ratio measures the extent of lateral contraction or expansion associated with axial deformation. These equations demonstrate that normal strains are not independent but are intrinsically coupled through the material's resistance to volumetric change.

The shear behaviour of an isotropic elastic solid is simpler in form. Shear stresses produce shear strains without inducing normal strains, and the relationship between them is governed by the shear modulus G. The engineering shear strains are related to the corresponding shear stresses by

$$\begin{Bmatrix} \gamma_{xy} \\ \gamma_{yz} \\ \gamma_{zx} \end{Bmatrix} = \begin{Bmatrix} \frac{1}{G}\tau_{xy} \\ \frac{1}{G}\tau_{yz} \\ \frac{1}{G}\tau_{zx} \end{Bmatrix} \tag{3.13}$$

The shear modulus represents the material's resistance to shape change at constant volume and is therefore associated with distortional deformation. In isotropic materials, the shear response is identical in all coordinate planes, reflecting the absence of preferred material directions.

Although several elastic constants are commonly used to describe material behaviour, they are not independent for an isotropic solid. Any two elastic constants uniquely determine all others. In particular, Young's modulus, Poisson's ratio, and the shear modulus are related by

$$G = \frac{E}{2(1 + \upsilon)} \tag{3.14}$$

3.4 Equilibrium equations

Equilibrium equations and compatibility equations. The stress components in a continuum must satisfy the differential equations of equilibrium, which are derived from force balance on an infinitesimal element. Using the reciprocity of shear stresses, the three-dimensional equilibrium equations can be expressed as:

Here, F_x,F_y, and F_z represent body force densities (e.g. gravitational or electromagnetic forces).

The local state of stress in a three-dimensional body is fully characterised by six independent stress components, namely the three normal stresses and the three shear stresses. However, the fundamental equations of equilibrium, expressed in the absence of body couples, provide only three independent scalar equations. These equilibrium relations, commonly written in differential form as

$$\frac{\partial \sigma_x}{\partial x} + \frac{\partial \sigma_x}{\partial x} + \frac{\partial \tau_{zx}}{\partial x} + F_x = 0 \tag{3.15}$$

$$\frac{\partial \sigma_y}{\partial y} + \frac{\partial \sigma_{yx}}{\partial x} + \frac{\partial \tau_{yz}}{\partial z} + F_y = 0 \tag{3.16}$$

$$\frac{\partial \sigma_z}{\partial z} + \frac{\partial \sigma_{zx}}{\partial x} + \frac{\partial \tau_{zy}}{\partial z} + F_z = 0 \tag{3.17}$$

Consequently, since equilibrium state supplies only three equations for six unknown stress components, the stress problem is underdetermined when equilibrium alone is considered. This situation is often described, in an informal sense, as statically indeterminate, although it is more precise to state that equilibrium conditions are insufficient to ensure a unique mechanical solution in a continuum setting.

To resolve this indeterminacy, additional physical constraints must be imposed. These constraints do not arise from statics, but from kinematics, specifically from the requirement that the deformation of the body be geometrically admissible. In physical terms, this means that the material must deform in a continuous manner, without gaps, overlaps, or interpenetration, and that the strain field must be derivable from a single-valued and continuous displacement field. Let the displacement vector be denoted by $\boldsymbol{u} = (u_1, u_2, u_3)$. Under the assumption of infinitesimal deformations, the components of the strain tensor are related to the displacement field through the linearised strain–displacement relations,

$$\varepsilon_{ij} = \frac{1}{2}\left(\frac{\partial u_i}{\partial x_j} + \frac{\partial u_j}{\partial x_i}\right), \ (i, j = 1, 2, 3) \tag{3.18}$$

While they ensure that a given displacement field produces a corresponding strain field, the inverse is not automatically guaranteed: not every arbitrary strain field corresponds to a physically admissible displacement field. This observation leads to the concept of strain compatibility.

Geometric compatibility requires that the strain field be consistent with a continuous displacement field throughout the body. Mathematically, this condition is enforced by eliminating the displacement components from the strain-displacement relations through successive differentiation and algebraic manipulation. By differentiating the strain components appropriately and combining the resulting expressions, one obtains a set of differential constraints involving only the strain components and their spatial derivatives. These constraints are known as the compatibility equations. For small strain theory in three dimensions, the compatibility conditions can be written in the general form

$$\frac{\partial^2 \varepsilon_{xx}}{\partial y^2} + \frac{\partial^2 \varepsilon_{yy}}{\partial x^2} - 2\frac{\partial^2 \varepsilon_{xy}}{\partial x \partial y} = 0 \tag{3.19}$$

$$\frac{\partial^2 \varepsilon_{yy}}{\partial z^2} + \frac{\partial^2 \varepsilon_{zz}}{\partial y^2} - 2\frac{\partial^2 \varepsilon_{yz}}{\partial y \partial z} = 0 \tag{3.20}$$

$$\frac{\partial^2 \varepsilon_{zz}}{\partial x^2} + \frac{\partial^2 \varepsilon_{xx}}{\partial z^2} - 2\frac{\partial^2 \varepsilon_{zx}}{\partial z \partial x} = 0 \tag{3.21}$$

These relations express the requirement that the second derivatives of the strain components satisfy specific symmetry conditions. Physically, they ensure that the strains do not imply contradictory deformations when integrated over space. If these equations are violated, the resulting strain field would correspond to a displacement field that is multivalued or discontinuous, which is physically inadmissible for a coherent solid.

The compatibility equations play a crucial role in closing the system of governing equations in elasticity. When combined with the equilibrium equations and an appropriate constitutive relation, such as Hooke's law for linear elastic materials,

$$\sigma_{ij} = C_{ijkl}\varepsilon_{kl} \tag{3.22}$$

where C_{ijkl} is the fourth order elasticity tensor, the full set of equations becomes sufficient to determine the stress, strain, and displacement fields uniquely, subject to boundary conditions. In this sense, compatibility does not directly provide additional equations for the stress components, but instead constrains the admissible strain fields, which subsequently restrict the stress field through the constitutive law.

It is important to emphasise that compatibility is a purely geometric requirement and is independent of material behaviour. Whether the material is elastic, viscoelastic, or plastic, the strain field must always satisfy compatibility if it is to correspond to a continuous displacement field. In linear elasticity, compatibility equations are particularly transparent and take the compact differential form shown above. In more advanced theories involving finite deformations, the compatibility conditions become more complex and are embedded within the nonlinear kinematic relations.

3.5 The stress function concept (Airy stress function)

In the framework of elastic stress analysis, one is confronted with a system of 15 unknown field quantities: the six independent components of the Cauchy stress tensor

σ_{ij}, the six components of the infinitesimal strain tensor ε_{ij}, and the three displacement components uiu_iui. Each of these is a function of the spatial coordinates (x,y,z) in Cartesian form or (r,θ,z) in cylindrical polar coordinates. The governing equations—comprising equilibrium, compatibility, and constitutive relations—are highly coupled and nonlinear in their most general form, making closed-form solutions tractable only for a restricted class of problems (Pister, 1954; Steinhardt & Chaudhari, 1981). To render these equations solvable, simplifying assumptions are often introduced based on the geometry, loading, and boundary conditions. For instance, in thin-walled pressure vessels, the plane stress approximation is sufficiently accurate, while in the case of a long, pressurised thick-walled cylinder, the plane strain condition is appropriate. Such assumptions reduce a three-dimensional elasticity problem to a two-dimensional formulation. Even then, the governing partial differential equations remain formidable. It was in this context that Sir George Biddell Airy introduced the *stress function*φ, a scalar potential from which the in-plane stress components can be obtained through the relations

$$\sigma_{xx} = \frac{\partial^2 \varphi}{\partial y^2} \tag{3.23}$$

$$\sigma_{yy} = \frac{\partial^2 \varphi}{\partial x^2} \tag{3.24}$$

$$\tau_{xy} = \frac{\partial^2 \varphi}{\partial x \partial y} \tag{3.25}$$

It should be emphasised that the Airy stress function φ not only satisfies the two-dimensional equilibrium equations

$$\frac{\partial \sigma_{xx}}{\partial x} + \frac{\partial \tau_{xy}}{\partial y} + F_x = 0 \tag{3.26}$$

$$\frac{\partial \tau_{xy}}{\partial x} + \frac{\partial \sigma_{yy}}{\partial y} + F_y = 0 \tag{3.27}$$

but must also satisfy the compatibility conditions, which ensure that the strain field corresponds to a continuous and single-valued displacement field. For two-dimensional elasticity, the compatibility equation reduces to

$$\frac{\partial^2 \varepsilon_{xx}}{\partial y^2} + \frac{\partial^2 \varepsilon_{yy}}{\partial x^2} - 2\frac{\partial^2 \gamma_{xy}}{\partial x \partial y} = 0 \tag{3.28}$$

Substituting strain state into the compatibility equation and combining with the equilibrium equations to eliminate the shear stress τ_{xy} gives the governing partial differential equation:

$$\frac{\partial^2}{\partial x^2}(\sigma_{xx} + \sigma_{yy}) + \frac{\partial^2}{\partial y^2}(\sigma_{xx} + \sigma_{yy}) = 0 \tag{3.29}$$

when body forces F_x and F_y are constant. A similar derivation holds for the plane stress case, leading to the same biharmonic-type equation for $\sigma_{xx} + \sigma_{yy}$. This formalism demonstrates that the stress function method elegantly reduces the problem to a single scalar equation, ensuring both equilibrium and compatibility, and thus providing a physically consistent stress field in two-dimensional elasticity.

This equation is commonly referred to as the *Laplace differential equation* or the *harmonic differential equation*. The function $(\sigma_{xx} + \sigma_{yy})$ is termed a *harmonic* function. It is important to note that the Laplace equation, which inherently includes the previous equations, does not explicitly depend on the elastic constants of the material. This observation carries significant implications for experimentalists: provided that geometric similarity, material isotropy, linearity, and similar applied loading exist between a model and its prototype, the stress distribution per unit load will be identical in both cases.

The stress function, as previously defined, must satisfy the *Laplace equation*:

$$\left(\frac{\partial^2}{\partial x^2} + \frac{\partial^2}{\partial y^2}\right)\left(\frac{\partial^2}{\partial x^2} + \frac{\partial^2}{\partial y^2}\right)\varphi = 0 \tag{3.30}$$

or equivalently:

$$\frac{\partial^4 \varphi}{\partial x^4} + 2\frac{\partial^4 \varphi}{\partial x^2 \partial y^2} + \frac{\partial^4 \varphi}{\partial y^4} = 0 \tag{3.31}$$

This can also be expressed in the more compact form:

$$\nabla^4 \varphi = 0 \tag{3.32}$$

which indicates that the stress function φ must be a biharmonic function. The state equation ($\nabla^4 \varphi = 0$) is frequently called the *biharmonic equation*, with φ often referred to as the *Airy stress function*.

At this stage, it is noteworthy that in-plane stress or plane strain problems, the task reduces to determining a biharmonic function φ such that the stress components satisfy the boundary conditions of the system.

Hence, if an appropriate polynomial expression in x and y (or in r and θ) is chosen for the stress function φ, both equilibrium and compatibility conditions are automatically satisfied. By considering the boundary conditions specific to the stress system, the coefficients of the polynomial can be determined, yielding a complete solution.

Importantly, the biharmonic equation itself does not 'fail' at singularities. Rather, it correctly predicts that no bounded stress solution exists under the imposed boundary conditions, which is a physically meaningful result. The singularity indicates a breakdown of the continuum description at atomic length scales, not a failure of the governing equation. Within the framework of linear elastic fracture mechanics, the biharmonic equation provides the exact asymptotic stress field near crack tips and remains valid up to distances where material nonlinearities or damage mechanisms become significant.

3.6 Virtual work

In solid mechanics, many physical laws can be expressed in alternative but equivalent forms, such as the principle of conservation of linear momentum, which states that the vector sum of all applied forces acting on a body must equal the total time rate of change of momentum, commonly known as Newton's second law, and which can also be derived from a variational principle. The traditional use of Newton's laws to formulate the governing equations of a structural problem requires isolating a typical volume element with its applied and reactive forces, a process that becomes increasingly cumbersome and intractable for complex systems, especially when the boundary conditions to be applied alongside the derived equations are unclear. In contrast, the variational approach provides governing equations through the principle of virtual displacements or by minimising the total potential energy of the system, an approach applicable to both linear and nonlinear theories, and particularly useful for deriving governing equations, boundary conditions, and approximate solutions via variational methods. Within this framework, the principle of virtual displacements is applied to develop the equations of motion of laminated plates, making it essential to review the concepts of virtual displacements and forces. From a geometrical standpoint, a mechanical system can adopt multiple possible configurations consistent with its geometric constraints, but only one corresponds to the actual equilibrium configuration satisfying Newton's second law. The complete set of configurations that satisfy geometric constraints, though not necessarily equilibrium, is known as the set of admissible configurations, which are obtained through infinitesimal variations of the true configuration, while keeping the geometric constraints and applied forces unchanged at their fixed values. When a system undergoes such hypothetical variations, it is said to experience virtual displacements, which are distinct from actual displacements caused by changes in loading; these variations are purely imagined but consistent with equilibrium requirements, and are therefore termed 'virtual'. The displacements at constrained boundaries are fixed and thus vanish, while within the system, the hypothetical displacements need not correspond to actual displacements. The work performed by real forces acting through such virtual displacements is called virtual work, and mathematically, the virtual work done by the actual forces $\boldsymbol{F}$ within a body Ω due to the virtual displacements $\delta\boldsymbol{u}$ is expressed as

$$\delta W = \int_{\Omega} \boldsymbol{F}\delta\boldsymbol{u}dV \tag{3.33}$$

In the material body Ω, where dV denotes the volume element, the external virtual work done by body forces $\boldsymbol{F}$ per unit volume and surface tractions $\boldsymbol{t}$ per unit area of the boundary Γ through virtual displacements $\delta\boldsymbol{u}$ is expressed as

$$\delta V = -\left(\int_{\Omega} \boldsymbol{F}\delta\boldsymbol{u}dV + \int_{\Gamma} \boldsymbol{t}\delta\boldsymbol{u}dS\right) \tag{3.34}$$

where dS denotes a surface element and Γ indicates the portion of the boundary where stresses are prescribed. The negative sign reflects that the work is performed on the body, and the displacements on the remaining portion of the boundary are fixed, making the virtual displacements vanish there regardless of whether the prescribed displacement is zero or nonzero. To compute the internal virtual work due to virtual displacement δu, consider an infinitesimal element of volume undergoing virtual strains $\delta \varepsilon_{ij}$ induced by displacements δu_i where

$$\delta \varepsilon_{ij} = \frac{1}{2}(\delta u_{i,j} + \delta u_{j,i}),\ \delta u_{i,j} = \frac{\partial \delta u_i}{\partial x_j} \tag{3.35}$$

For example, the work done by stress σ_{11} during virtual $\delta u_1 = \delta \varepsilon_{11} dx_1$ is given by $\sigma_{11} dx_2 dx_3 \cdot \delta \varepsilon_{11} dx_1 = \sigma_{11} \delta \varepsilon_{11} dV$, and similarly, the work done by stress σ_{12} corresponds to $\tau_{12} dx_2 dx_3 \cdot 2\gamma \varepsilon_{11} dx_1 = \tau_{12} 2\delta \varepsilon_{12} dV$. Therefore the total internal virtual work done by all stresses within the element as they move through their corresponding virtual strains is expressed as

$$\int (\sigma_{11}\delta\varepsilon_{11} + \sigma_{22}\delta\varepsilon_{22} + \cdots + 2\sigma_{12}\delta\gamma_{12})dV = \int \sigma_{ij}\delta\varepsilon_{ij} dV \tag{3.36}$$

The total internal virtual work can be obtained by integrating the expression for stress and virtual strain throughout the body, giving

$$\delta U = \int_{\Omega} \sigma_{ij}\delta\varepsilon_{ij} dV \tag{3.37}$$

This formulation is universally valid for any deformable body, regardless of its constitutive behaviour, and the quantity is referred to as the *virtual strain energy*. Alternatively, the concept of virtual work can also be expressed in terms of virtual stresses $\delta\sigma_{ij}$ acting through the actual strains ε_{ij}, leading to:

$$\delta U^* = \int_{\Omega} \varepsilon_{ij}\delta\sigma_{ij} dV \tag{3.38}$$

which is identified as the *virtual complementary strain energy*. In this case, the variations in forces and tractions (δf_i, δt_i) and stresses $\delta\sigma_{ij}$ must satisfy both the equilibrium equations within the body and the prescribed boundary conditions on the stress surface, expressed as

$$\delta\sigma_{ij}|_j + \delta f_i = 0 on \Omega \tag{3.39}$$

$$\delta t_i = \delta\sigma_{ji} n_i = 0 on \Gamma \tag{3.40}$$

However, in the present context, the analysis does not make use of complementary energy principles.

The concept of a variational operator is now introduced. When applied to virtual displacements and forces, the operator – denoted by the symbol δ – represents a variation or change in a quantity. For example, δu signifies a virtual or imagined change in the displacement u. In this sense, the variational operator functions in a manner similar to the differential operator d, which represents an infinitesimal change in a variable such as x. The variational operator is particularly useful in the derivation of governing equations and in reformulating principles of virtual work into precise mathematical expressions.

To illustrate, consider a function F that depends on a variable u and its $u' = \frac{du}{dx}$. For a fixed x, the total differential of F can be expressed as

$$dF = \frac{\partial F}{\partial x}dx + \frac{\partial F}{\partial u}du + \frac{\partial F}{\partial u'}du' \tag{3.41}$$

The first variation of F, on the other hand, is written as:

$$\delta F = \frac{\delta F}{\delta u}\cdot\delta u + \frac{\delta F}{\delta u'}\cdot\delta u' + \frac{1}{2}\frac{\delta^2 F}{\delta u^2}\cdot(\delta u)^2 + \frac{1}{2}\frac{\delta^2 F}{\delta u'^2}\cdot(\delta u')^2 + . \tag{3.42}$$

Because δu represents an infinitesimal variation, all higher-order terms involving squares or higher powers of δu may be neglected. Thus the first variation of F reduces to

$$\delta F = \frac{\partial F}{\partial u}\delta u + \frac{\partial F}{\partial u'}\delta u' \tag{3.43}$$

Since x is held fixed during the variation of u to $u + \delta u$, we have dx . Comparing δF with dF reveals an important point: the variational operator δ is essentially a differential operator, but it acts with respect to the dependent variable, u. Consequently, the rules governing variations of sums, products, ratios, powers, and similar forms mirror the usual rules of differentiation. The following identities illustrate the fundamental properties of the variational operator:

$$\delta(\nabla u) = \nabla(\delta u) \tag{3.44}$$

$$\delta\left(\int_\Omega u d\Omega\right) = \int_\Omega \delta u d\Omega \tag{3.45}$$

$$\delta(F_1 \mp F_2) = \delta F_1 \mp \delta F_2 \tag{3.46}$$

3.7 Principle of virtual displacements

The concept of *virtual displacement* refers to an imagined infinitesimal change in the configuration of a body that is fully consistent with the imposed geometric constraints, but does not correspond to an actual motion in time. The principle of virtual displacements states that the external work done by the applied forces during such virtual displacements must balance the internal virtual work of the system, provided the body is in equilibrium.

To illustrate, consider a rigid body subjected to a system of applied forces $F_1, \ldots, F_n$. Let the points of application of these forces undergo small, arbitrary, but geometrically admissible displacements $\delta u_1, \ldots, \delta u_n$ The external virtual work corresponding to these displacements is expressed as

$$\delta\Theta = -\sum_{i=1}^{n} F_i \delta u_i \tag{3.47}$$

Since a perfectly rigid body cannot undergo deformation, the internal virtual work δU vanishes, because no virtual strains are generated. Furthermore, in the case of a rigid body, all virtual displacements are identical, i.e., $\delta u_1 = \delta u_2 = \ldots = \delta u_n = \delta u$. Substituting this into the above relation yields

$$\delta\Theta = -\left(\sum_{i=1}^{n} F_i\right)\cdot\delta u, \ \delta U = 0 \tag{3.48}$$

According to Newton's second law, the vector sum of all forces acting on a body in static equilibrium must be zero. Consequently, the total virtual work, given by the sum of external and internal contributions $\delta\Theta + \delta U$, must vanish. Therefore for a body in equilibrium, the *total virtual work due to any admissible virtual displacement is equal to zero*.

The condition that the total virtual work associated with virtual displacements vanishes is referred to as the principle of virtual displacements. This principle is not restricted to rigid systems; it extends naturally to continuous, deformable media, where in general $\delta U \neq 0$. Its importance lies in its central role in the formulation of structural and continuum theories (such as plate and shell theories) and in the development of variational approaches for approximate solutions.

Consider a continuous body B in equilibrium under the combined influence of body forces f and prescribed surface tractions t. Let the undeformed reference configuration be denoted by C^0, with volume Ω_0. Assume that the total boundary Γ of Ω_0 is partitioned into two disjoint subsets:

- On Γ_u, the displacements are prescribed as $\overline{u}$,
- On Γ_t, the tractions are specified as $\overline{t}$.

Thus the complete boundary is written as $\Gamma = \Gamma_u \cup \Gamma_t$ with $\Gamma_u \cap \Gamma_t = \varnothing$.

Let u represent the displacement field of the equilibrium configuration, while σ and ε denote the corresponding stress and strain tensors. The admissible displacement fields are defined as sufficiently smooth functions satisfying the geometric boundary condition

$$u = \overline{u} on \Gamma_u \tag{3.49}$$

If equilibrium holds, then among all admissible configurations, the actual displacement field minimises the virtual work, resulting in the total virtual work being identically zero. To derive the governing equations, we allow the body to undergo a virtual displacement δu from the equilibrium configuration C. These variations δu are arbitrary continuous functions, subject only to the homogeneous geometric boundary condition $\delta u = 0$ on Γ_u.

The principle of virtual displacements can be concisely expressed as:

$$\delta W = \delta U + \delta\Theta = 0 \tag{3.50}$$

That is for a continuous body in equilibrium, the total virtual work of all internal and external forces corresponding to any admissible virtual displacement must vanish.

This statement is completely general: it applies to elastic as well as inelastic continua, and to both linear and nonlinear material behaviour.

For a solid body, the external and internal virtual work contributions may be written as:

$$\int_{\Omega_0} \sigma : \delta\varepsilon dV - \int_{\Omega_0} f : \delta u dV - \int_{\Gamma_t} t : \delta u dS = 0 \tag{3.51}$$

Here:
$\sigma : \delta\varepsilon$ denotes the double contraction (or inner product) of the stress tensor with the virtual strain tensor, and Ω_0 is the volume of the undeformed body.

In Cartesian tensor notation, the above equation can be expressed as:

$$\int_{\Omega_0} (\sigma_{ij}\delta\varepsilon_{ij} - f_i\delta u_i)dV - \int_{\Gamma_t} t_i\delta u_i dS \tag{3.52}$$

with the usual summation convention over repeated indices. The Euler–Lagrange equations derived correspond directly to the classical equilibrium equations of continuum mechanics. Hence, the principle of virtual displacements provides a unifying variational framework for obtaining the governing equations of equilibrium.

Within the framework of three-dimensional elasticity, the strain–displacement relations provide the link between the virtual displacements δu_i and the corresponding virtual strains $\delta\varepsilon_{ij}$. These are given by

$$\delta\varepsilon_{ij} = \frac{1}{2}(\delta u_{i,j} + \delta u_{j,i}),\ \delta u_{i,j} = \frac{\partial \delta u_i}{\partial x_j} \tag{3.53}$$

Substituting the above expression for $\delta\varepsilon_{ij}$ into the Cartesian tensor notation, and applying the divergence theorem to transfer derivatives from δu_i to their coefficients, one obtains (with considering Principal stress state at which $\sigma_{ij} = \sigma_{ji}$):

$$\int_{\Omega_0} \left(\frac{1}{2}\sigma_{ij}(\delta u_{i,j} + \delta u_{j,i}) - f_i\delta u_i\right)dV - \int_{\Gamma_\sigma} t_i\delta u_i dS = 0 \tag{3.54}$$

Since the total boundary $\Gamma = \Gamma_u \cup \Gamma_a$ and $\delta u_i = 0$ on Γ_u, the expression simplifies to:

$$\int_{\Omega_0} (\sigma_{ij,j} + f_i)\delta u_i dV + \int_{\Gamma_0} (\sigma_{ij}n_j + t_i)\delta u_i dS = 0 \tag{3.55}$$

Because the virtual displacements δu_i are arbitrary within Ω_0 and on Γ_σ, it leads directly to the following relations:

$$\sigma_{ij,j} + f_i = 0 in \Omega_0 \tag{3.56}$$

$$\sigma_{ij}n_j - t_i = 0 in \Gamma_\sigma \tag{3.57}$$

Eqs. (3.56) and (3.57) represent the Euler–Lagrange equations corresponding to the principle of virtual displacements for a body undergoing infinitesimal deformation. The boundary conditions are the natural boundary conditions. Importantly, the validity of the virtual displacement principle is completely general: it applies to continuous bodies regardless of the specific constitutive law, whether the material behaviour is elastic, inelastic, linear, or nonlinear.

3.8 Principle of minimum total potential energy

An important special case of the principle of virtual displacements, particularly for elastic as well as nonlinear elastic bodies, is known as the principle of minimum total potential energy. For an elastic body (assuming no thermal effects), one can define a *strain energy density function* U_0 such that

$$\sigma = \frac{\partial U_0}{\partial \varepsilon} \tag{3.58}$$

This relation constitutes the constitutive law of an elastic material. The function U_0 is a scalar-valued quantity depending only on the strain state at a point, and is assumed to be positive definite.

By substituting the strain energy density into the expression of the virtual work principle, the governing relation can be expressed in terms of U_0

$$\int_{\Omega_0} \frac{\partial U_0}{\partial \varepsilon} : \delta\varepsilon dV - \int_{\Omega_0} f \cdot \delta u dV - \int_{\Omega_0} t \cdot \delta u dS = 0 \tag{3.59}$$

The first integral corresponds to:

$$\delta U = \int_{\Omega_0} \delta U_0 dV \tag{3.60}$$

where U is the internal strain energy functional, defined as:

$$U = \int_{\Omega_0} U_0 dV \tag{3.61}$$

Thus the principle of minimum total potential energy states that, for an elastic system in equilibrium, the total potential energy functional – comprising the internal strain energy and the potential of the external forces – assumes a minimum value.

Suppose there exists a potential function Θ whose first variation is expressed as

$$\delta\Theta = -\int_{\Omega_0} f_i \delta u_i dV - \int_{\Gamma_\sigma} t_i \delta u_i dS \tag{3.62}$$

In this form, the principle of virtual work can be rewritten as

$$\delta\Pi = \delta U + \delta\Theta = 0 \tag{3.63}$$

Here, the functional $\delta\Pi$ is defined as the total potential energy of the elastic body. This expression is commonly known as the principle of minimum total potential energy.

This principle states that among all admissible displacement fields, the true equilibrium displacement u is the one that minimises the total potential energy functional. In mathematical terms, this can be written as

$$\Pi(u) \le \Pi(\overline{u}) \tag{3.64}$$

where $\overline{u}$ represents any admissible (kinematically possible) displacement field. Equality holds only when $\overline{u} - u = 0$.

3.9 Hamilton's principle

Hamilton's principle can be regarded as a generalisation of the principle of virtual displacements, extended to dynamical systems. The principle states that when a system evolves in time, its motion is governed by the balance between two fundamental energy contributions: the kinetic energy $\Bbbk$ and the potential energy Π. For deformable continua, both energy terms can be expressed in terms of the problem's dependent variables, which are functions of both spatial position and time. In this way, Hamilton's principle can be viewed as the dynamic analogue of the virtual work principle.

Newton's second law of motion, when applied to deformable bodies, provides the global balance of linear momentum. While it represents the conservation of linear momentum for a continuous medium, it is insufficient on its own to fully describe the body's motion, because additional constitutive equations and kinematic relations are required.

In general form, Newton's second law for a continuous body may be expressed as

$$\boldsymbol{F} = m\boldsymbol{a} \tag{3.65}$$

where mmm denotes the mass, $\boldsymbol{a}$ the acceleration vector, and $\boldsymbol{F}$ the resultant of all external forces acting on the body of mass m. The actual trajectory of a material point in the body is represented as $u = u(\boldsymbol{x}, t)$, which is consistent with prescribed kinematic (essential) boundary conditions. To apply Hamilton's principle, we consider a variation of this path, $u + \delta u$, where δu represents an admissible (virtual) displacement.

The admissible variation δu is assumed to coincide with the actual path at the time interval, $t \in [t_1, t_2]$. Thus the conditions imposed on δu are:

$$\forall\, t,\ \delta u = 0 on S \tag{3.66}$$

$$\forall\, \boldsymbol{x},\ \delta u(\boldsymbol{x}, t_1) = \delta u(\boldsymbol{x}, t_2) = 0 \tag{3.67}$$

These conditions ensure that variations vanish at the temporal boundaries, while remaining arbitrary within the time interval (t_1, t_2). Consequently, Hamilton's principle provides a variational framework that combines both kinetic and potential energies to describe the true motion of the system.

Here, S_1 denotes the portion of the boundary where the displacement vector u is prescribed. Notice that the scalar product with δu corresponds to the virtual work at a point x, since $\boldsymbol{F}$, a, and u are vector quantities, while the resulting product is scalar. By integrating this contribution over the body volume (and surface), we obtain the total virtual work performed by all points of the medium.

The virtual work done on the body at a given time t by the resultant forces acting through the virtual displacement δu\delta $u\delta u$ can be written as

$$\int_V f.\, \delta u dV + \int_{S_2} \hat{t}.\, \delta u dS - \int_V \sigma\colon \delta\varepsilon dV \tag{3.68}$$

where $\boldsymbol{f}$ is the body force per unit volume, $\hat{\boldsymbol{t}}$ is the prescribed traction vector, and $\boldsymbol{\sigma}$, and $\boldsymbol{\varepsilon}$ are the stress and strain tensors, respectively. The final term represents the virtual work of internal stresses stored in the body. The virtual strains $\boldsymbol{\delta\varepsilon}$ are assumed to satisfy the compatibility conditions implied by the strain–displacement relations.

The virtual work associated with the inertial forces $\boldsymbol{ma}$ acting through the displacement δu is expressed as

$$\int_V \rho \frac{\partial^2 u}{\partial t^2} \cdot \delta u dV \tag{3.69}$$

where ρ denotes the mass density of the medium (which may vary with position).

Combining these expressions over the time interval (t_1, t_2), we obtain

$$\int_{t_1}^{t_2}\left\{\int_V \rho\frac{\partial^2 u}{\partial t^2}\cdot\delta u dV - \left(\int_V (f\cdot\delta u - \sigma: \delta\varepsilon)dV + \int_{S_2} \hat{t}.\ \delta u dS\right)\right\} = 0 \tag{3.70}$$

The above equations provide the general expression of Hamilton's principle for a continuum medium, applicable irrespective of whether the system is conservative or nonconservative, and valid for both elastic and inelastic materials. In variational form, Hamilton's principle states that the actual motion of the system renders the action integral – which consists of the kinetic energy $\Bbbk$, potential of external forces Θ, and strain energy U – stationary.

$$\int_{t_1}^{t_2} \{\Bbbk - (\Theta + U)\}dt = 0 \tag{3.71}$$

3.10 Vibration of discrete systems

In engineering practice, the process of building a mathematical model for a mechanical system is often closer to an art of approximation than an exact science. Real-world systems are typically so complex that representing every detail with absolute accuracy is not feasible, and more importantly, not necessary. The goal of modelling is usually to capture the essential features that determine the system's performance or safety. In preliminary design stages, engineers are mainly concerned with verifying whether a proposed system can meet specific performance requirements. In other situations, only certain aspects of the system's response need to be predicted, so the physical system is represented by a simplified model tailored for that purpose.

A model can therefore be regarded as an approximation of the physical reality, and its usefulness depends on how well it preserves the fundamental dynamic properties of the real system. The central idea is that the mathematical behaviour of the model should mimic, at least to a reasonable extent, the observed response of the actual system under comparable conditions.

When focusing on vibrations, mechanical systems can generally be divided into two categories: lumped-parameter models and distributed-parameter models. In lumped-parameter systems, the physical quantities such as mass, stiffness, and damping are assumed to be concentrated at discrete points, and the equations of motion are expressed using a finite number of coordinates. Distributed-parameter models, in contrast, treat the system as continuous, where properties like mass and elasticity are spread throughout the body, leading to partial differential equations. In practice, the same physical system can sometimes be represented in either form depending on the purpose of the analysis, meaning that the classification is not rigid but often subjective.

To illustrate, consider the case of a fuel tank mounted to an automobile body through rubber isolators, modelled as a lumped-parameter system with base excitation from the chassis. Let the tank (treated as a rigid body in vertical translation) have mass m_t, and let the N rubber mounts be idealised by linear springs k_i and viscous dampers

c_i acting in parallel at the attachment points. For pure vertical motion, the mounts combine to equivalent parameters

$$k_{eq} = \sum_{i=1}^{N} k_i \tag{3.72}$$

And

$$c_{eq} = \sum_{i=1}^{N} c_i \tag{3.73}$$

Denote by $z(t)$ the vertical displacement of the tank's centre of mass and by $y(t)$ the imposed vertical motion (road/chassis input) at the mount bases. The equation of motion with base excitation is

$$m_t z'' + c_{eq}(z' - y') + k_{eq}(z - y) = 0 \tag{3.74}$$

which is the canonical single degree of freedom isolator form. Define the natural frequency:

$$\omega_n = \sqrt{\frac{k_{eq}}{m_t}} \tag{3.75}$$

and damping ratio:

$$\varsigma = \frac{c_{eq}}{\sqrt{m_t k_{eq}}} \tag{3.76}$$

For harmonic base motion:

$$y(t) = Ye^{i\omega t} \tag{3.77}$$

The steady state complex amplitude ratio, commonly referred to as displacement transmissibility, is defined as the ratio between the amplitude of the system response and the amplitude of the imposed base excitation under harmonic steady state conditions.

$$\frac{Z}{Y} = \frac{1 + i2\varsigma r}{1 - r^2 + i2\varsigma r} \tag{3.78}$$

where

$$r = \frac{\omega}{\omega_n} \tag{3.79}$$

and the magnitude is

$$\left|\frac{Z}{Y}\right| = \sqrt{\frac{1 + (2\varsigma r)^2}{(1 - r^2)^2 + (2\varsigma r)^2}} \tag{3.80}$$

The force transmitted to the body is

$$F_T k_{eq}(z - y) + c_{eq}(z' - y') \tag{3.81}$$

its amplitude relative to the static load $k_{eq}Y$ is the classical force transmissibility

$$T_F = \left|\frac{F_T}{k_{eq}Y}\right| = \sqrt{\frac{(1 - r^2)^2 + 4\varsigma^2 r^2}{1 + 4\varsigma^2 r^2}} \tag{3.82}$$

Isolation in the high-frequency regime ($r \gg 1$) is improved by lowering ω_n (softer mounts or larger m_t) while maintaining adequate ς to control resonance near $r \approx 1$. To capture fluid–structure interaction from fuel slosh, include a sloshing degree of freedom $x(t)$ representing the dominant lateral free-surface mode, modelled as an internal mass m_s connected to the tank via a spring–damper pair (k_s, c_s) (these arise from hydrostatic restoring and hydrodynamic dissipation and can be estimated from tank geometry and fill level). The minimal two-degree-of-freedom model then reads

$$\begin{cases} m_t z'' c_{eq}(z' - y') + k_{eq}(z - y) + c_s(z' - x') + k_s(z - x) = 0 \\ m_s x'' c_s(x' - z') + k_s(x - z) = 0 \end{cases} \tag{3.83}$$

or, in matrix form with $\boldsymbol{q} = [zx]^T$ and base input y,

$$\boldsymbol{M}\boldsymbol{q}'' + \boldsymbol{C}\boldsymbol{q}' + \boldsymbol{K}\boldsymbol{q} = \begin{bmatrix} c_{eq} \\ 0 \end{bmatrix} y' + \begin{bmatrix} k_{eq} \\ 0 \end{bmatrix} y \tag{3.84}$$

where M and C are presented in Eqs. (3.83), and (3.84), respectively.

$$\boldsymbol{M} = \begin{bmatrix} m_t & 0 \\ 0 & m_s \end{bmatrix} \tag{3.85}$$

$$\boldsymbol{C} = \begin{bmatrix} c_{eq} + c_s & -c_s \\ -c_s & c_s \end{bmatrix} \tag{3.86}$$

The coupled eigenproblem $\det(\boldsymbol{K} - \omega^2\boldsymbol{M}) = 0$ yields two natural frequencies: a rigid-like tank bounce dominated by m_t and k_{eq}, and a sloshing-dominated mode governed by m_s and k_s. When the road/engine excitation frequency ω approaches either eigenfrequency, the corresponding mode is strongly excited; proper selection of isolator stiffness m_t and k_s and damping ς shifts the bounce resonance away from dominant input spectra, while (k_s, c_s) depend on fill height and tank geometry, often varying during operation (illustrated in Fig. 3.4).

3.11 Linear vibration of plates

In the preceding sections, vibration problems were formulated by modelling mechanical systems as finite, discrete assemblies of masses and stiffness elements. Such discretisation provides an efficient and accurate means of analysing a wide range of practical problems while keeping mathematical complexity to a minimum. However, for certain classes of structures, the assumption of discreteness is no longer adequate, and the inherent continuity of the material must be taken into account. A prominent example is the vibration of plates, where deformation and inertia are distributed continuously over the structural domain. In applications, such as fuel

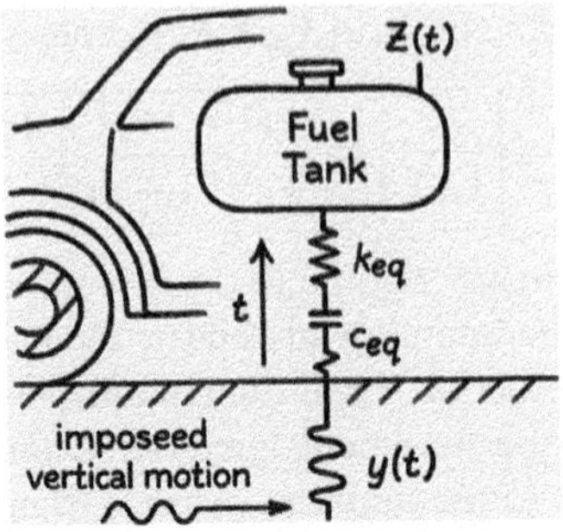

Figure 3.4 Equivalent lumped parameter spring–damper model of a fuel tank subjected to vertical base excitation, illustrating mass, stiffness, damping, and sloshing degree of freedom representation.

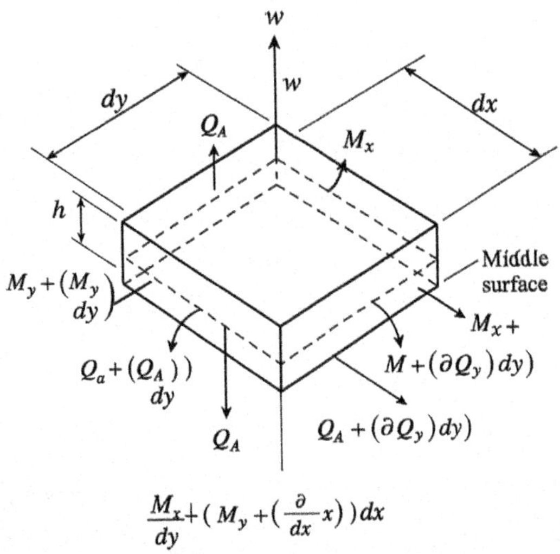

Figure 3.5 Infinitesimal plate element of a thin elastic plate showing bending moments, twisting moment, and transverse shear forces acting on the mid-surface according to classical plate theory.

tanks, containment vessels, and thin-walled containers, the use of continuum plate and shell theories becomes essential for capturing the correct dynamic behaviour. Accordingly, this section provides a brief introduction to the vibration of continuous plate structures and the theoretical framework used in their analysis (Ambartsumian et al., 1966; Chauhan & Ashwell, 1969; Gol'denveizer, 1966; Lindholm & Hu, 1966; Pan, 1969; Sinitsyn, 1969; Yu, 1965).

In solid mechanics, the connection between stresses and strains is described by generalised Hooke's law. This relation can be expressed compactly using tensor notation. For a three-dimensional material, the stresses (σ_x, σ_y, σ_z) represent the normal stresses, which act perpendicular to the coordinate planes, while (τ_{xy}, τ_{yz}, τ_{zx}) represent the shear stresses, acting tangentially to those planes (illustrated in Fig. 3.5).

The corresponding strains are (ε_x, ε_y, ε_z) for the normal strains, which measure the relative elongation or contraction along each axis, and (γ_{xy}, γ_{yz}, γ_{zx}) for the shear strains, which quantify angular distortion.

The stress–strain relation in its full tensor form is written as:

$$\begin{Bmatrix} \sigma_x \\ \sigma_y \\ \sigma_z \\ \tau_{xy} \\ \tau_{yz} \\ \tau_{zx} \end{Bmatrix} = \begin{bmatrix} c_{11} & c_{12} & c_{13} & c_{14} & c_{15} & c_{16} \\ c_{21} & c_{22} & c_{23} & c_{24} & c_{25} & c_{26} \\ c_{31} & c_{32} & c_{33} & c_{34} & c_{35} & c_{36} \\ c_{41} & c_{42} & c_{43} & c_{44} & c_{45} & c_{46} \\ c_{51} & c_{52} & c_{53} & c_{54} & c_{55} & c_{56} \\ c_{61} & c_{62} & c_{63} & c_{64} & c_{65} & c_{66} \end{bmatrix} \begin{Bmatrix} \varepsilon_x \\ \varepsilon_y \\ \varepsilon_z \\ \gamma_{xy} \\ \gamma_{yz} \\ \gamma_{zx} \end{Bmatrix} \tag{3.87}$$

Here, the coefficients c_{ij} are the elastic stiffness constants. They characterise how the material responds to strain by developing stresses. In general, there are 21 independent constants for a fully an isotropic elastic material, though the number reduces significantly if the material exhibits additional symmetry (e.g. isotropic or orthotropic materials).

For simplicity, these relations are often rewritten in contracted notation using Voigt's representation, which shortens the six-component stress and strain vectors while retaining the same physical meaning.

Overview of Solid Mechanics in Sloshing Applications

Tensor formulation can be compactly expressed as:

$$\{\sigma\} = [C]\{\varepsilon\} \tag{3.88}$$

Here, $\{\sigma\}$ is the stress vector, $\{\varepsilon\}$ is the strain vector, and $[C]$ is the stiffness matrix. In Cartesian tensor notation, this may be written as:

$$\sigma_i = c_{ij}\varepsilon_j i, j = 1, 2, 3, .,6 \tag{3.89}$$

The matrix $[C]$ contains the elastic stiffness coefficients, which relate applied strain to the resulting stress. The inverse of this relation is expressed through the compliance matrix $[S]$:

$$\{\varepsilon\} = [S]\{\sigma\} \tag{3.90}$$

or equivalently, in tensor form,

$$\varepsilon_i = s_{ij}\sigma_j i, j = 1, 2, 3, .,6 \tag{3.91}$$

Both $[C]$ and $[S]$ are symmetric due to the symmetry of the stress and strain tensors. In the most general anisotropic case, there are 36 coefficients in these matrices, but this number reduces to 21 independent elastic constants because $c_{ij} = c_{ji}$ and $s_{ij} = s_{ji}$

The strains that develop due to these displacements are related through the following expressions:

$$\begin{Bmatrix} \varepsilon_x \\ \varepsilon_y \\ \varepsilon_z \end{Bmatrix} = \begin{Bmatrix} \dfrac{\partial u}{\partial x} \\ \dfrac{\partial v}{\partial y} \\ \dfrac{\partial w}{\partial z} \end{Bmatrix} \tag{3.92}$$

The shear strains are expressed as

$$\begin{Bmatrix} \gamma_{xy} \\ \gamma_{yz} \\ \gamma_{zx} \end{Bmatrix} = \begin{Bmatrix} \frac{1}{2}\left(\frac{\partial u}{\partial y} + \frac{\partial v}{\partial x}\right) \\ \frac{1}{2}\left(\frac{\partial v}{\partial z} + \frac{\partial w}{\partial y}\right) \\ \frac{1}{2}\left(\frac{\partial w}{\partial x} + \frac{\partial u}{\partial z}\right) \end{Bmatrix} \tag{3.93}$$

The compatibility conditions ensure that the displacement field of an elastic body is continuous and single-valued, meaning there are no gaps or overlaps in the deformed configuration. While many vibration problems focus only on displacement expressions, for completeness, the compatibility equations can be written as follows:

$$\begin{Bmatrix} \frac{\partial^2 \varepsilon_x}{\partial y \partial z} \\ \frac{\partial^2 \varepsilon_y}{\partial z \partial x} \\ \frac{\partial^2 \varepsilon_z}{\partial x \partial y} \end{Bmatrix} = \begin{Bmatrix} \frac{\partial}{\partial y}\left(\frac{\partial \gamma_{xz}}{\partial y} + \frac{\partial \gamma_{xy}}{\partial z} + \frac{\partial \gamma_{yz}}{\partial x}\right) \\ \frac{\partial}{\partial y}\left(\frac{\partial \gamma_{yz}}{\partial x} + \frac{\partial \gamma_{xy}}{\partial z} + \frac{\partial \gamma_{xz}}{\partial y}\right) \\ \frac{\partial}{\partial z}\left(\frac{\partial \gamma_{zx}}{\partial y} + \frac{\partial \gamma_{yx}}{\partial x} + \frac{\partial \gamma_{xy}}{\partial z}\right) \end{Bmatrix} \tag{3.94}$$

Additionally, the following relations hold:

$$\begin{Bmatrix} 2\frac{\partial^2 \gamma_{xy}}{\partial x \partial y} \\ 2\frac{\partial^2 \gamma_{yz}}{\partial y \partial z} \\ 2\frac{\partial^2 \gamma_{zx}}{\partial z \partial x} \end{Bmatrix} = \begin{Bmatrix} \frac{\partial^2 \varepsilon_x}{\partial y^2} + \frac{\partial^2 \varepsilon_y}{\partial x^2} \\ \frac{\partial^2 \varepsilon_y}{\partial z^2} + \frac{\partial^2 \varepsilon_z}{\partial y^2} \\ \frac{\partial^2 \varepsilon_z}{\partial x^2} + \frac{\partial^2 \varepsilon_x}{\partial z^2} \end{Bmatrix} \tag{3.95}$$

To study the deformation of a plate, it is convenient to examine the motion of its mid-surface. Consider a plate of thickness h, with its middle surface lying in the xy-plane when the plate is undeformed. Suppose a distributed load $q(x, y)$ acts on the top surface, while shear forces Q and bending moments M act along the boundary edges ∂R of the plate.

The deformation can be separated into two components:

1. Stretching deformations, caused by in-plane forces applied along the boundary.
2. Bending deformations, resulting from transverse loads and bending moments.

Let u_s and v_s represent the horizontal displacement components of a material point (x, y, z) in the plate. Since these displacements are assumed to be the same through the thickness (i.e., independent of z, they can be written as):

$$u_s(x, y, z) = u_s(x, y, 0) = u_s(x, y) \tag{3.96}$$

$$v_s(x, y, z) = v_s(x, y, 0) = v_s(x, y) \tag{3.97}$$

This implies that any line originally perpendicular to the xy-plane and connecting the top and bottom surfaces of the plate translates horizontally without changing orientation, due to the stretching action.

In addition to stretching, the plate also deforms due to bending. According to the classical plate theory, the normals to the undeformed mid-surface remain straight and perpendicular to the mid-surface even after bending. This introduces a rotation similar to what occurs in rigid body motion, along with vertical displacement.

If u_b and v_b are the in-plane displacement components associated with bending, they can be expressed as:

$$u_b(x, y, z) = -z\frac{\partial w(x, y)}{\partial x} \tag{3.98}$$

$$v_b(x, y, z) = -z\frac{\partial w(x, y)}{\partial y} \tag{3.99}$$

Here, $w(x, y)$ represents the transverse displacement of the mid-surface.

The bending contribution introduces additional displacements beyond the membrane behaviour. According to classical plate theory, lines initially perpendicular to the mid-surface remain straight and normal to the deformed mid-surface, although they may rotate. Consequently, bending produces both in-plane displacements and transverse deflection.

The displacements in the x- and y-directions caused by bending are denoted as u_b and v_b Due to the bending action, a rotation similar to that of rigid body elements occurs, leading also to a vertical translation. These in-plane displacements can be expressed as:

$$u_b(x, y, z) = -z\frac{\partial w(x, y)}{\partial x} \tag{3.100}$$

$$v_b(x, y, z) = -z\frac{\partial w(x, y)}{\partial y} \tag{3.101}$$

In classical thin-plate theory, the deformation field is characterised by the transverse displacement $w(x,y)$. To capture bending effects, we define bending and twisting moments per unit length through integration of the stress components across the plate thickness. For example, the bending moments about the xx- and y-axes are written as

$$M_x = \int_{-h/2}^{h/2} \sigma_x z dz \tag{3.102}$$

$$M_y = \int_{-h/2}^{h/2} \sigma_y z dz \tag{3.103}$$

and the twisting moment is

$$M_{xy} = \int_{-h/2}^{h/2} \tau_{xy} z dz \tag{3.104}$$

where h is the plate thickness and z the distance from the mid-plane. These definitions inherently account for the linear stress distribution along the thickness predicted by Kirchhoff's assumptions.

By using Hooke's law for plane stress and the kinematic relations between strains and the curvature of the plate surface, the moments simplify to

$$M_z = D\left(\frac{\partial^2 \omega}{\partial x^2} + \upsilon\frac{\partial^2 \omega}{\partial y^2}\right) \tag{3.105}$$

$$M_z = D\left(\frac{\partial^2 \omega}{\partial y^2} + \upsilon\frac{\partial^2 \omega}{\partial x^2}\right) \tag{3.106}$$

$$M_{xy} = D(1-\upsilon)\frac{\partial^2 \omega}{\partial x \partial y} \tag{3.107}$$

where the constant D can be described as

$$D = \frac{Eh^3}{12(1-\upsilon^2)} \tag{3.108}$$

is the *flexural rigidity*, dependent on Young's modulus E, plate thickness h, and Poisson's ratio υ.

Equilibrium of an infinitesimal plate element requires that the transverse shear resultants, Q_x and Q_y, satisfy

$$Q_x = \frac{\partial M_x}{\partial x} + \frac{\partial M_{xy}}{\partial y} \tag{3.109}$$

$$Q_y = \frac{\partial M_y}{\partial y} + \frac{\partial M_{xy}}{\partial x} \tag{3.110}$$

Further enforcing equilibrium in the transverse direction, including a distributed normal load $q(x, y)$, leads to

$$\frac{\partial Q_x}{\partial x} + \frac{\partial Q_y}{\partial y} + q(x, y) = 0 \tag{3.111}$$

Substituting the expressions for, Q_x and Q_y, yields the governing *plate equation*:

$$\frac{\partial^4 \omega}{\partial x^4} + 2\frac{\partial^4 \omega}{\partial x^2 \partial y^2} + \frac{\partial^4 \omega}{\partial y^4} = -\frac{q(x, y)}{D} \tag{3.112}$$

This compact form is expressed as

$$\nabla^4 \omega = \frac{q}{D} \tag{3.113}$$

This is known as the biharmonic equation. The biharmonic equation can also be derived variationally. The total potential energy of the plate comprises the strain energy stored in bending and the work done by external loads. The bending strain energy is

$$U = \frac{D}{2} \iint_R \left[\left(\frac{\partial^2 \omega}{\partial x^2} \right)^2 + \left(\frac{\partial^2 \omega}{\partial y^2} \right)^2 + 2\upsilon \frac{\partial^2 \omega}{\partial x^2} \frac{\partial^2 \omega}{\partial y^2} + 2(1 - \upsilon) \left(\frac{\partial^2 \omega}{\partial x \partial y} \right)^2 \right] dxdy \tag{3.114}$$

while the potential of external transverse loading is

$$V = \iint_R q(x, y)\omega(x, y)dxdy \tag{3.115}$$

Thus the total potential energy functional becomes

$$\Pi = U - V \tag{3.116}$$

Applying the calculus of variations, the stationary condition $\delta\Pi = 0$ recovers the governing differential equation

$$\nabla^4 \omega = \frac{q}{D} \tag{3.117}$$

This energy approach not only confirms the biharmonic equation but also directly produces boundary terms that reveal the natural boundary conditions of the system.

For dynamic problems, inertia must be included. The kinetic energy of a vibrating plate is

$$T = \frac{1}{2} \iint_R \rho h \dot{\omega}^2 dxdy \tag{3.118}$$

where ρ is the material density. Using Hamilton's principle, $\delta \int_{t_1}^{t_2} (T - \Pi)dt = 0$, one obtains the dynamic equilibrium:

$$D\nabla^4 \omega + \rho h \ddot{\omega} = q(x, y, t) \tag{3.119}$$

This is the fundamental equation of motion for thin elastic plates, combining flexural stiffness, inertia, and external excitation.

If the plate thickness (or material properties) vary spatially, the flexural rigidity $D = D(x,y)$ is no longer constant. In this case, equilibrium leads to a generalised equation:

$$\rho h \ddot{\omega} = \nabla^2 (D \nabla^2 \omega) - (1 - \nu) \left[\frac{\partial^2 D}{\partial y^2} \frac{\partial^2 \omega}{\partial x^2} - 2 \frac{\partial^2 D}{\partial x \partial y} \frac{\partial^2 \omega}{\partial x \partial y} + \frac{\partial^2 D}{\partial x^2} \frac{\partial^2 \omega}{\partial y^2} \right] \tag{3.120}$$

This more complex form governs plates of nonuniform thickness or graded materials, which are increasingly relevant in advanced aerospace and mechanical applications.

The variational derivation also highlights the boundary terms, which dictate the admissible boundary conditions. For a boundary curve C with outward normal direction j and tangential direction hh, two essential types of conditions emerge:

1. Kinematic (displacement-type) conditions: Either the transverse deflection ww is specified (clamped or simply supported case), or the slope normal to the boundary, $\frac{\partial \omega}{\partial j}$, is prescribed.
2. Natural (force/moment-type) conditions: Either the bending moment M_j vanishes along the boundary, or the effective shear force is zero.

These conditions encapsulate all common physical constraints, such as clamped, simply supported, or free edges, and provide the closure needed to solve the biharmonic plate equation for real structures.

3.12 Concluding remarks

This chapter has established the structural mechanics foundation necessary for a physically consistent treatment of sloshing systems. While sloshing is initiated and sustained by fluid motion, the enclosing structure defines the mechanical environment within which that motion evolves. Structural stiffness, inertia, damping, and boundary constraints influence pressure distribution, energy transfer, and resonance characteristics. In practical systems, the distinction between fluid dynamics and structural response is often artificial; both participate in a coupled dynamic process. The development began with the fundamental measures of stress and strain and proceeded through constitutive relations for linear elastic materials. The equilibrium and compatibility conditions clarified that force balance alone is insufficient to define a mechanical state. Geometric admissibility and material behaviour must also be satisfied to obtain a physically meaningful solution. Variational principles were introduced not merely as mathematical tools, but as a unifying framework capable of deriving governing equations, natural boundary conditions, and energy-based interpretations of structural response. These principles form the conceptual bridge to modern computational approaches and reduced order modelling strategies used in fluid-structure interaction problems. Both discrete and continuous structural representations were examined. Lumped parameter vibration models provide transparency and engineering intuition, particularly for global tank motion and isolation behaviour. In contrast, plate theory captures distributed deformation effects that become important in thin-walled containers and in resonance scenarios where structural modes interact with free surface oscillations. With this structural framework in place, the discussion now turns to the fluid domain. The following chapter introduces the mathematical modelling of sloshing dynamics, where the governing equations of free surface motion are formulated and analytical and reduced order representations of sloshing behaviour are developed.

References

Ambartsumian, S. A., Bagdasarian, G. E., Durgarian, S. M., & Gnuny, V. T. (1966). Some problems of vibration and stability of shells and plates. *International Journal of Solids and Structures, 2*(1), 59–81. https://doi.org/10.1016/0020-7683(66)90007-2.

Babu, S. S., & Bhattacharyya, S. K. (1996). Finite element analysis of fluid-structure interaction effect on liquid retaining structures due to sloshing. *Computers and Structures, 59*(6), 1165–1171. https://doi.org/10.1016/0045-7949(95)00271-5.

Calayir, Y., & Dumanoğlu, A. A. (1993). Static and dynamic analysis of fluid and fluid-structure systems by the Lagrangian method. *Computers and Structures, 49*(4), 625–632. https://doi.org/10.1016/0045-7949(93)90067-N.

Chauhan, A. P., & Ashwell, D. G. (1969). Small- and large-amplitude free vibrations of square shallow shells. *International Journal of Mechanical Sciences, 11*(4), 337–349. https://doi.org/10.1016/0020-7403(69)90068-X.

Gantenbein, F., & Buland, P. (1991). DEMT experimental and analytical studies on seismic isolation. *Nuclear Engineering and Design, 127*(3), 409–418. https://doi.org/10.1016/0029-5493(91)90064-O.

Gol'denveizer, A. L. (1966). Qualitative analysis of free vibrations of an elastic thin shell. *Journal of Applied Mathematics and Mechanics, 30*(1), 110–127. https://doi.org/10.1016/0021-8928(66)90061-X.

Hung, R. J., Lee, C. C., & Leslie, F. W. (1990). Effect of g-jitters on the stability of rotating bubble under microgravity environment. *Acta Astronautica, 21*(5), 309–321. https://doi.org/10.1016/0094-5765(90)90091-X.

Hung, R. J., Lee, C. C., & Leslie, F. W. (1992). Similarity rules in gravity jitterrelated spacecraft liquid propellant slosh waves excitation. *Journal of Fluids and Structures, 6*(4), 493–522. https://doi.org/10.1016/0889-9746(92)90028-2.

Hung, R. J., & Long, Y. T. (1996). Response of lateral impulse on liquid helium sloshing with baffle effect in microgravity. *International Journal of Mechanical Sciences, 38*(8–9), 951–965. https://doi.org/10.1016/0020-7403(95)00089-5.

Hung, R. J., & Pan, H. L. (1994). Gravity gradient or gravity jitter induced viscous stress and moment fluctuations in microgravity. *Fluid Dynamics Research, 14*(1), 29–51. https://doi.org/10.1016/0169-5983(94)90020-5.

Hung, R. J., & Pan, H. L. (1995a). Combined gravity gradient and jitter accelerations acting on liquid-vapor interface oscillations in reduced gravity. *International Journal of Mechanical Sciences, 37*(11), 1203–1221. https://doi.org/10.1016/0020-7403(95)00012-M.

Hung, R. J., & Pan, H. L. (1995b). Mathematical model of bubble sloshing dynamics for cryogenic liquid helium in orbital spacecraft dewar container. *Applied Mathematical Modelling, 19*(8), 483–498. https://doi.org/10.1016/0307-904X(95)00033-G.

Hung, R. J., & Pan, H. L. (1996). Modelling of sloshing modulated angular momentum fluctuations actuated by gravity gradient associated with spacecraft slew motion. *Applied Mathematical Modelling, 20*(5), 399–409. https://doi.org/10.1016/0307-904X(95)00159-H.

Lindholm, U. S., & Hu, W. C. L. (1966). Non-symmetric transverse vibrations of truncated conical shells. *International Journal of Mechanical Sciences, 8*(9), 561–579. https://doi.org/10.1016/0020-7403(66)90078-6.

Pan, H. H. (1969). Axisymmetrical vibrations of a circular sandwich shell with a viscoelastic core layer. *Journal of Sound and Vibration, 9*(2), 338–348. https://doi.org/10.1016/0022-460X(69)90038-8.

Pister, K. S. (1954). The airy stress function in curvilinear coordinates with application to the uniform flexure of a naturally curved spiral beam. *Journal of the Franklin Institute, 257*(1), 25–36.

Schulkes, R. M. S. (1992). Interactions of an elastic solid with a viscous fluid: Eigenmode analysis. *Journal of Computational Physics, 100*(2), 270–283. https://doi.org/10.1016/0021-9991(92)90234-P.

Sinitsyn, A. P. (1969). Vibrations and stability of shells due to loads and temperature. *International Journal of Solids and Structures, 5*(12), 1333–1337. https://doi.org/10.1016/0020-7683(69)90075-4.

Steinhardt, P. J., & Chaudhari, P. (1981). Airy stress function for atomic models. *Journal of Computational Physics, 42*(2), 266–276. https://doi.org/10.1016/0021-9991(81)90244-8.

Yu, Y.-Y. (1965). On linear equations of isotropic elastic plates and shells. *Journal of the Franklin Institute, 280*(5), 395–416. https://doi.org/10.1016/0016-0032(65)90530-2.

Mathematical modeling of sloshing dynamics

4

4.1 Introduction

Sloshing refers to the oscillatory motion of a liquid contained within a vessel when the container is subjected to external disturbances. This behaviour arises in partially filled containers exposed to motions such as horizontal or vertical translation, rotation, or mechanical vibration.

Well-known examples include the movement of liquid in large storage tanks during seismic events, the oscillation of propellant within rocket and spacecraft fuel tanks, and the motion of liquids in ship tanks induced by wave action. In these situations, sloshing can produce considerable transient pressures on the container walls and can significantly affect the overall stability and structural integrity of the system.

From a physical perspective, sloshing is governed by the coupling between the liquid free surface and the imposed motion of the container. When the frequency of excitation approaches one of the natural frequencies of the liquid system, resonant conditions may develop. Under resonance, free surface displacements can grow markedly, leading to amplified forces and enhanced dynamic effects. This resonance behaviour is therefore of particular relevance in engineering design and safety assessment (Debongnie, 1986; Hashimoto & Sudo, 1988; Modi & Welt, 1988; Sakai et al., 1984; Shemer & Kit, 1988; Sinai, 1985).

The scientific investigation of sloshing began largely in the context of naval engineering and liquid storage facilities. Early research relied mainly on experimental observations, while subsequent developments introduced analytical descriptions based on linear wave theory. These studies demonstrated that sloshing dynamics are strongly influenced by factors such as container shape, fill level, and the nature of the imposed excitation.

More recently, sloshing has gained increased attention due to its importance in aerospace applications, the seismic design of storage tanks, and fluid behaviour under reduced gravity conditions. In all these fields, reliable prediction of sloshing-induced forces and free surface motion is essential for safe and efficient system design.

This chapter introduces the core principles of sloshing dynamics and outlines its fundamental characteristics, establishing the conceptual foundation for the analytical and mathematical modelling approaches developed herein. Numerical methods commonly employed for the simulation of sloshing phenomena are addressed separately and are presented in Chapter 5, where fully discretised computational techniques are discussed in detail.

Slosh Dynamics. DOI: https://doi.org/10.1016/B978-0-443-44330-5.00007-4

4.2 Overview of sloshing

When sloshing takes place, the liquid free surface can respond in a range of distinct motion patterns, depending on the characteristics of the imposed excitation. In many practical systems, the motion applied to a container is not purely unidirectional. Instead, it often consists of a combination of translational movements and rotations. This combined forcing leads to complex liquid responses, in which nonlinear behaviour commonly becomes evident.

As the amplitude of excitation increases, the assumptions underlying linear theories progressively lose validity. Under such conditions, pronounced free surface deformation may develop, accompanied by phenomena such as wave breaking, direct impacts of the liquid on the container walls, and enhanced energy dissipation associated with turbulent flow. These nonlinear effects have a decisive influence on the resulting sloshing loads and therefore play a critical role in engineering analysis and design.

The dynamic response of sloshing is also highly sensitive to the geometry of the container. Tanks with cylindrical, rectangular, or spherical shapes exhibit distinct sloshing modes and characteristic natural frequencies. Moreover, even within a given container geometry, variations in the liquid fill level can lead to substantial changes in free surface dynamics and resonance behaviour.

From an engineering standpoint, sloshing should be regarded not merely as a problem of fluid motion, but as a coupled fluid–structure interaction. The unsteady pressures generated by the moving liquid act on the container walls and can induce structural vibrations. In certain cases, resonance may arise from the interaction between the sloshing motion and the structural response, potentially amplifying stresses and displacements.

4.2.1 *Fundamental categories of sloshing*

Sloshing phenomena may be broadly categorised according to the nature of the excitation and the resulting liquid motion. Common classifications include planar

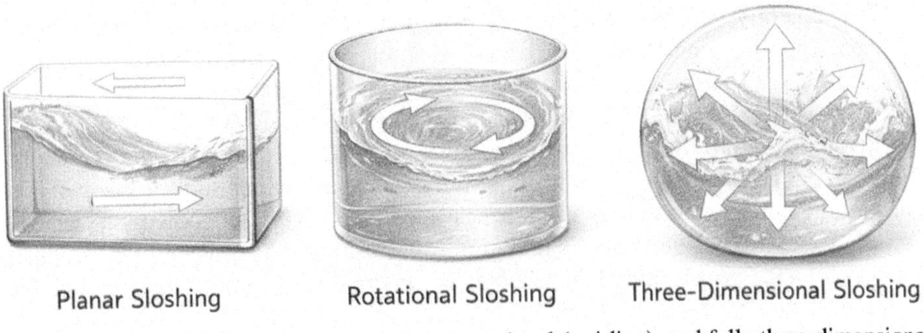

Figure 4.1 Schematic representation of planar, rotational (swirling), and fully three-dimensional sloshing modes in partially filled containers, illustrating the qualitative differences in free surface motion patterns.

sloshing, rotational sloshing, and fully three-dimensional sloshing (illustrated in Fig. 4.1). In planar sloshing, the free surface motion is largely confined to a single plane, with oscillations occurring predominantly in one direction. Rotational sloshing (Swirling), by contrast, is characterised by a swirling or circumferential motion of the free surface. Additional sloshing patterns, particularly those observed in circular containers, exhibit more complex three-dimensional behaviour.

Depending on the nature and intensity of the imposed excitation, sloshing may involve more than simple oscillations of the liquid free surface. Under certain conditions, additional phenomena such as wave breaking, hydraulic impacts against the container walls, and sloshing-induced vibrations can arise. In extreme situations, the liquid motion may lose regularity and evolve into a highly irregular or chaotic response. From an engineering viewpoint, sloshing behaviour is often distinguished according to whether the free surface motion remains weakly disturbed or enters a strongly nonlinear regime. When the wave height is small relative to the liquid depth, the governing equations may be linearised, and linear sloshing theory provides an adequate description. As the amplitude of motion increases, however, nonlinear effects become increasingly important, and linear assumptions cease to be valid.

4.2.2 Natural frequency of sloshing

The natural frequency is a key parameter characterising sloshing motion in partially filled containers. For the idealised case of small-amplitude oscillations, the sloshing frequency can be obtained by assuming inviscid flow and negligible free surface deformation. For a two-dimensional rectangular container with width L and liquid depth h, the natural frequency of the fundamental sloshing mode is expressed as:

$$f_n = \frac{1}{2\pi}\sqrt{gk\tanh(kh)} \tag{4.1}$$

where g denotes the gravitational acceleration, and k is the wave number, defined by:

$$k = \frac{n\pi}{L},\ (n = 1, 2, 3, .) \tag{4.2}$$

In most practical applications, the fundamental mode corresponding to n=1 dominates the sloshing response.

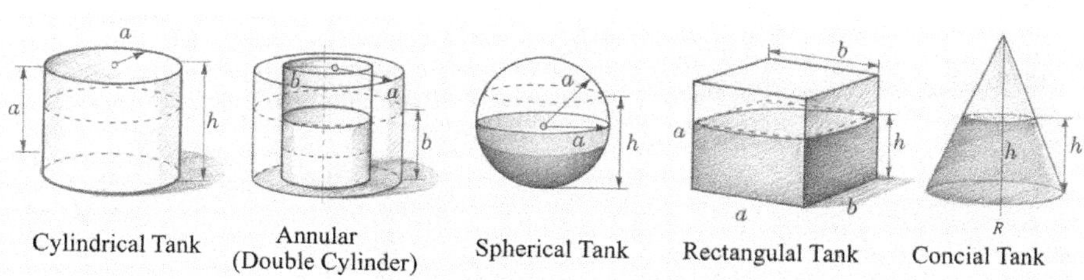

Figure 4.2 Summarises representative container shapes, while their corresponding natural angular frequencies are listed in Table 4.1.

Table 4.1 Natural angular frequencies of the fundamental sloshing modes for the representative container geometries are shown.

Shape	Natural angular frequency (ω) equation	Parameters/Coefficients
Cylindrical tank	$\omega^2 = \xi \frac{g}{a} \tanh\left(\xi \frac{h}{a}\right)$	$\xi_{11} = 1.8412$, $\xi_{12} = 5.3314$
Annular (double cylinder)	$\omega^2 = \xi \frac{g}{a} \tanh\left(\xi \frac{h}{a}\right)$	Inner radius$b$, Outer radius$a$
Spherical tank	$\frac{\omega^2 a}{g} = \frac{1}{3}\left\{\Xi \cdot \sin\left(\frac{\pi}{4}(2-\kappa)\left(1+\frac{\kappa}{3}\right)\right)\right\}$	$\Gamma = \frac{240-220\kappa+72\kappa^2-9\kappa^3}{80-100\kappa+44\kappa^2-9\kappa^3+\kappa^4}$, $\kappa = \frac{h}{a}$
Conical tank	$\omega^2 = C\frac{g}{R_b}$	$R_b$bottom radius, $c$dimensionless factor
Rectangular tank	$\omega_{nm}^2 = g\lambda_{nm} \tanh(\lambda_{nm} h)$	$\lambda_{nm}^2 = \left(\frac{(2n-1)\pi}{a}\right)^2 + \left(\frac{(2m-1)\pi}{b}\right)^2$

In the limit of sufficiently deep liquid, namely when $kh \gg 1$, the hyperbolic tangent term approaches unity. Under this condition, the natural frequency may be approximated by

$$f_n = \frac{1}{2\pi}\sqrt{gk} \tag{4.3}$$

Extending the discussion from the two-dimensional rectangular configuration to simple three-dimensional tank geometries, as illustrated in Fig. 4.2 and Table 4.1, leads to the classical sloshing regime commonly referred to as lateral sloshing.

An important outcome of this analysis is that, provided the liquid layer is not extremely shallow, the natural sloshing frequency f is governed primarily by the characteristic horizontal dimension of the container, namely the tank diameter ($D=2a$) for cylindrical vessels, or the side length in the case of rectangular tanks. For a standard cylindrical container, the first mode natural frequency f_{11} may be estimated by assuming a gravitational acceleration of ($g=9.81$ m/s^2). When the tank radius a is expressed in metres, the frequency is given by

$$f_{11} = \frac{1}{2\pi}\sqrt{\frac{1.8412g}{a}} = \frac{0.676469}{\sqrt{a}} \approx \frac{1}{\sqrt{D}} \tag{4.4}$$

This relation provides a convenient physical interpretation: a cylindrical tank with a diameter of 10 metre exhibits a fundamental sloshing frequency of approximately 10 Hz, corresponding to an oscillation period of 10 s.

4.3 Basic types of sloshing

In real engineering systems, sloshing rarely appears as an isolated motion of the liquid free surface. Instead, it is closely associated with the dynamic interaction between the liquid and the container structure. When the container is subjected to external excitation, such as earthquake motion, ship motion, or mechanical vibration, the oscillating liquid generates time-dependent pressure on the container walls. These pressures may excite structural vibrations, which in turn influence the liquid motion. This mutual dependence constitutes a typical fluid–structure interaction problem.

If the container is sufficiently rigid, structural deformation can be neglected, and the sloshing behaviour is governed mainly by gravity, liquid depth, and container geometry. However, in large-scale or thin-walled containers, structural flexibility becomes significant. Under such conditions, sloshing can induce bulging deformation, in which parts of the container wall deform outward in response to fluctuating hydrodynamic pressure. This deformation modifies the boundary conditions of the liquid and alters the characteristics of the sloshing motion.

Another important structural response associated with sloshing is breathing vibration. Breathing vibration refers to an axisymmetric mode in which the container expands and contracts uniformly in the radial direction. Unlike bulging, which is often localised, breathing vibration involves the global deformation of the container shell. When the natural frequency of this mode approaches that of the sloshing motion, strong coupling may occur, leading to significant amplification of both liquid motion and structural response.

From a theoretical standpoint, these coupled phenomena cannot be described adequately using free surface sloshing theory alone. Instead, it is necessary to introduce simplified models that represent both the liquid dynamics and the structural response. Such approaches are essential for practical engineering design, particularly in seismic safety assessment and vibration analysis.

4.4 Mechanical models of sloshing

To analyse sloshing phenomena in an engineering framework, it is common to employ mechanical analogue models that represent the complex fluid motion using simplified dynamic elements. One widely used approach is the equivalent mass model, in which the liquid is conceptually divided into different components according to their dynamic behaviour.

In this representation, part of the liquid moves together with the container and is treated as a rigidly attached mass. The remaining portion, which participates in the oscillatory free surface motion, is modelled as a sloshing mass connected to the container by a spring. The stiffness of this spring is determined from the natural frequency of the sloshing mode. This simple model captures the essential characteristics of sloshing while allowing the system to be analysed using classical vibration theory.

When structural flexibility is taken into account, additional degrees of freedom are introduced into the model. Springs and masses are added to represent the bulging modes and breathing vibration of the container wall. In this way, the interaction between liquid motion and structural deformation can be analysed within a unified mechanical framework. Such models are particularly useful for predicting resonance phenomena and evaluating dynamic loads acting on the container.

Although mechanical analogue models involve strong simplifications, they provide valuable physical insight and are widely used in preliminary design and safety evaluation. Their applicability, however, is limited when sloshing amplitudes become large, nonlinear effects dominate, or complex three-dimensional container geometries are involved. In such cases, more advanced numerical or experimental methods are required.

4.5 Equivalent mechanical representation of sloshing

To enable practical engineering analysis, sloshing motion is often represented by a mechanical analogue consisting of discrete masses and springs. This approach replaces the distributed liquid motion with a simplified dynamic system that retains the essential physical characteristics of sloshing (see Fig. 4.3).

In the most common representation, the total liquid mass M is divided into two components:

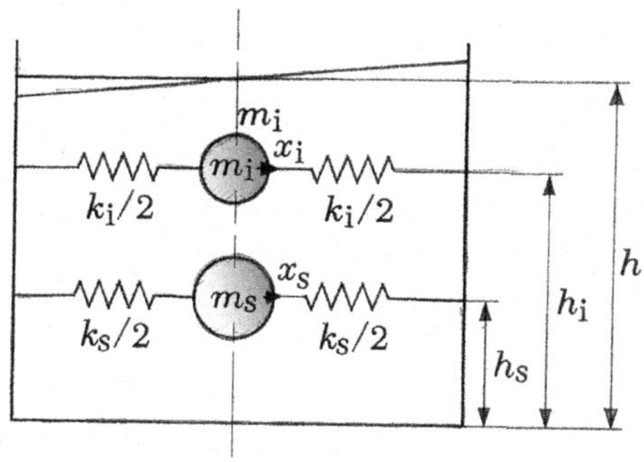

Figure 4.3 Equivalent mass spring representation of liquid sloshing dynamics, showing the impulsive mass moving with the container and the sloshing mass connected through an elastic restoring element.

1. Sloshing mass M_s, which participates in oscillatory motion relative to the container.
2. Impulsive mass M_i, which moves rigidly together with the container, and

The sloshing mass M_s is modelled as a concentrated mass connected to the container by a linear spring of stiffness k_s. The equation of motion for the sloshing mass relative to the container can be written as:

$$M_s\ddot{x}_s + k_s x_s = -M_s\ddot{x}_c \tag{4.5}$$

where x_s is the relative displacement of the sloshing mass, and x_c is the displacement of the container.

k_s is chosen such that the natural frequency of the mechanical system matches the natural frequency of the sloshing mode obtained from fluid mechanics:

$$k_s = M_s\omega_s^2 \tag{4.6}$$

Here, ω_s is the angular natural frequency of sloshing, previously derived from the free-surface wave analysis. This relationship ensures that the simplified mechanical model reproduces the correct dynamic behaviour of the liquid free surface. The impulsive mass M_i represents the portion of the liquid that moves rigidly with the container and therefore does not introduce an additional degree of freedom. Its motion is fully constrained by the container motion, such that

$$x_i(t) = x_c(t) \tag{4.7}$$

As a consequence, the impulsive mass contributes only to the inertial force acting on the structure, given by

$$F_i(t) = M_i\ddot{x}_c(t) \tag{4.8}$$

The impulsive mass does not require a separate equation of motion, nor is it associated with a restoring spring. Instead, it is incorporated directly into the inertia of the container in the global dynamic formulation.

When structural flexibility is taken into account, additional degrees of freedom must be introduced to represent vibration modes of the container, such as bulging and breathing vibration. These modes may be modelled using structural masses M_b, M_{br} and corresponding stiffnesses k_b, k_{br}.

The equations of motion of the coupled liquid structure system may then be written in matrix form as:

$$\boldsymbol{M}\ddot{\boldsymbol{x}} + \boldsymbol{K}\boldsymbol{x} = \boldsymbol{F}(t) \tag{4.9}$$

where M is the total mass matrix representing both liquid and structural inertia, K is the total stiffness matrix representing sloshing and structural restoring forces, and $F(t)$ is the total excitation force induced by container motion. Although these mechanical models are highly idealised, they provide a powerful conceptual and computational tool. They allow resonance conditions, dynamic amplification, and force transmission to be analysed efficiently, particularly in seismic design and preliminary safety assessment. It must be emphasised, however, that these models are valid primarily under the assumptions of small amplitude motion, linear behaviour, and simple container geometry. When sloshing becomes strongly nonlinear, when wave breaking occurs, or when geometric complexity is significant, the accuracy of the mechanical analogue decreases, and more advanced numerical or experimental methods are required.

4.6 Swirling

Swirling sloshing represents a class of free-surface motion in partially filled containers, in which the liquid surface no longer oscillates back and forth along a fixed plane but instead develops a rotational or circumferential travelling wave. In contrast to planar sloshing, where nodal lines remain stationary, and the free surface motion is described by standing waves, swirling sloshing is characterised by a continuous phase progression in the azimuthal direction, resulting in a rotating crest that travels around the container axis. This behaviour is most naturally observed in containers with rotational symmetry, such as cylindrical or spherical tanks, where the underlying geometry supports degenerate sloshing modes with identical natural frequencies. The physical origin of swirling sloshing lies in the simultaneous excitation of two orthogonal planar sloshing modes with a finite phase difference, typically close to $\pi/2$, which causes their superposition to manifest as a rotating wave rather than a standing oscillation.

From a kinematic perspective, planar sloshing in an idealised cylindrical tank may be decomposed into two independent orthogonal modes aligned, for example, with the x and y axes of the horizontal plane. In linear theory, these modes are uncoupled and may be expressed as

$$\begin{cases} \eta_x(r, \theta, t) = A(r)\cos(\theta)\cos(\omega t) \\ \eta_y(r, \theta, t) = A(r)\sin(\theta)\sin(\omega t) \end{cases} \tag{4.10}$$

where η_x, and η_y denote the free surface elevation in x and y directions respectively, r and θ are the radial and azimuthal coordinates. A is a characteristic wave amplitude. When these two motions are excited simultaneously with a quarter-period phase shift, their superposition yields

$$\eta(r, \theta, t) = A(r)\cos(\theta - \omega t) \tag{4.11}$$

which represents a travelling wave propagating continuously in the circumferential direction. This expression makes explicit the defining feature of swirling sloshing: the free surface pattern rotates rigidly about the container axis with angular velocity ω, while maintaining a nearly constant spatial shape. Unlike planar sloshing, where energy oscillates locally between kinetic and potential forms within a fixed region, swirling sloshing redistributes energy azimuthally around the entire container perimeter. The physical conditions required for the emergence of swirling sloshing are readily encountered in practical systems. Any excitation that contains two horizontal components with a phase lag, such as combined translational and rotational motion, oblique seismic input, or off-axis mechanical forcing, can trigger the simultaneous activation of orthogonal sloshing modes. Even in nominally unidirectional excitation, weak nonlinearities, geometric imperfections, or asymmetries in boundary conditions may break the ideal independence of the modes and allow energy transfer between them. Once a finite phase difference develops, the system naturally evolves towards a rotating free-surface response. This sensitivity to small imperfections explains why swirling sloshing is frequently observed in experiments and real structures, even when the excitation appears simple. The dynamic importance of swirling sloshing stems from the fundamentally different hydrodynamic forces it generates compared with planar sloshing. In planar motion, the dominant loads act alternately on opposing container walls and are primarily normal to the boundary. By contrast, swirling sloshing produces substantial tangential velocity components and circumferential pressure gradients, giving rise to torsional moments and azimuthal shear forces. The instantaneous pressure field $p(r,\theta,z,t)$, obtained from the unsteady Bernoulli equation

$$p = -\rho\frac{\partial\phi}{\partial t} - \frac{1}{2}\rho\,|\nabla\phi|^2 - \rho g z \tag{4.12}$$

where ϕ is the velocity potential and ρ the liquid density, rotates with the free surface and continuously loads different regions of the container wall. As a consequence, the structural response may involve modes that are not activated by purely planar sloshing, including torsional deformation, circumferential shell vibration, and coupled bending–twisting behaviour in flexible tanks. This is particularly critical in thin-walled cylindrical vessels, spacecraft propellant tanks, and large storage tanks subjected to seismic excitation. From an energetic standpoint, swirling sloshing alters the balance between local dissipation and global momentum transfer. Because the wave crest is not repeatedly reflected at fixed boundaries, local wave breaking and impact may be less severe than in high-amplitude planar sloshing. However, the continuous circulation of liquid mass generates a persistent angular momentum, leading to significant global loads and prolonged interaction with the structure. The total kinetic energy of the swirling motion,

$$U_k = \frac{1}{2} \int_V \rho \, |\mathbf{u}|^2 dV \tag{4.13}$$

remains distributed around the circumference and can couple efficiently with low-frequency structural modes, especially when the container exhibits near-axisymmetric flexibility. This coupling mechanism explains why swirling sloshing is often associated with long-duration vibrations and fatigue-critical loading scenarios. The transition from planar to swirling sloshing cannot be fully captured within strictly linear theory, even though the final swirling motion itself may often be approximated as linear in amplitude. In ideal linear systems, orthogonal modes remain independent, and no phase locking occurs. In real systems, however, weak nonlinear terms arising from free-surface boundary conditions, finite wave steepness, or viscous effects introduce modal coupling. When two degenerate modes exchange energy, the system naturally favours a travelling wave configuration, as this minimises temporal variations in local energy density. This behaviour is analogous to mode locking in coupled oscillators and is a robust feature of rotationally symmetric fluid systems. Mathematically, swirling sloshing corresponds to a travelling wave solution rather than a standing wave solution of the free-surface boundary value problem. This distinction has important implications for pressure distribution, phase relationships, and resonance behaviour. In planar sloshing, resonance occurs when the excitation frequency matches the natural frequency of a standing mode, leading to large oscillation amplitudes at fixed locations. In swirling sloshing, resonance may instead manifest as a synchronisation between excitation and rotational wave speed, resulting in sustained angular momentum transfer rather than purely vertical free-surface growth. Consequently, conventional design criteria based solely on maximum wave height or wall impact pressure may underestimate the true structural demand in systems susceptible to swirling motion.

4.7 Nonlinear sloshing

In practical engineering systems, sloshing motion almost inevitably enters the nonlinear regime once the amplitude of the free-surface displacement exceeds a modest threshold. While linear sloshing theory provides valuable insight into the fundamental modes and resonance characteristics of liquid motion, it relies on restrictive assumptions, most notably infinitesimal wave height, small surface slopes, and weak coupling between modes. When these assumptions are violated, nonlinear sloshing behaviour emerges, and the liquid response can no longer be represented accurately by linear potential theory alone. This nonlinear regime is particularly relevant under strong external excitation, near-resonant forcing, or long-duration dynamic loading, where the resulting hydrodynamic forces may be substantially larger and more complex than those predicted by linear models (Akkaoui et al., 2020; Alemi Ardakani, 2016; Ali Goudarzi & Reza Sabbagh-Yazdi, 2012; Du et al., 2019;

Gurusamy & Kumar, 2020; Hara & Watanabe, 2016; Huang et al., 2018; Ikeda et al., 2016; Love & Haskett, 2018; Moslemi et al., 2019; Strelnikova et al., 2020; Xin et al., 2020; Zhang et al., 2015; Zhao et al., 2018). The mathematical origin of nonlinear sloshing lies primarily in the free-surface boundary conditions, which are intrinsically nonlinear even when the bulk flow is assumed to be inviscid and irrotational. The velocity field may still be expressed in terms of a scalar potential ϕ, satisfying Laplace's equation within the liquid domain,

$$\nabla^2\phi = 0 \tag{4.14}$$

but the kinematic and dynamic conditions imposed at the moving free surface introduce nonlinearities that dominate the system response at finite amplitudes. The kinematic free-surface condition, enforcing that the free surface moves with the fluid, is given by

$$\frac{\partial\eta}{\partial t} + \nabla\phi.\ \nabla\eta = \frac{\partial\phi}{\partial z} \tag{4.15}$$

The presence of the nonlinear convective term $\nabla\phi\cdot\nabla\eta$ implies that wave steepness directly influences the evolution of the free surface.

Similarly, the dynamic boundary condition obtained from Bernoulli's equation introduces quadratic velocity terms,

$$\frac{\partial\phi}{\partial t} + \frac{1}{2}\,|\nabla\phi|^2 + g\eta = 0 \tag{4.16}$$

In a linear perspective, the quadratic term is neglected, leading to a linear relationship between pressure and surface elevation. In nonlinear sloshing, however, this term becomes significant and gives rise to amplitude-dependent dynamics and asymmetric wave shapes.

One of the most important consequences of nonlinear sloshing is the dependence of the natural frequency on oscillation amplitude. In the linear regime, the natural angular frequency ω_n is a fixed system parameter determined by geometry and liquid depth. In the nonlinear regime, the frequency shifts as the free-surface amplitude increases, a phenomenon often referred to as nonlinear frequency modulation. This effect may be represented schematically as

$$\omega = \omega_n + \alpha A^2 \tag{4.17}$$

where A is a characteristic wave amplitude and α is a nonlinear coefficient that depends on container geometry and boundary conditions. This amplitude–frequency coupling leads to resonance detuning and may result in jump phenomena, hysteresis, and saturation of the sloshing response under harmonic excitation.

Nonlinear sloshing is also characterised by distortion of the free-surface waveform and the generation of higher harmonics. While linear sloshing produces nearly sinusoidal oscillations, finite-amplitude motion results in steepened crests and flattened troughs. The free-surface elevation may therefore be expanded as a harmonic series,

$$\eta(t) = A_1 \cos(\omega t) + A_2 \cos(2\omega t) + A_3 \cos(3\omega t) + \cdots \tag{4.18}$$

The emergence of higher-order harmonics A_2,A_3,...is a clear indicator of nonlinear behaviour and contributes to increased peak pressures and nonuniform loading on the container walls.

At higher excitation levels, nonlinear sloshing may evolve into strongly dissipative phenomena such as wave breaking and hydraulic jumps. Wave breaking occurs when the local wave steepness exceeds a critical value, leading to overturning of the crest and rapid conversion of organised wave energy into turbulence and heat. In confined containers, breaking waves frequently impact the tank walls or roof, producing short-duration impulsive pressures. These impact loads are often modelled using momentum balance arguments, where the impulse III exerted on the wall may be expressed as

$$I = \int_{t_1}^{t_2} F(t)dt \tag{4.19}$$

with $F(t)$ denoting the transient hydrodynamic force. Such impulsive events are entirely absent from linear sloshing models but often govern structural damage in severe loading scenarios.

Another defining feature of nonlinear sloshing is the coupling between different sloshing modes. In linear theory, orthogonality ensures that modes evolve independently. Nonlinearities, however, allow energy exchange between modes, particularly when internal resonance conditions are satisfied. A common internal resonance condition takes the form

$$\omega_2 \approx 2\omega_1 \tag{4.20}$$

Under such conditions, energy initially contained in the fundamental mode may be transferred to higher modes, resulting in complex multifrequency responses and spatially intricate free-surface patterns.

Fig. 4.4A and B illustrate typical nonlinear free-surface configurations under planar and rotational excitations. In nonlinear planar sloshing, the liquid oscillates predominantly along a single horizontal direction, but the wave shape becomes increasingly asymmetric with amplitude growth, leading to uneven pressure

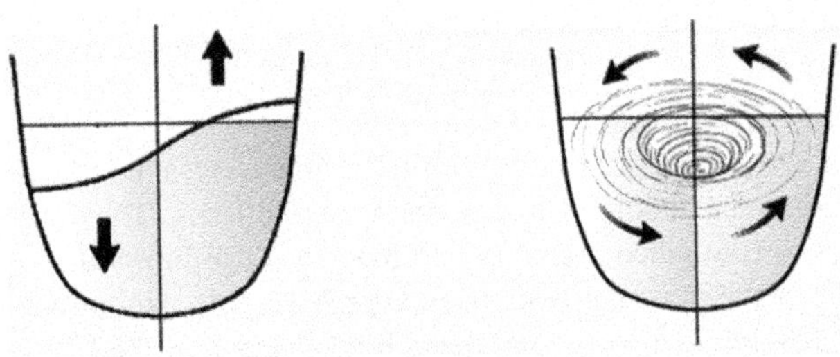

(A) Planar Motion (Lateral Sloshing) (B) Rotational Motion (Swirl)

Figure 4.4 Nonlinear free surface responses under (A) planar excitation and (B) rotational excitation, demonstrating amplitude-dependent deformation and asymmetric wave development.

distributions and localised impact zones. In contrast, nonlinear rotational sloshing exhibits a circumferential travelling-wave pattern, where nonlinear effects manifest as amplitude-dependent wave speed, azimuthal modulation of wave height, and enhanced tangential momentum transfer. While rotational motion may reduce localised breaking, it significantly increases global torsional loading and angular momentum exchange between the liquid and the structure.

4.7.1 Transition from linear to nonlinear sloshing

Transient sloshing refers to the time-dependent evolution of free surface motion from an initial quiescent or disturbed state towards a developed oscillatory response and represents a physically rich regime in which inertia, gravity, and dissipation interact in a strongly nonstationary manner. Unlike steady state or purely periodic sloshing, transient behaviour is governed by the balance between energy input, storage, redistribution, and loss, and therefore cannot be characterised by a single frequency or amplitude. From a continuum mechanics perspective, the liquid motion during transients is still governed by the conservation laws of mass and momentum, which for an incompressible Newtonian fluid that is presented in Eqs. 4.21 and 4.22

$$\nabla \boldsymbol{u} = 0 \tag{4.21}$$

$$\rho\left(\frac{\partial \boldsymbol{u}}{\partial t} + \boldsymbol{u}\cdot\nabla \boldsymbol{u}\right) = -\nabla p + \rho \boldsymbol{g} + \mu \nabla^2 \boldsymbol{u} \tag{4.22}$$

The defining feature of the transient regime is the explicit role of time derivatives, which dominate the early stages of motion and determine how the system departs from its initial condition. When a container is suddenly accelerated, decelerated, or subjected to a time-varying excitation, the free surface initially responds through impulsive acceleration of the liquid mass, generating inertial waves whose spatial structure is dictated by container geometry. During this initial phase, the free surface elevation η often grows rapidly.

The transient dynamics are then governed by a set of coupled ordinary differential equations of the form

$$\ddot{q} + 2\zeta_n \omega_n \dot{q}_n + \omega_n^2 q_n = F(t) + N_n(q) \tag{4.23}$$

where ζ_n represents modal damping, $F(t)$ the time-dependent forcing, and N_n nonlinear interaction terms. Physically, the transient response is characterised by an initial growth phase, during which energy supplied by the excitation is stored primarily in the lowest modes, followed by a redistribution phase in which energy is transferred among modes and partially dissipated through viscous effects and wave breaking. The rate at which the transient grows depends strongly on the characteristic time scales of the system. A key timescale is the gravitational wave time.

$$T_g \sim \sqrt{\frac{h}{g}} \tag{4.24}$$

which sets the natural response speed of the free surface, while another is the excitation time scale $T=1/\Omega$, where Ω is the forcing frequency. When these time scales are comparable, resonance-like growth may occur even before a steady state is reached, leading to large transient amplitudes that can exceed the eventual periodic response. From an energy viewpoint, the transient regime can be understood by examining the evolution of the total mechanical energy

$$E(t) = \int_V \left(\frac{1}{2}\rho\ |\boldsymbol{u}|^2 + \rho g z\right) dV \tag{4.25}$$

The time derivative of this energy satisfies

$$\frac{dE}{dt} = P_m - P_{\text{dissp}} \tag{4.26}$$

where P_m is the power injected by container motion and P_{dissp} represents viscous dissipation and energy loss due to turbulence or wave breaking. During the early transient, P_m typically exceeds P_{dissp}, resulting in rapid energy growth. As the motion develops, dissipation increases, particularly as velocity gradients intensify near walls and at the free surface, eventually leading to a balance that defines the long-term response. A distinctive feature of transient sloshing is the strong sensitivity to initial conditions. Small differences in initial free surface shape or velocity can lead to markedly different transient trajectories, especially in regimes where nonlinear effects are present. This sensitivity is often observed experimentally as run-to-run variability in peak impact loads or maximum wave heights, even under nominally identical excitation. Another important aspect is the role of damping. While viscous effects are often weak in steady sloshing analyses, they play a decisive role in transients by controlling the rate at which oscillations build up and decay. The effective damping ratio ζ is not constant but may evolve in time as flow regimes change, for example transitioning from laminar boundary layers to turbulent motion with enhanced dissipation. Dimensionless parameters again provide a compact description of transient sloshing behaviour. The Froude number is the characteristic acceleration amplitude, governs the relative importance of inertial forcing to gravity and thus the initial growth rate of the free surface.

$$Fr = \frac{a}{g} \tag{4.27}$$

where a is the characteristics acceleration.

The Reynolds number, on the other hand, controls the relative strength of inertia to viscous dissipation and therefore influences how rapidly transient oscillations decay or saturate.

The relative wave amplitude η/h evolves dynamically during the transient and serves as an indicator of when nonlinear effects become dominant. In shallow or intermediate depth containers, transient sloshing may also involve the formation of bores or hydraulic jumps, which act as highly effective energy dissipators and can abruptly limit further growth. From a structural and engineering perspective, transient sloshing is often more critical than steady state motion, because the largest forces on container walls and internal structures frequently occur during the build-up phase rather than after the response has settled. The impulsive pressures associated with rapidly accelerating liquid masses can be estimated by

$$p_{imp} \sim \rho h a_{eff} \tag{4.28}$$

where a_{eff} is an effective transient acceleration that may significantly exceed the nominal excitation amplitude due to modal superposition. These transient loads are particularly relevant in applications involving start-up, shutdown, seismic excitation, or manoeuvring, where the forcing is inherently time-limited and nonperiodic.

4.8 Mass spring modelling of propellant sloshing as a dynamical system

Liquid sloshing arises whenever a partially filled container undergoes translational or rotational acceleration. In spacecraft, where propellant mass fractions are often large,and manoeuvres induce sustained accelerations, the resulting motion of the free surface can couple strongly with vehicle dynamics. This coupling manifests through disturbance forces, torque oscillations, and effective migration of the centre of mass, all of which can degrade pointing accuracy or, in extreme cases, compromise stability.

In the early phases of a mission, when propellant levels are high, slosh effects are typically dominant. However, even during later operational phases, residual sloshing can remain relevant when attitude jitter budgets are tight. The central engineering problem is therefore not merely to predict fluid motion, but to represent its dominant dynamical effects in a form compatible with spacecraft guidance, navigation, and control analysis.

This article develops sloshing dynamics as an **equivalent mechanical system**, replacing the distributed fluid motion with a small number of lumped degrees of freedom. Such representations allow the essential physics of slosh to be embedded directly within rigid-body spacecraft simulations and control design environments.

4.8.1 *Equivalent mechanical representations of slosh*

In equivalent modelling, the total propellant mass is partitioned into two components:

- a rigidly attached mass, which moves with the tank and contributes directly to the spacecraft's inertia, and
- a sloshing mass, which represents the dominant free surface mode and moves relative to the tank.

The sloshing mass is constrained by mechanical elements that mimic restoring forces and energy dissipation within the fluid.

Two principal representations are employed: the mass spring damper model as illustrated in Fig. 4.5.

The validity of any equivalent model depends on the relative magnitudes of inertial, gravitational, and capillary forces. These are quantified using classical dimensionless numbers, which are Bond number, Weber number and Froude number that have been explained in Chapter 2. For typical spacecraft propellants under

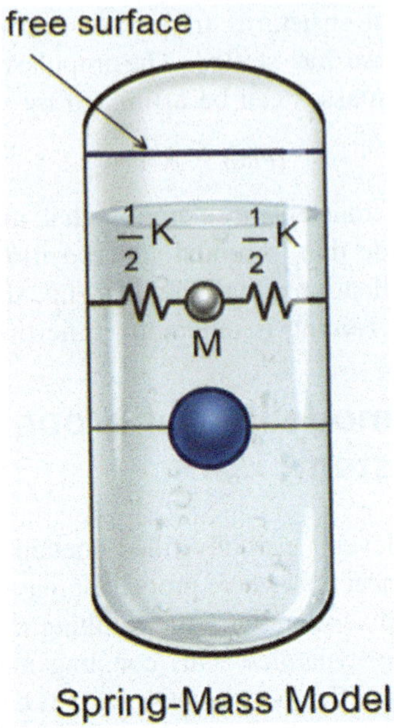

Figure 4.5 Equivalent mass spring damper model for propellant sloshing, illustrating the decomposition of total liquid mass into rigidly attached and dynamically active components.

manoeuvre accelerations, gravitational and inertial effects dominate, allowing surface tension to be neglected in first-order models.

The model can be summarised as the system of rigidly attached or impulsive component m_0 (moves with the tank), with an associated rigid inertia I_0. The mass spring damper model represents lateral sloshing as a linear oscillator attached to the tank wall. The schematic is shown in Fig. 4.6. The sloshing mass (m_1) is connected to the tank through a spring of stiffness (k_1) and a damper with coefficient (c_1). This model is valid for small amplitude motions and is particularly suitable for terrestrial applications or spacecraft manoeuvres with modest accelerations.

Figure Linear mass–spring–damper model of lateral sloshing, showing impulsive and sloshing mass components coupled to the tank.

Let x_1 denote the displacement of the sloshing mass relative to the tank, and x_0 the translational motion of the tank. The equation of motion is

$$m_1(\ddot{x}_0 + \ddot{x}_1 + H_1\ddot{\alpha} - g\alpha) + c_1\dot{x}_1 + k_1x_1 = 0 \tag{4.29}$$

where H_1 is the vertical offset of the sloshing mass and α is the small angular rotation of the tank. The force exerted by the fluid on the tank is

$$F = -m_1(\ddot{x}_1 + \ddot{x}_0) - m_0\ddot{x}_0 \tag{4.30}$$

and the associated moment is

$$M = -(I_0 + m_0H_0)\ddot{\alpha} - m_1H_1(\ddot{x}_1 + H_1\ddot{\alpha}) + gm_1x_1 \tag{4.31}$$

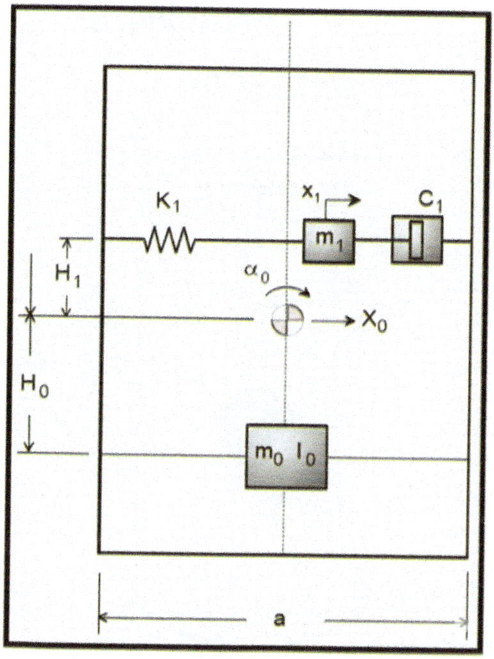

Figure 4.6 Linear mass spring damper model of lateral sloshing showing coupling between sloshing mass, tank motion, restoring stiffness, and damping elements.

These expressions show explicitly how slosh dynamics couple into spacecraft translational and rotational motion. Because the restoring stiffness is constant, large excitations can produce unphysical displacements exceeding the tank dimensions. Consequently, this model loses validity for large amplitude or low-gravity sloshing, motivating the use of pendulum-based representations.

4.9 Pendulum slosh model of propellant sloshing as a dynamical system

The pendulum model constrains the sloshing mass to move along a curved path, ensuring bounded motion even under sustained acceleration (as shown in Figs. 4.7 and 4.8). This is particularly appropriate for spacecraft tanks, where propellant tends to settle opposite the direction of acceleration.

In the pendulum-based representation of propellant sloshing, the total liquid mass is decomposed into a rigidly attached (impulsive) component and a dynamically active sloshing component. The impulsive mass moves together with the tank and contributes directly to the rigid-body inertia of the system, while the sloshing mass represents the dominant free-surface mode that responds dynamically to external excitation. The pendulum model constrains the sloshing mass to move along a curved

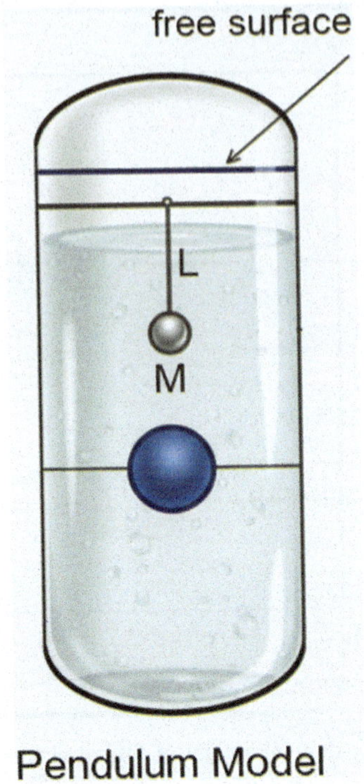

Figure 4.7 Pendulum based equivalent model of propellant sloshing, representing bounded rotational motion of the sloshing mass under sustained acceleration.

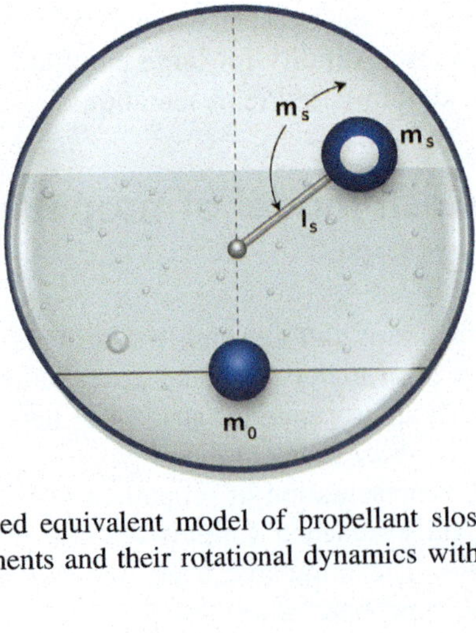

Figure 4.8 Pendulum-based equivalent model of propellant sloshing, illustrating impulsive and sloshing mass components and their rotational dynamics within the tank.

trajectory about a pivot point within the tank. This geometric constraint ensures bounded motion, even under sustained or steady accelerations, and provides a physically consistent description of large-scale liquid motion. Such behaviour is particularly relevant in spacecraft propellant tanks, where the liquid tends to accumulate opposite to the direction of acceleration and undergoes rotational or pendulum-like oscillations rather than unbounded lateral translation. Energy dissipation mechanisms associated with viscous effects, wave breaking, and internal fluid friction are represented through torsional damping at the pivot. Within its range of validity, the pendulum slosh model offers a compact and physically interpretable approximation of sloshing-induced forces, moments, and angular momentum exchange between the propellant and the tank structure. It is therefore widely employed in spacecraft dynamics, attitude control analysis, and guidance system design.

Let θ denote the pendulum angle, L the effective pendulum length, and m_p the sloshing mass. The governing equation is:

$$m_p L^2 \ddot{\theta} + c_\theta \dot{\theta} - m_p \alpha_T L \sin\theta - m_p \alpha_A L \cos\theta = 0 \tag{4.32}$$

where α_A and α_T are axial and transverse accelerations respectively.

The transverse force on the tank is:

$$F_T = m_p(\alpha_T - L\dot{\theta}^2 + L\ddot{\theta}\sin\theta) \tag{4.33}$$

and the axial force is

$$F_A = -m_p(L\dot{\theta}^2 \sin\theta + L\ddot{\theta}\cos\theta) \tag{4.34}$$

These forces represent the tangential and normal components of the sloshing motion. Unlike the mass spring model, the pendulum formulation naturally accommodates varying acceleration levels and large angular displacements. It is therefore the preferred choice for spacecraft propellant slosh analysis.

4.10 Nonlinear two-dimensional sloshing example using discretized concentrated mass approach

Consider a rigid rectangular tank of horizontal length (L_x) and liquid depth (Lz), partially filled with an incompressible Newtonian liquid. The tank is subjected to prescribed horizontal excitation, representative of seismic or mechanical forcing. The objective is to compute the nonlinear sloshing response, including free-surface elevation and internal fluid motion, using a concentrated mass model rather than a full continuum or finite-element formulation.

The physical configuration corresponds to Fig. 4.9A of the reference paper, where the liquid occupies a two-dimensional domain and undergoes finite-amplitude oscillations without wave breaking. Surface tension is neglected, and gravity is assumed constant.

The liquid domain is discretised into Nx elements in the horizontal direction and Ny elements in the vertical direction, as schematically illustrated in

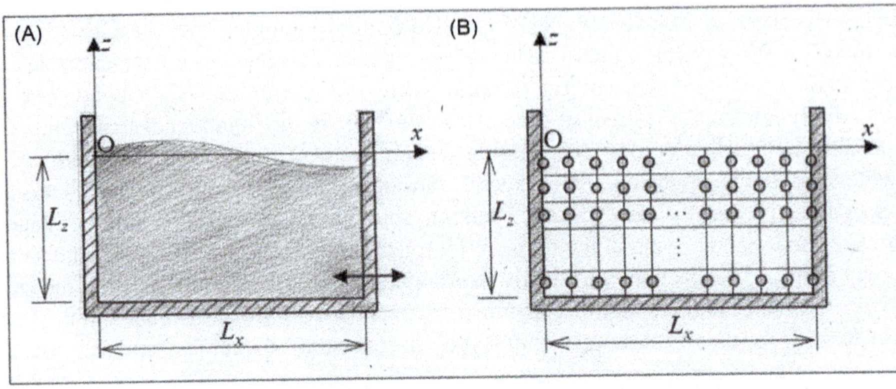

Figure 4.9 Two-dimensional sloshing configuration in a rigid rectangular tank: (A) physical liquid domain under horizontal excitation and (B) discretised representation using concentrated mass elements.

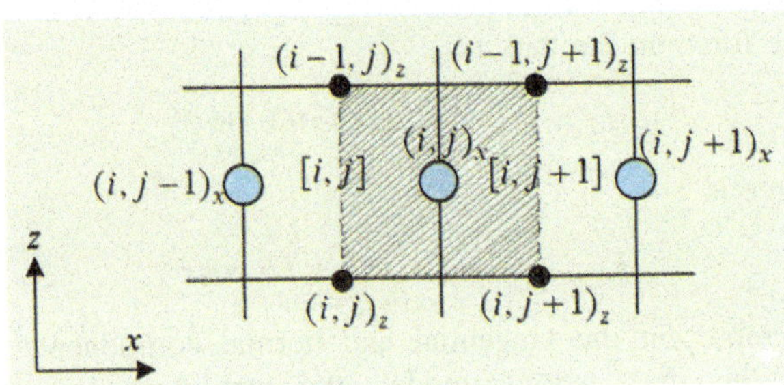

Figure 4.10 Structured discretisation of the liquid domain into horizontal and vertical elements for the nonlinear concentrated mass formulation.

Figs. 4.9B and 4.10 The adopted discretisation directly defines the number of concentrated mass units used to represent the liquid system.

Each unit mass corresponds to a finite portion of the fluid domain and therefore embodies the local inertia and kinematic response of that region. Increasing the number of discretisation units increases the spatial resolution of the model, allowing a more accurate representation of local sloshing characteristics such as spatial variation of velocity, deformation of the free surface, and interaction between different regions of the liquid. In the limit of a sufficiently fine discretisation, the concentrated mass formulation approaches the behaviour of a continuous fluid description, while for coarser discretisations it provides a reduced order but still physically meaningful approximation of the global sloshing dynamics.

This structured subdivision transforms the continuous fluid domain into a finite set of representative subdomains, each of which is characterised by uniform geometric and inertial properties. The resulting elemental dimensions are:

$$\begin{cases} l_x = \frac{L_x}{N_x} \\ l_z = \frac{L_z}{N_z} \end{cases} \tag{4.35}$$

so that the discretisation spans the full tank length and liquid depth in a consistent and physically interpretable manner. The mass of liquid contained within each element is assumed to be lumped at discrete mass points located on the vertical faces of the elements. This concentrated mass representation reflects the fact that, for horizontally excited sloshing, the dominant fluid motion is primarily translational in the horizontal direction, while vertical velocities remain comparatively small except in the immediate vicinity of the free surface. Accordingly, each mass point is restricted to move only in the horizontal direction, an assumption that preserves the essential sloshing kinematics while significantly reducing the number of degrees of freedom. The mass associated with each concentrated point is therefore given by:

$$m = \rho l_x l_z \tag{4.36}$$

is the liquid density. Vertical deformation of the liquid column and the associated free surface evolution are accounted for indirectly through nodal points positioned at the interfaces between elements in the vertical direction. These nodal points do not carry mass but act as kinematic constraints that enforce incompressibility and geometric compatibility between adjacent elements. Through their relative motion, they capture changes in liquid column height and free surface shape without introducing explicit vertical momentum equations. This discretisation strategy thus provides a balanced compromise between physical fidelity and computational efficiency, enabling both linear and nonlinear sloshing behaviour to be modelled within a reduced-order framework while retaining a clear connection to the underlying fluid mechanics.

4.10.1 Incompressibility and nonlinear geometric coupling

The incompressibility of the liquid imposes a strict kinematic constraint linking horizontal and vertical motions, such that any horizontal displacement of the concentrated masses must be accompanied by compatible vertical displacements in order to preserve local volume. Let $\delta x_{i,j}$ denote the horizontal displacement of the concentrated mass located in row i and column j, and let $\delta z_{i,j}$ denote the vertical displacement of the corresponding nodal point at the interface between elements. Because the liquid within each column behaves as a deformable but incompressible column, relative horizontal displacements between adjacent columns necessarily induce changes in column width, which must be compensated by vertical deformation. Enforcing volume conservation within a representative column leads to the nonlinear geometric relation

$$\delta z_{i-1,j} - \delta z_{i,j} = \frac{l_z\left(\delta x_{i,j} - \delta x_{i,j-1}\right)}{l_x + \delta x_j - \delta x_{j-1}} \tag{4.37}$$

where l_x and l_z are the elemental dimensions defined previously. The term δ_{x_j} epresents the average horizontal displacement of all mass points within column j, introduced to ensure geometric compatibility between neighbouring columns and defined as

$$\delta_{x_j} = \frac{1}{N_z} \sum_{i=1}^{N_i} \delta x_{i,j} \tag{4.38}$$

By specifying the tank bottom as a fixed reference, the vertical displacement at the lowest nodal point is set to

$$\delta z N_z j = 0 \tag{4.39}$$

which provides a boundary condition for the reconstruction of the vertical displacement field. Summing the relative vertical displacements upward from the bottom yields the vertical displacement of an arbitrary nodal point within column j as

$$\delta z_{i,j} = \frac{l_z}{l_x + \delta x_j - \delta x_{j-1}} \sum_{k=i+1}^{N_z} \delta x_{k,j} - \delta x_{k,j-1} \tag{4.40}$$

This expression makes explicit the nonlinear geometric coupling inherent in the formulation: vertical motion arises entirely from horizontal mass displacements through volume preservation, and the nonlinearity enters through the displacement-dependent denominator. As a result, even though individual masses are constrained to move horizontally, the model is capable of capturing finite-amplitude free surface deformation and nonlinear sloshing behaviour in a physically consistent manner.

4.10.2 Free-surface elevation

The free-surface elevation in column j, denoted η_j, corresponds to the displacement of the top nodal point:

$$\eta_j = \delta z_{0,j} = \frac{l_z(\delta x_j - \delta x_{j-1})}{l_x + \delta x_j - \delta x_{j-1}} \tag{4.41}$$

The restoring force acting on each mass point arises from hydrostatic and hydrodynamic pressure contributions. The vertical momentum balance in a liquid column is governed by Euler's equation,

$$\frac{\partial w_j(z,t)}{\partial t} = \frac{1}{\rho} \frac{\partial p_j(z,t)}{\partial z} - g \tag{4.42}$$

where w_j is the vertical velocity and p_j the pressure.

Integrating from the free surface $z = \eta_j$ down to the centre of element i,j (as presented in Fig. 4.11) yields the pressure at that element.

This pressure is decomposed as

$$p_{i,j} = p_{i,j}^{(s)} + p_{i,j}^{(d)} \tag{4.43}$$

where the static component is

$$p_{i,j}^{(s)} = \rho g \left(\eta_j + (i-1) l_z + \frac{l_z}{2} \right) \tag{4.44}$$

and the dynamic component accounts for vertical acceleration of the liquid.

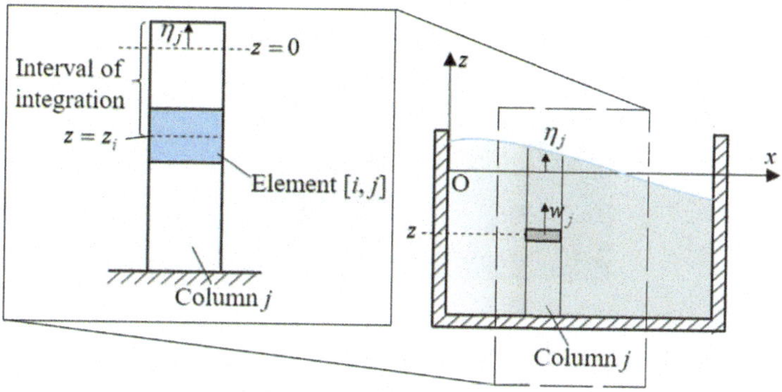

Figure 4.11 Column-wise-discretisation and free surface elevation definition used for pressure integration and evaluation of hydrostatic and hydrodynamic contributions.

The resulting horizontal restoring force on mass point i, j is

$$f_{i,j} = \alpha_{i,j} + \beta_{i,j} \tag{4.45}$$

with

$$\begin{cases} \alpha_{i,j} = \rho g l_z \eta_j + \frac{1}{2}\rho g \eta_j^2 ifi = 1 \\ \alpha_{i,j} = \rho g l_z \eta_j otherwise \end{cases} \tag{4.46}$$

and $\beta_{i,j}$ containing contributions from vertical accelerations.

The nonlinear equation of motion for each mass point is obtained by balancing inertia, restoring forces, damping, and external excitation:

$$m\ddot{\delta}x_{i,j} = \alpha_{i,j} + \beta_{i,j} - \alpha_{i,j+1} - \beta_{i,j+1} - c\dot{\delta}x_{i,j} + f_{i,j}^{(e)}(t) \tag{4.47}$$

The external excitation due to horizontal tank motion of amplitude Δ and angular frequency Ω is

$$f_{i,j}^{(e)}(t) = m\Delta\Omega^2 \sin(\Omega t) \tag{4.48}$$

Collecting all degrees of freedom into a vector $\boldsymbol{x}$, the governing system assumes the compact form

$$\boldsymbol{M}\ddot{\boldsymbol{x}} + \boldsymbol{C}\dot{\boldsymbol{x}} + \boldsymbol{K}\boldsymbol{x} = \boldsymbol{G} \tag{4.49}$$

Here, $\boldsymbol{M}$ and $\boldsymbol{C}$ are state-dependent, reflecting nonlinear inertia and stiffness.

The numerical results presented in this study are obtained using a concentrated mass discretisation in which the liquid domain is divided into N_x=30 elements in the horizontal direction and N_z=10 elements in the vertical direction. The geometrical and physical parameters adopted for the representative case are a tank length L_x=0.5 m, an equilibrium liquid depth L_z=0.25 m, and the liquid density ρ=1000 kg/m^3. Within this configuration, the reduced-degree-of-freedom formulation described in the section on degree-of-freedom reduction enables an accurate and physically consistent

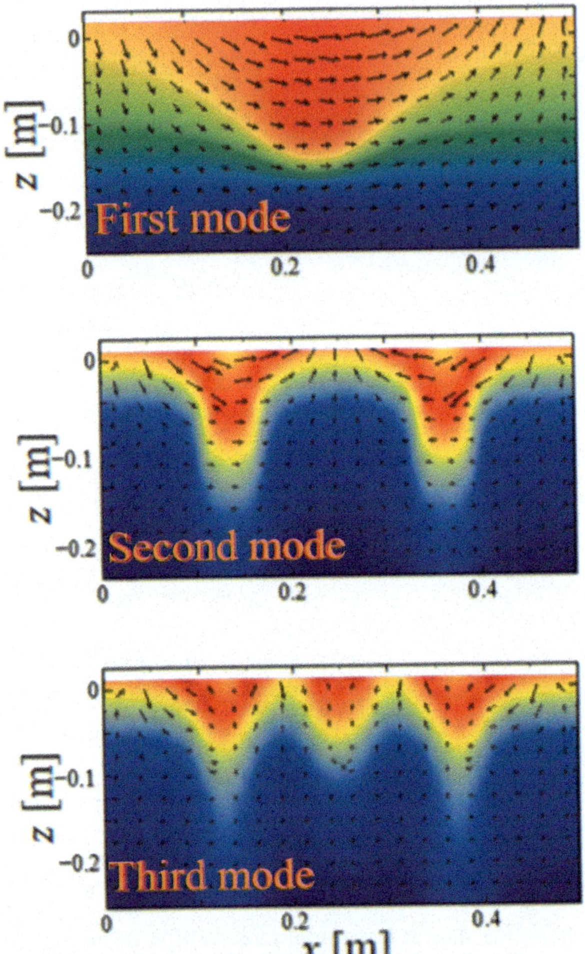

Figure 4.12 Time evolution of the nonlinear free surface profile obtained using the reduced degree of freedom concentrated mass sloshing model.

representation of the free-surface motion, as illustrated by the detailed evolution of the liquid surface in Fig. 4.12.

The ability of the model to reproduce the free-surface dynamics with high fidelity highlights the fundamental role of discretisation in modern computational engineering. Discretisation represents the process by which a continuous physical domain governed by partial differential equations is transformed into a finite set of algebraic equations that can be solved numerically. This concept underpins virtually all contemporary numerical methods used in fluid and structural mechanics, including the finite-element method, the finite volume method, and finite difference formulations. In addition, discretisation forms the conceptual foundation of so-called meshless approaches, such as smoothed particle hydrodynamics, as well as more recent data-driven and hybrid techniques that incorporate machine learning and neural networks.

Although these methods differ significantly in their mathematical formulation and numerical implementation, they all share a common objective: to approximate the governing conservation laws of mass, momentum, and energy in a form that is computationally tractable. Each method provides its own interpretation of the underlying physics, whether through nodal interpolation functions, control-volume flux balances, particle interactions, or learned surrogate representations. In this sense, discretisation does not alter the physics itself but rather defines how the physics is represented, resolved, and ultimately interpreted in numerical form. A more detailed and systematic discussion of discretisation strategies, their theoretical foundations, and their implications for accuracy and stability is therefore deferred to Chapter 5.

From a historical perspective, discretisation may be regarded as one of the most transformative innovations in twentieth-century engineering and applied science. Its development enabled engineers and scientists to move beyond purely analytical solutions, which are limited to highly idealised geometries and boundary conditions, and to address realistic, industrially relevant problems. The earliest numerical solutions based on discretised formulations were carried out using semimechanical computational devices and tabulating machines. These calculations often required one or more days to complete and relied heavily on manual preparation, including the physical loading of punched-card decks to encode the governing equations and input data.

Despite these severe limitations in computational capability, such early discretised models played a decisive role in some of the most significant technological achievements of the modern era. They contributed directly to the development of jet propulsion systems, the structural and aerodynamic design of early aircraft, and the analysis that ultimately enabled the launch of the first artificial satellites and rockets into space. In many cases, these pioneering calculations represented the first practical means of predicting complex coupled fluid–structure phenomena, including sloshing, vibration, and dynamic instability.

In contrast, contemporary numerical simulations operate on an entirely different computational scale. Modern discretisation-based methods, particularly finite volume formulations widely used in industrial fluid dynamics, can be executed on high-performance computing platforms comprising hundreds of physical processing cores and several terabytes of shared memory. For example, simulations of sloshing in automotive fuel tanks or washer reservoirs during vehicle manoeuvres may be performed on systems with approximately 512 physical cores and more than 2000 GB of random-access memory, with typical runtimes ranging from eight to ten hours. Each processor may sustain computational throughputs on the order of 10 teraflops, enabling the resolution of highly nonlinear, three-dimensional, transient flow phenomena with remarkable detail.

Within this broader historical and technological context, the concentrated mass approach presented in this chapter occupies a distinctive and valuable position. By combining physically motivated discretisation with a reduced number of degrees of freedom, it offers a computationally efficient alternative to large-scale continuum simulations while retaining essential nonlinear characteristics of sloshing dynamics.

4.11 Sloshing under reduced gravity conditions

Sloshing under reduced gravity conditions represents a profound shift in the governing physics of free surface motion, because the classical balance between inertia and gravity that underpins terrestrial sloshing theory is fundamentally weakened, allowing surface tension to emerge as a leading restoring mechanism, particularly in spacecraft, satellites, orbital platforms, and other low acceleration environments (Hung et al., 1991, 1992, 1996; Hung & Pan, 1994, 1995, 1996; Hung & Shyu, 1992, 1995; Kimura et al., 1995; Snyder, 1999; Veldman & Vogels, 1984). When the effective gravitational acceleration g is reduced by several orders of magnitude, the familiar gravity wave picture, in which the free surface oscillates due to the competition between inertia and weight, becomes inadequate, and the liquid interface instead behaves as a deformable membrane whose dynamics are controlled by capillary forces. The relative importance of gravity and surface tension is most naturally quantified by the Bond number

$$Bo = \frac{\rho g L^2}{\sigma} \tag{4.50}$$

L a characteristic length scale such as the tank radius or liquid depth, and σ the surface tension coefficient. This dimensionless group measures the ratio of gravitational body forces to capillary forces acting at the interface. In conventional terrestrial tanks, $Bo \gg 1$, meaning that gravity dominates and surface tension only influences small-scale phenomena such as wave breaking or contact line behaviour. In contrast, under reduced gravity conditions, g becomes small, and Bo may approach unity or fall well below one, at which point surface tension controls the shape, stability, and dynamics of the free surface. Physically, this transition has striking consequences. Instead of forming nearly planar gravity waves that oscillate about a horizontal mean level, the free surface in microgravity often adopts curved, axisymmetric, or container conforming shapes that minimise interfacial energy rather than gravitational potential energy. The equilibrium configuration is determined by a balance between capillary pressure

$$\Delta p = \sigma \kappa \tag{4.51}$$

where κ is the interface curvature, and any residual hydrostatic pressure gradient. As a result, liquids may wet container walls, form central domes or depressions, or even completely coat the interior surface, depending on the wetting properties and contact angle. From a dynamical perspective, sloshing in this regime can no longer be interpreted as a gravity wave phenomenon but must be analysed as a capillary wave problem. The governing equations of motion remain those of incompressible fluid flow, but the free surface boundary conditions acquire a dominant curvature-dependent term through the Young-Laplace relation, such that the dynamic condition at the interface, which is given by:

$$p = p_{ext} + \sigma\kappa \tag{4.52}$$

Linearisation about a curved equilibrium surface leads to oscillatory modes whose restoring force is proportional to surface tension rather than weight. This shift is reflected directly in the scaling of the natural frequency. While classical gravity driven sloshing frequencies scale as ($\omega \sim g/L$), capillary dominated oscillations scale instead as

$$\omega \sim \sqrt{\frac{\sigma}{\rho L^3}} \tag{4.53}$$

which highlighting the absence of gravity from the leading order balance. More generally, in the intermediate regime where both gravity and surface tension contribute, the natural frequency may be expressed in the form

$$\omega = \sqrt{\frac{\sigma L^3}{\rho}} f(Bo) \tag{4.54}$$

where the function $f(Bo)$ accounts for the combined influence of residual gravity and capillarity. In the limit Bo≫1, this expression recovers the gravity wave result, whereas for Bo≪1, it reduces to the pure capillary scaling (i.e. $\omega = \sqrt{\frac{\sigma L^3}{\rho}}$). The physical interpretation of this result is central to understanding sloshing in space systems: the characteristic time scale of free surface motion is no longer set by how fast a liquid parcel falls under gravity, but by how rapidly surface tension can restore curvature disturbances. This has important consequences for transient response, damping, and stability. Capillary waves typically have higher frequencies than their gravity-driven counterparts for small length scales, leading to fast oscillations that can be easily excited by mechanical disturbances or residual accelerations known as g jitter. At the same time, viscous dissipation becomes more influential because the energy stored in capillary deformation is often small compared with the energy dissipated in thin boundary layers near the interface and container walls. Another distinctive feature of reduced gravity sloshing is the strong coupling between interface shape and container geometry. In cylindrical or spherical tanks, the natural modes are often axisymmetric, and their spatial structure reflects the curvature of the equilibrium free surface. Small perturbations can propagate as travelling or standing capillary waves along the interface, while larger disturbances may lead to global rearrangement of the liquid mass, including sudden reorientation of the free surface or relocation of the liquid to different regions of the container.

Below is a fully rewritten and substantially enriched section suitable for direct inclusion in a scientific book or review article. The tone is British scientific English, technically rigorous but intentionally clear and readable.

4.12 Damping mechanisms in sloshing and the role of baffles

Sloshing motion is inherently dissipative. Whenever a liquid undergoes oscillatory motion relative to its container, part of the mechanical energy supplied by the excitation is irreversibly converted into heat through a variety of physical mechanisms. This energy loss manifests macroscopically as damping, which governs the decay rate of free oscillations, limits resonance amplitudes, and strongly influences the dynamic loads transmitted to the tank structure. Accurate representation of damping is therefore essential for realistic prediction of sloshing responses in engineering systems ranging from fuel tanks and cargo vessels to seismic storage tanks and spacecraft propellant systems.

Unlike stiffness or inertia, which can often be derived directly from geometry and mass distribution, sloshing damping is not associated with a single physical process. Instead, it arises from the combined action of several mechanisms operating simultaneously over different spatial and temporal scales. The dominant mechanism depends on the excitation amplitude, fluid properties, container geometry, and gravitational environment.

4.12.1 *Principal damping mechanisms in sloshing*

The most important sources of energy dissipation in sloshing flows may be classified as follows.

1. **Viscous dissipation within the liquid bulk:** Even in nominally inviscid sloshing theories, real liquids possess finite viscosity. Velocity gradients within the oscillating flow field lead to viscous shear stresses, which continuously convert kinetic energy into heat. For small amplitude sloshing, this bulk viscous dissipation is typically weak but becomes increasingly relevant for highly viscous liquids or at low excitation frequencies where shear layers persist over long times.
2. **Boundary layer losses at tank walls and bottom:** As the liquid oscillates relative to the container, thin viscous boundary layers develop along the tank walls and base. Within these layers, velocity gradients are large, and viscous dissipation is concentrated. For rigid tanks with smooth walls, wall boundary layer damping often dominates over bulk dissipation at small amplitudes. Its magnitude depends strongly on wall roughness, liquid viscosity, and the characteristic sloshing frequency.
3. **Wave breaking and turbulence at large amplitudes:** When the free surface steepens sufficiently, nonlinear effects become pronounced, and wave breaking may occur. Breaking waves generate localised jets, splashes, and entrained air bubbles, leading to intense turbulence and rapid energy loss. In this regime, damping increases dramatically and is highly nonlinear. It cannot be represented accurately by simple linear models and must usually be determined experimentally or through high-fidelity numerical simulations.
4. **Contact line motion and surface contamination effects:** At the intersection between the free surface and the tank wall, the contact line undergoes periodic motion. This motion is associated with additional dissipation due to microscopic wetting dynamics, surface roughness, and the presence of contaminants or surfactants. Although often neglected in simplified models, contact line damping can be significant in small-scale tanks, low-gravity environments, or systems with high surface-to-volume ratios.

These mechanisms act concurrently, and their relative importance varies across operating regimes. Table 4.1 summarises representative damping formulations commonly adopted in sloshing analysis for different container geometries and excitation conditions.

4.12.2 Linear damping representation for small amplitude sloshing

For sufficiently small amplitude oscillations, sloshing motion remains weakly nonlinear, and the various dissipation mechanisms can often be lumped into an equivalent linear viscous damping term. In this case, the damping force acting on the generalised sloshing coordinate x may be written as:

$$F_d = c\dot{x} \tag{4.55}$$

where c is an equivalent damping coefficient. This coefficient encapsulates the combined effects of bulk viscosity, wall boundary layers, and mild free surface deformation.

Within the framework of reduced order models, such as the classical concentrated mass or pendulum analogies, this linear damping term leads to equations of motion of the form

$$m\ddot{x} + c\dot{x} + kx = f(t) \tag{4.56}$$

where m is the effective sloshing mass, k is the equivalent stiffness, and f(t) represents external excitation. Although this representation is phenomenological, it is often adequate for engineering design under low excitation levels and has the advantage of mathematical simplicity.

However, it must be emphasised that the coefficient c is not a material constant. It depends on liquid properties, tank geometry, fill level, and excitation frequency. In practice, it is frequently obtained from experimental decay tests or calibrated numerical simulations.

4.12.3 Nonlinear and amplitude-dependent damping

As the sloshing amplitude increases, the assumption of linear damping rapidly loses validity. So its representation assumes that the resisting force is directly proportional to the velocity and always acts in opposition to the motion. Such a linear damping model is only valid when velocity amplitudes remain small, the flow field stays laminar or only weakly disturbed, and free-surface deformations are mild, so that phenomena such as flow separation, wave breaking, and turbulence do not occur. Under these conditions, energy dissipation is governed primarily by viscous shear and can be reasonably approximated as a linear process. However, as sloshing amplitudes increase, these assumptions progressively fail. The flow becomes increasingly nonlinear, with the emergence of separation, turbulence, and violent free-surface motion, causing the rate of energy dissipation to depart from a simple linear dependence on velocity. In such conditions, the damping force is more appropriately expressed as a nonlinear function of velocity and displacement, for example

$$F_d = c_1\dot{x} + c_2\ |\dot{x}|\ \dot{x} \tag{4.57}$$

As such, the damping force is expressed explicitly as a function of velocity. The linear term $c_1\dot{x}$ represents dissipation mechanisms that remain approximately proportional to velocity, such as viscous shear in the liquid bulk and laminar boundary layers at the tank walls. The quadratic term $c_2\ |\dot{x}|\ \dot{x}$ is introduced to capture inertial and turbulence-dominated losses, including flow separation, vortex shedding, and wave breaking, for which energy dissipation scales with the square of the velocity. This form is particularly effective when damping is controlled primarily by the intensity of the flow, and when the free surface geometry does not vary drastically within a sloshing cycle. It provides a practical compromise between physical realism and mathematical simplicity and is therefore widely used in engineering reduced-order models.

However, in many sloshing situations, the level of dissipation cannot be determined by velocity alone. As such the damping force may take the form:

$$F_d = c(x,\ \dot{x})\dot{x} \tag{4.58}$$

where the effective damping coefficient itself depends on the instantaneous state of motion.

In reduced gravity environments, nonlinear damping effects are particularly pronounced. When gravitational restoring forces are weak, surface tension and viscous forces play a comparatively larger role, and energy dissipation is strongly influenced by free surface curvature, contact line dynamics, and interfacial phenomena. Under such conditions, damping must usually be characterised through dedicated experiments or high-resolution multiphase simulations.

4.13 Baffles in sloshing tanks: purpose and physical principles

A baffle is an internal structural element installed within a tank for the explicit purpose of modifying and controlling liquid motion. Baffles represent one of the most effective and widely adopted passive means of sloshing mitigation and are routinely employed in road tankers, marine vessels, launch vehicles, and stationary storage tanks. Their primary function is to dissipate sloshing energy and to disrupt the development of large-scale, coherent free surface waves that would otherwise form under external excitation. By partially obstructing the flow and introducing additional wetted surfaces within the liquid domain, baffles increase viscous dissipation, promote the early onset of turbulence even at moderate excitation levels, and effectively reduce the characteristic sloshing length over which liquid momentum can accumulate. From a physical standpoint, baffles act by redistributing momentum within the tank, interrupting the predominantly horizontal motion of the liquid and forcing it to change direction, fragment, and reorganise into smaller-scale flow structures, thereby lowering peak velocities associated with global sloshing modes. At the same time, the presence of baffle surfaces generates additional boundary layers and regions of strong shear, while flow separation at baffle edges further enhances turbulent energy loss. By subdividing the fluid domain, baffles also modify the effective sloshing geometry, altering the natural frequencies of the system and shifting them away from dominant excitation frequencies. Finally, by limiting wave growth and steepening, baffles can suppress or delay the onset of violent wave breaking, thereby reducing impact loads on tank walls and roofs and contributing to improved structural safety (238. The use of wire gauze baffles in the vacuum columns for distilling, 1967; BERLAMONT & VANDERSTAPPEN, 1980; Collias & Prud'homme, 1985; Holmes et al., 1964; Idem et al., 1987; Kolarik & Pipkin, 1983; Lewis & Davidson, 1983; Nakli & Welsby, 1978; Nienow, 1968; Okuyama et al., 1986; Prakash et al., 1983; SHASHKOV, 1968; Voncken et al., 1964; Yuu & Oda, 1980).

4.13.1 Mathematical representation of baffle effects

In reduced order models, the influence of baffles is commonly introduced through modified damping and stiffness terms. A simple approach is to augment the damping coefficient,

$$c = c_0 + c_b \tag{4.59}$$

where c_0 represents the intrinsic damping of the unbaffled tank, and c_b is an additional contribution due to the baffle.

More refined models account for the fact that baffle-induced damping is often nonlinear and velocity dependent, leading to expressions such as

$$F_b = c_b \ |\dot{x}| \ \dot{x} \tag{4.60}$$

In continuum-based formulations, baffles appear explicitly as internal boundaries in the governing Navier-Stokes equations, enforcing no-slip conditions and inducing flow separation and recirculation zones. The resulting increase in dissipation emerges naturally from the resolved velocity gradients.

4.13.2 Perforated baffles and the role of holes

In many applications, baffles are intentionally perforated, containing holes or slots that allow controlled fluid exchange between adjacent compartments. The inclusion of holes serves several important purposes.

First, perforations prevent excessive pressure differences across the baffle, which could otherwise lead to high structural loads or flow-induced vibration. Second, they allow partial mass transfer, which smooths the liquid motion while still providing significant damping.

From a physical standpoint, flow through baffle holes generates localised jets and vortices. These features are highly dissipative and contribute strongly to energy loss. The associated pressure drop across a perforated baffle may be approximated by relations of the form

$$\Delta p = \frac{1}{2}\rho C_d\left(\frac{Q}{A_h}\right)^2 \tag{4.61}$$

where Q is the volumetric flow rate through the holes, A_h is the total open area, and C_d is a discharge coefficient dependent on hole geometry.

The size, number, and distribution of holes therefore play a critical role in tuning the damping characteristics. Small holes increase dissipation but may restrict flow excessively, while large holes reduce damping effectiveness. Optimal designs balance these competing effects.

4.13.3 Under baffled and over baffled tanks

The terms *under baffled* and *over baffled* are commonly used in engineering practice, and their conceptual meaning is physically sound.

An under baffled tank is one in which the baffle arrangement is insufficient to suppress dominant sloshing modes. In such cases, large-scale oscillations persist, damping remains low, and dynamic loads may exceed acceptable limits. This situation often arises when baffles are too few, too small, or poorly positioned relative to the dominant mode shapes.

An over baffled tank, by contrast, contains an excessive number of baffles or overly restrictive perforations. While sloshing amplitudes are strongly reduced, overbaffling may introduce undesirable consequences, such as increased pressure losses during filling and draining, excessive structural loads on baffles, or trapped liquid volumes. In propulsion systems, overbaffling can impair propellant management and lead to uneven mass distribution.

From a modelling perspective, overbaffling manifests as excessively high effective damping, potentially leading to unrealistic suppression of dynamic response. Therefore optimal baffle design seeks a balance rather than maximal damping.

4.13.4 Nondimensional regime classification of sloshing dynamics

Direct experimental investigation of sloshing under realistic flight conditions is rarely feasible. Full-scale tanks, real propellants, and flight-like acceleration environments are costly, hazardous, and often inaccessible. As a result, engineers and researchers rely heavily on subscale experiments, ground testing, and reduced-gravity platforms. The central scientific challenge is therefore not how to perform slosh experiments, but how to ensure that the results obtained from such experiments are physically representative of the real system.

The cornerstone of this representativeness is nondimensional similarity analysis. Rather than attempting to replicate absolute dimensions, forces, or motions, similarity analysis seeks to preserve the *relative balance of physical mechanisms* governing the flow. When this balance is preserved, two systems are said to be fluid dynamically similar, and their sloshing behaviour can be meaningfully compared despite differences in size, fluid, or acceleration environment.

This article presents a rigorous and unified treatment of nondimensional sloshing analysis. Emphasis is placed on the physical interpretation of the governing dimensionless numbers, their interdependence, the regimes they define, and their application to practical slosh testing and modelling (Debongnie, 1986; Demirbilek, 1983).

4.13.4.1 Physical description of sloshing

Sloshing, from a physical standpoint, is governed by the interaction of four fundamental mechanisms that compete and cooperate to shape the observed dynamics. Inertial effects arise from the need to accelerate the liquid mass and tend to amplify motion once it has been initiated. Body acceleration effects, which may be due to gravity, thrust, or manoeuvre-induced forces, act to orient the free surface and provide a dominant restoring influence in many terrestrial and high-acceleration environments. Viscous effects are responsible for the dissipation of kinetic energy through shear and boundary layer development, thereby controlling damping rates and the decay of oscillations over time. Surface tension acts at the liquid-gas interface and resists deformation by penalising curvature, playing a particularly important role in small-scale systems and reduced gravity environments where it may dominate over both inertia and body acceleration. The sloshing regime that ultimately emerges is determined not by the absolute magnitude of any one of these effects, but by their relative importance, and nondimensional analysis provides a powerful and systematic framework for assessing this balance in a way that is independent of scale. For isothermal sloshing, the behaviour of the system may be expressed in terms of a small set of characteristic quantities describing the geometry, motion, and fluid properties, namely a representative length scale such as the tank diameter, a characteristic liquid velocity associated with the sloshing motion, the applied body acceleration, and the density, viscosity, and surface tension of the liquid. Although these quantities possess physical dimensions, they may be combined into a reduced set of dimensionless parameters that encapsulate the essential physics of the problem. Among these, several nondimensional numbers have become standard tools for interpreting and classifying sloshing behaviour because of their clear physical meaning. The Reynolds number compares inertial effects with viscous resistance and governs the development of boundary layers along container walls, the efficiency of viscous damping, and whether the flow remains laminar or becomes turbulent during a sloshing cycle. High Reynolds number sloshing is typically associated with thin boundary layers and weak viscous attenuation, whereas low Reynolds number motion is heavily damped and may suppress large-scale oscillations altogether. The Weber number measures the relative importance of inertia and surface tension and therefore controls the degree of free surface deformation, the smoothness of wave motion, and the likelihood of wave breaking or droplet formation. In low Weber number conditions, the interface remains smooth and resistant to distortion, while at high Weber number inertia, it overwhelms surface tension, and the free surface becomes highly mobile and unstable. The Bond number compares body acceleration effects with surface tension and determines whether the liquid configuration is primarily shaped by gravity or imposed acceleration, or instead by capillary forces. This distinction is of central importance in reduced gravity and microgravity environments, where even modest surface tension can dominate the behaviour of the liquid and produce configurations that are counterintuitive from a terrestrial perspective. These nondimensional measures are not independent of one another, and their ratios give rise to additional parameters such as the Froude number, which characterises the relationship between inertial motion and body acceleration and is closely linked to wave propagation speed, the Capillary number, which reflects the balance between viscous stresses and surface tension and is particularly relevant to wetting and thin film dynamics, and the Galileo number, which compares body acceleration to viscous diffusion and is useful for describing acceleration driven bulk motion and settling processes. In an idealised sense, two sloshing systems would be exactly similar if all relevant nondimensional parameters were identical, but in practice this is rarely achievable due to constraints on fluid selection, tank size, acceleration environment, and safety considerations. Consequently, the concept of regime similarity is adopted, whereby systems are considered comparable if they occupy the same region of nondimensional parameter space and are governed by the same dominant physical mechanisms, even if the numerical values of the parameters differ. A particularly insightful way of visualising this concept is through the combined consideration of Weber and Bond numbers, which together delineate distinct physical regimes ranging from capillary-dominated motion, through inertia-dominated behaviour, to body acceleration-dominated sloshing. Experience shows that surface tension effects weaken progressively as the influence of body acceleration increases and become negligible once acceleration forces overwhelmingly dominate, while inertia-controlled regimes are associated with energetic and often nonlinear free surface motion. An additional complexity unique to sloshing is the absence of a single natural velocity scale, since the liquid motion is generated by the excitation itself rather than imposed externally, and the appropriate velocity must therefore be inferred from the underlying physical mechanism, whether it be wave motion associated with natural sloshing modes or bulk acceleration-driven transport as encountered during propellant settling manoeuvres. The role of viscosity within this framework is strongly context dependent, and while exact matching of viscous effects between experimental models and real systems is often impractical, meaningful similarity may still be achieved provided that both operate within the same qualitative flow regime, such as laminar or turbulent, or when other dissipation mechanisms dominate the dynamics. In reduced gravity and microgravity conditions, the dramatic reduction of body acceleration shifts the balance decisively towards surface tension, making wetting properties, contact angle, and interfacial stability central to the behaviour of the liquid, and leading to sloshing modes and frequencies that differ fundamentally from those observed under terrestrial gravity. In such environments, sloshing may be more appropriately interpreted as a capillary-driven oscillatory system, with surface tension providing the primary restoring influence rather than gravity.

4.13.4.2 Limitations and assumptions

Nondimensional sloshing analysis is unavoidably built upon a set of simplifying assumptions, and the scientific validity of any resulting conclusions depends on how clearly these assumptions are articulated and how well they reflect the physical reality of the application under consideration. Typical assumptions adopted in sloshing studies include treating the flow as isothermal, prescribing a fixed fill level, idealising the container walls as smooth and rigid, neglecting mass transfer between the liquid and the surrounding gas, and restricting attention to small or moderate free surface amplitudes for which extreme nonlinear effects remain limited. Such assumptions are not merely mathematical conveniences; they define the boundaries within which the analysis is applicable, and if they are violated in practice, similarity arguments may become misleading or entirely invalid. Although nondimensional analysis does not remove the inherent complexity of sloshing experiments, it offers a coherent and physically grounded framework that guides critical decisions, such as the selection of suitable test fluids, the choice of tank scale, the design of excitation manoeuvres, the interpretation of experimental measurements, and the validation of analytical or numerical models. Most importantly, it acts as a safeguard against drawing erroneous conclusions from experiments that may appear similar in a superficial sense but are in fact governed by different physical mechanisms. Sloshing is fundamentally a multiscale and multiphysics phenomenon, and its dominant behaviour is controlled not by absolute size, velocity, or acceleration, but by the relative balance between inertia, body acceleration, viscosity, and surface tension. Nondimensional similarity analysis provides the conceptual structure required to identify and manage this balance, allowing engineers and researchers to focus on the governing regimes rather than attempting to achieve unattainable exact parameter matching. This regime-based philosophy enables the design of meaningful and physically relevant experiments under severe practical constraints and forms the foundation of modern slosh testing, modelling, and simulation practices, remaining indispensable for the safe and efficient design of liquid-containing systems

References

Akkaoui, Q., Capiez-Lernout, E., Soize, C., & Ohayon, R. (2020). Uncertainty quantification for dynamics of geometrically nonlinear structures coupled with internal acoustic fluids in presence of sloshing and capillarity. *Journal of Fluids and Structures, 94*, 102966. https://doi.org/10.1016/j.jfluidstructs.2020.102966.

Alemi Ardakani, H. (2016). A symplectic integrator for dynamic coupling between nonlinear vessel motion with variable cross-section and bottom topography and interior shallow-water sloshing. *Journal of Fluids and Structures, 65*, 30–43. https://doi.org/10.1016/j.jfluidstructs.2016.03.013.

Ali Goudarzi, M., & Reza Sabbagh-Yazdi, S. (2012). Investigation of nonlinear sloshing effects in seismically excited tanks. *Soil Dynamics and Earthquake Engineering, 43*, 355–365. https://doi.org/10.1016/j.soildyn.2012.08.001.

BERLAMONT, J., & VANDERSTAPPEN, N. (1980). On the effect of baffle walls in a water tower tank under wind action. In J. E. CERMAK (Ed.). *Wind Engineering: Proceedings of the Fifth International Conference, Fort Collins, Colorado, USA, July 1979* (pp. 1195–1201). Pergamon. https://doi.org/10.1016/B978-1-4832-8367-8.50110-8.

Classified abstract 1–266 (1967). 238. The use of wire gauze baffles in the vacuum columns for distilling: M A Mamadovet al, Neftepererabotka i neftekhim, no 4, 1966, 34–36 (in Russian). *Vacuum, 17*(1), 67. https://doi.org/10.1016/0042-207X(67)92770-4.

Collias, D. J., & Prud'homme, R. K. (1985). The effect of fluid elasticity on power consumption and mixing times in stirred tanks. *Chemical Engineering Science, 40*(8), 1495–1505. https://doi.org/10.1016/0009-2509(85)80091-9.

Debongnie, J. F. (1986). On a purely lagrangian formulation of sloshing and fluid-induced vibrations of tanks. *Computer Methods in Applied Mechanics and Engineering, 58*(1), 1–18. https://doi.org/10.1016/0045-7825(86)90075-7.

Demirbilek, Z. (1983). Energy dissipation in sloshing waves in a rolling rectangular tank—II. Solution method and analysis of numerical technique. *Ocean Engineering, 10*(5), 359–374. https://doi.org/10.1016/0029-8018(83)90005-7.

Du, Y., Wang, C., & Zhang, N. (2019). Numerical simulation on coupled ship motions with nonlinear sloshing. *Ocean Engineering, 178*, 493–500. https://doi.org/10.1016/j.oceaneng.2019.02.057.

Gurusamy, S., & Kumar, D. (2020). Experimental study on nonlinear sloshing frequency in shallow water tanks under the effects of excitation amplitude and dispersion parameter. *Ocean Engineering, 213*, 107761. https://doi.org/10.1016/j.oceaneng.2020.107761.

Hara, K., & Watanabe, M. (2016). Formulation of the nonlinear sloshing-structure coupled problem based on the Hamiltonian mechanics for constraint systems. *Journal of Fluids and Structures, 62*, 104–124. https://doi.org/10.1016/j.jfluidstructs.2015.12.011.

Hashimoto, H., & Sudo, S. (1988). Violent liquid sloshing in vertically excited cylindrical containers. *Experimental Thermal and Fluid Science, 1*(2), 159–169. https://doi.org/10.1016/0894-1777(88)90033-7.

Holmes, D. B., Voncken, R. M., & Dekker, J. A. (1964). Fluid flow in turbine-stirred, baffled tanks—I: Circulation time. *Chemical Engineering Science, 19*(3), 201–208. https://doi.org/10.1016/0009-2509(64)85030-2.

Huang, S., Duan, W., Han, X., Nicoll, R., You, Y., & Sheng, S. (2018). Nonlinear analysis of sloshing and floating body coupled motion in the time-domain. *Ocean Engineering, 164*, 350–366. https://doi.org/10.1016/j.oceaneng.2018.06.003.

Hung, R. J., & Shyu, K. L. (1992). Excitation of slosh waves associated with low frequency impulsive reverse gravity acceleration of geyser initiation. *Acta Astronautica, 26*(6), 425–433. https://doi.org/10.1016/0094-5765(92)90071-P.

Hung, R. J., & Pan, H. L. (1994). Simulation of sloshing dynamics induced forces and torques actuated on Dewar container driven by gravity gradient and jitter accelerations in microgravity. *Simulation Practice and Theory,* 2(3), 91–120. https://doi.org/10.1016/0928-4869(94)90007-8.

Hung, R. J., & Pan, H. L. (1995). Combined gravity gradient and jitter accelerations acting on liquid-vapor interface oscillations in reduced gravity. *International Journal of Mechanical Sciences, 37*(11), 1203–1221. https://doi.org/10.1016/0020-7403(95)00012-M.

Hung, R. J., & Shyu, K. L. (1995). Slosh wave and geyser excitations due to liquid hydrogen shut-off during draining in microgravity. *Acta Astronautica, 35*(8), 509–523. https://doi.org/10.1016/0094-5765(94)00291-S.

Hung, R. J., & Pan, H. L. (1996). Modelling of sloshing modulated angular momentum fluctuations actuated by gravity gradient associated with spacecraft slew motion. *Applied Mathematical Modelling, 20*(5), 399–409. https://doi.org/10.1016/0307-904X(95)00159-H.

Hung, R. J., Lee, C. C., & Leslie, F. W. (1991). Slosh wave excitation in a partially filled rotating tank due to gravity jitters in a microgravity environment. *Acta Astronautica, 25*(8), 523–551. https://doi.org/10.1016/0094-5765(91)90035-4.

Hung, R. J., Lee, C. C., & Leslie, F. W. (1992). Similarity rules in gravity jitterrelated spacecraft liquid propellant slosh waves excitation. *Journal of Fluids and Structures, 6*(4), 493–522. https://doi.org/10.1016/0889-9746(92)90028-2.

Hung, R. J., Long, Y. T., & Zu, G. J. (1996). Sloshing of cryogenic helium driven by lateral impulse/gravity gradient-dominated/or g-jitter-dominated accelerations and orbital dynamics. *Cryogenics, 36*(10), 829–841. https://doi.org/10.1016/0011-2275(96)00046-X.

Idem, S. A., Jung, C., Gonzalez, G. J., & Goldschmidt, V. W. (1987). Performance of air-to-water copper finned-tube heat exchangers at moderately low air-side Reynolds numbers, including effects of baffles. *International Journal of Heat and Mass Transfer, 30*(8), 1733–1741. https://doi.org/10.1016/0017-9310(87)90320-6.

Ikeda, T., Harata, Y., & Osasa, T. (2016). Internal resonance of nonlinear sloshing in rectangular liquid tanks subjected to obliquely horizontal excitation. *Journal of Sound and Vibration, 361*, 210–225. https://doi.org/10.1016/j.jsv.2015.09.032.

Kimura, K., Ogura, K., Mieda, T., Yamamoto, K., Eguchi, Y., Moriya, S., Hagiwara, Y., Takakuwa, M., Kodama, T., & Koike, K. (1995). Experimental and analytical studies on the multi-surface sloshing characteristics of a top entry loop type FBR. *Nuclear Engineering and Design, 157*(1), 49–63. https://doi.org/10.1016/0029-5493(95)00982-I.

Kolarik, Z., & Pipkin, N. (1983). Drop size distribution in stirred tanks using the uranyl nitrate/tributyl phosphate extraction system. *Hydrometallurgy, 9*(3), 371–379. https://doi.org/10.1016/0304-386X(83)90031-2.

Lewis, D. A., & Davidson, J. F. (1983). Bubble sizes produced by shear and turbulence in a bubble column. *Chemical Engineering Science, 38*(1), 161–167. https://doi.org/10.1016/0009-2509(83)80147-X.

Love, J. S., & Haskett, T. C. (2018). Nonlinear modelling of tuned sloshing dampers with large internal obstructions: Damping and frequency effects. *Journal of Fluids and Structures, 79*, 1–13. https://doi.org/10.1016/j.jfluidstructs.2018.01.012.

Modi, V. J., & Welt, F. (1988). Damping of wind induced oscillations through liquid sloshing. *Journal of Wind Engineering and Industrial Aerodynamics, 30*(1), 85–94. https://doi.org/10.1016/0167-6105(88)90074-8.

Moslemi, M., Farzin, A., & Kianoush, M. R. (2019). Nonlinear sloshing response of liquid-filled rectangular concrete tanks under seismic excitation. *Engineering Structures, 188*, 564–577. https://doi.org/10.1016/j.engstruct.2019.03.037.

Nakli, N., & Welsby, V. G. (1978). The use of geometrical diffraction theory to predict the directional response of acoustic arrays with side baffles. *Journal of Sound and Vibration, 60*(4), 565–585. https://doi.org/10.1016/S0022-460X(78)80093-5.

Nienow, A. W. (1968). Suspension of solid particles in turbine agitated baffled vessels. *Chemical Engineering Science, 23*(12), 1453–1459. https://doi.org/10.1016/0009-2509(68)89055-4.

Okuyama, K., Kousaka, Y., Yamamoto, S., & Hosokawa, T. (1986). Particle loss of aerosols with particle diameters between 6 and 2000 nm in stirred tank. *Journal of Colloid and Interface Science, 110*(1), 214–223. https://doi.org/10.1016/0021-9797(86)90370-X.

Prakash, J., Garg, H. P., & Datta, G. (1983). Effect of baffle plate on the performance of built-in storage type solar water heater. *Energy, 8*(5), 381–387. https://doi.org/10.1016/0360-5442(83)90088-9.

Sakai, F., Nishimura, M., & Ogawa, H. (1984). Sloshing behavior of floating-roof oil storage tanks. *Computers & Structures, 19*(1), 183–192. https://doi.org/10.1016/0045-7949(84)90217-7.

SHASHKOV, A. G. (1968). THEORY OF CONTROL DEVICES OF THE "NOZZLE-BAFFLE" TYPE, WORKING WITH OIL. In M. A. AIZERMAN (Ed.). *Pneumatic and Hydraulic Control Systems: Volume 1: Seminar on Pneumohydraulic Automation (First Session)* (pp. 285–308). Pergamon. https://doi.org/10.1016/B978-0-08-001103-5.50028-1.

Shemer, L., & Kit, E. (1988). Study of the role of dissipation in evolution of nonlinear sloshing waves in a rectangular channel. *Fluid Dynamics Research, 4*(2), 89–105. https://doi.org/10.1016/0169-5983(88)90017-2.

Sinai, Y. L. (1985). Fundamental sloshing frequencies of stratified two-fluid systems in closed prismatic tanks. *International Journal of Heat and Fluid Flow, 6*(2), 142–144. https://doi.org/10.1016/0142-727X(85)90052-9.

Snyder, H. A. (1999). Sloshing in microgravity. *Cryogenics, 39*(12), 1047–1055. https://doi.org/10.1016/S0011-2275(99)00120-4.

Strelnikova, E., Kriutchenko, D., Gnitko, V., & Degtyarev, K. (2020). Boundary element method in nonlinear sloshing analysis for shells of revolution under longitudinal excitations. *Engineering Analysis with Boundary Elements, 111*, 78–87. https://doi.org/10.1016/j.enganabound.2019.10.008.

Veldman, A. E. P., & Vogels, M. E. S. (1984). Axisymmetric liquid sloshing under low-gravity conditions. *Acta Astronautica, 11*(10), 641–649. https://doi.org/10.1016/0094-5765(84)90048-1.

Voncken, R. M., Holmes, D. B., & den Hartog, H. W. (1964). Fluid flow in turbine-stirred, baffled tanks—III: Dispersion during circulation. *Chemical Engineering Science, 19*(3), 209–213. https://doi.org/10.1016/0009-2509(64)85031-4.

Xin, J., Chen, Z., Shi, F., Shi, F., & Jin, Q. (2020). Numerical simulation of nonlinear sloshing in a prismatic tank by a Cartesian grid based three-dimensional multiphase flow model. *Ocean Engineering, 213*, 107629. https://doi.org/10.1016/j.oceaneng.2020.107629.

Yuu, S., & Oda, T. (1980). Measurement of turbulence parameters in a non-baffled stirred tank with high rotation speeds. *The Chemical Engineering Journal, 20*(1), 35–42. https://doi.org/10.1016/0300-9467(80)85004-0.

Zhang, C., Li, Y., & Meng, Q. (2015). Fully nonlinear analysis of second-order sloshing resonance in a three-dimensional tank. *Computers & Fluids, 116*, 88–104. https://doi.org/10.1016/j.compfluid.2015.04.016.

Zhao, D., Hu, Z., Chen, G., Lim, S., & Wang, S. (2018). Nonlinear sloshing in rectangular tanks under forced excitation. *International Journal of Naval Architecture and Ocean Engineering, 10*(5), 545–565. https://doi.org/10.1016/j.ijnaoe.2017.10.005.

Numerical methods for sloshing simulations

5

5.1 Introduction

Sloshing dynamics represents one of the most challenging and practically important problems in fluid mechanics, owing to the inherently nonlinear behaviour of free surface flows subjected to external excitation. In its simplest form, sloshing refers to the oscillatory motion of a liquid within a partially filled container. However, in realistic engineering systems, sloshing rarely remains simple. Large amplitude motion, wave breaking, impact loads, air entrapment, multiphase interactions, and strong coupling with the surrounding structure often occur simultaneously. These features introduce pronounced nonlinearity, geometric complexity, and sensitivity to operating conditions, making sloshing both difficult to predict and critical to control. The importance of accurately modelling sloshing dynamics extends across a wide range of high-consequence engineering applications. In aerospace engineering, liquid propellant sloshing in spacecraft and launch vehicles directly affects attitude stability, control accuracy, and structural integrity. Even relatively small sloshing-induced forces can couple with vehicle dynamics, leading to instability or excessive control effort. In the automotive sector, the rapid advancement of electric vehicles and hybrid power systems has increased the prevalence of liquid cooling systems, fuel tanks, and battery thermal management units, where sloshing can influence performance, noise, and durability. Similarly, in high-speed food and pharmaceutical manufacturing, containers are often transported, filled, and sealed at high accelerations, where uncontrolled sloshing can result in spillage, contamination, or inconsistent product quality. At larger scales, sloshing plays a decisive role in marine and offshore engineering. Oil tankers, liquefied natural gas carriers, and floating production systems routinely transport large volumes of liquid under wave-induced motion. Sloshing loads can produce severe impact pressures on tank walls and internal structures, potentially leading to fatigue damage or catastrophic failure. On land, large storage tanks for water, chemicals, and fuels are vulnerable to sloshing during seismic events, where resonance effects may amplify free surface motion and compromise structural safety. In high-risk facilities such as nuclear reactors, where coolant behaviour under extreme conditions is of paramount importance, the robustness of multiphase fluid systems is directly linked to operational safety. In such contexts, inaccurate prediction of sloshing phenomena can have serious economic, environmental, and human consequences. Given this breadth of applications and the severity of potential failure modes, the need for reliable and predictive numerical tools for sloshing analysis is evident. Experimental investigations remain indispensable, particularly for validation and physical insight, but they are often costly, time-consuming, and limited in parameter space. As a result, numerical simulation has become an essential component of modern sloshing research and

Slosh Dynamics. DOI: https://doi.org/10.1016/B978-0-443-44330-5.00012-8

engineering design. However, the numerical modelling of sloshing is far from trivial. The presence of a moving free surface, large deformations, and frequent topological changes challenges conventional computational fluid dynamics (CFD) techniques and places stringent demands on numerical stability, accuracy, and robustness. This chapter is therefore devoted to the numerical methods most commonly employed for the simulation of free surface flows in sloshing dynamics. The focus is placed on three major computational frameworks that have shaped both academic research and industrial practice: the Finite Element Method (FEM), the Finite Volume Method (FVM), and Smoothed Particle Hydrodynamics (SPH). Each of these methods approaches the governing equations of fluid motion from a distinct conceptual standpoint, leading to different strengths, limitations, and areas of applicability. By presenting them within a unified narrative, this chapter aims to provide the reader with a clear understanding of how these methods relate to one another and how they may be selected or combined for specific sloshing problems. The chapter begins with the FEM, which is formulated through the weak or variational form of the governing equations. FEM is widely regarded as one of the most mathematically rigorous numerical methods in computational mechanics. Its foundation in functional analysis allows for systematic error estimation, convergence analysis, and the natural treatment of complex boundary conditions. In the context of fluid dynamics, FEM offers considerable flexibility in representing irregular geometries and coupling fluid motion with structural deformation, making it particularly attractive for fluid–structure interaction problems associated with sloshing. Within this book, the FEM is presented with emphasis on spatial discretisation through shape functions, nodal-based communication, and the construction of global system matrices. Both incompressible and compressible flow formulations are discussed, along with the challenges associated with enforcing incompressibility constraints. Special attention is given to stabilisation techniques required for convection-dominated flows, such as those encountered in violent sloshing, where numerical oscillations may arise if standard Galerkin formulations are used without modification. Through this discussion, the reader gains insight into why FEM remains a powerful and widely used tool, while also appreciating the numerical care required when applying it to strongly nonlinear free surface problems. The chapter then introduces the FVM as an alternative numerical framework rooted in the integral form of the conservation laws. In FVM, the computational domain is divided into a collection of control volumes, and conservation of mass, momentum, and energy is enforced by balancing fluxes across the faces of each volume. This face-based formulation ensures strict local and global conservation, a property that is particularly valuable for transient flows characterised by strong advection and sharp gradients, both of which are common in sloshing phenomena. The FVM has achieved widespread adoption in industrial CFD, largely due to its robustness, conservation properties, and compatibility with complex geometries. In this chapter, FVM is presented in a manner that highlights its conceptual differences from FEM, including face-centred communication, flux evaluation strategies, and time marching schemes. The discussion clarifies why FVM has become the method of choice for many large-scale engineering simulations, while also addressing the challenges it faces when dealing with moving free surfaces and interface capturing. These challenges often necessitate additional modelling techniques, which can increase numerical complexity. The final part of the chapter is devoted to SPH, a fundamentally different approach that abandons fixed spatial discretisation altogether. SPH is a mesh-free, fully Lagrangian method in which the fluid is represented by a set of moving particles. Each particle carries physical quantities such as mass, velocity, density, and energy, and the governing equations are approximated using kernel-based interpolation over neighbouring particles. Because the particles move with the flow, large deformations of the free surface are captured naturally, without the need for mesh deformation, re-meshing, or explicit interface tracking. This property makes SPH particularly well suited to sloshing problems involving wave breaking, fragmentation, and violent impact. In this chapter, SPH is introduced from its fundamental principles, with emphasis on kernel approximation, particle interaction, and the Lagrangian formulation of the governing equations. At the same time, the numerical challenges associated with SPH are discussed in detail. These include pressure stability, boundary representation, particle disorder, and computational cost. The chapter also highlights why SPH has benefited significantly from advances in high-performance computing, especially graphics processing unit (GPU) acceleration, as its local interaction structure maps efficiently onto massively parallel architectures. It is important to recognise that FEM, FVM, and SPH do not exhaust the range of numerical methods available for sloshing and free surface flows. Other approaches, such as the Lattice Boltzmann Method, Marker-and-Cell techniques, and Volume of Fluid formulations, have also been developed and applied with considerable success, particularly in academic research. These methods offer alternative ways of representing interfaces and modelling multiphase flows, often with impressive accuracy under specific conditions. However, they also introduce their own modelling assumptions, numerical complexities, and implementation challenges. The primary objective of this book is not to provide an exhaustive survey of all available numerical techniques, but rather to focus on the methods that have achieved the widest adoption and practical relevance in engineering applications. For this reason, advanced interface capturing techniques such as Volume of Fluid, Marker-and-Cell, and related methods are not treated in detail. Interested readers are encouraged to consult the extensive literature dedicated to these approaches, which is referenced where appropriate. This book aims to bridge the gap between academic theory and engineering practice. The emphasis is placed on physical interpretation, numerical robustness, and practical applicability, rather than on methodological novelty alone. Through this approach, the chapter provides a solid foundation for understanding how modern numerical methods are used to analyse sloshing dynamics in systems ranging from laboratory-scale experiments to large-scale industrial and safety-critical applications.

5.2 Finite element method

The FEM is first discussed as a numerical framework for the simulation of sloshing phenomena. FEM is based on the weak, or variational, formulation of the governing equations, in which the computational domain is discretised into a finite number of elements. Within each element, the flow variables are approximated using shape functions, and the unknowns are associated with nodal degrees of freedom. Communication between elements takes place through shared nodes, ensuring continuity of the solution over the domain.

The method allows a consistent treatment of incompressible and compressible flows, provided that appropriate pressure–velocity coupling and stabilisation techniques are employed. These aspects are particularly important for accurately resolving free surface motion and transient pressure fields (Connelly & Kokini, 2004; Kallivokas & Bielak, 1993; Kalumuck et al., 1995; Kokkinos & Reddy, 1995; Lavooij & Tusseling, 1991; Luo, 1998; Pramila, 1986; Rajagopalan et al., 1993; Shimizu & Ohtani, 1996; Wang & Tan, 1998).

5.2.1 A general algorithm for compressible and incompressible fluid flow

The numerical simulation of fluid flow presents a fundamental challenge due to the wide diversity of physical regimes encountered in practice. These range from incompressible viscous flows at low Mach numbers to highly compressible, inviscid or weakly viscous flows involving shock waves and strong nonlinearities. A truly general computational framework must therefore accommodate both compressible and incompressible behaviour without artificial switching of governing equations or numerical strategy. The algorithm introduced here aims to provide such a unified formulation, grounded in the conservation laws of continuum mechanics and implemented within a finite element framework.

At the core of fluid dynamics lie the conservation equations of mass, momentum, and energy. For a Newtonian fluid, these may be expressed in conservative form as

$$\frac{\partial \rho}{\partial t} + \nabla \cdot (\rho \boldsymbol{u}) = 0 \tag{5.1}$$

$$\frac{\partial (\rho \boldsymbol{u})}{\partial t} + \nabla \cdot (\rho \boldsymbol{u} \otimes \boldsymbol{u}) + \nabla p - \nabla \cdot \boldsymbol{\tau} = \rho f \tag{5.2}$$

$$\frac{\partial (\rho e)}{\partial t} + \nabla \cdot [(\rho e + p)\boldsymbol{u}] - \nabla \cdot (\boldsymbol{\tau} \boldsymbol{u}) - \nabla \cdot (k \nabla T) = \rho r \tag{5.3}$$

where e is the total specific energy, $\boldsymbol{\tau}$ the viscous stress tensor, k the thermal conductivity, T the temperature, and $\mathbf{f}$ and r represent body forces and volumetric heat sources, respectively.

A key observation is that incompressible flow is not governed by a fundamentally different set of equations, but rather emerges as a limiting case of the compressible formulation when density variations become negligible. In this limit, density is treated as constant, and the continuity equation reduces to the solenoidal constraint

$$\nabla . \boldsymbol{u} = 0 \tag{5.4}$$

The principal numerical difficulty then lies in the treatment of pressure, which no longer follows from an equation of state but instead acts as a Lagrange multiplier enforcing incompressibility.

The algorithm under consideration avoids introducing separate formulations for compressible and incompressible regimes. Instead, it retains the full conservation form of the equations and allows the numerical solution process to naturally adapt to the governing physics through appropriate scaling and discretisation. This approach is particularly advantageous in problems where regions of differing compressibility coexist or where flow transitions occur during the simulation.

To achieve numerical robustness, the equations are first cast into nondimensional form. Let L be a characteristic length, U a characteristic velocity, and (\rho_0) a reference density. Introducing dimensionless variables (denoted by overbars),

$$\begin{cases} x = L\overline{x} \\ t = \frac{L}{U}\overline{t} \\ \boldsymbol{u} = U\overline{\boldsymbol{u}} \\ \rho = \rho_0\overline{\rho} \end{cases} \tag{5.5}$$

A defining feature of the method is its use of a characteristic-based time integration strategy. Rather than treating convection explicitly as a purely Eulerian process, the formulation accounts for transport along characteristic directions associated with the hyperbolic nature of the governing equations. This significantly enhances stability, particularly in advection-dominated and high-speed flows.

Time discretisation proceeds by expressing the governing equations in incremental form over a time step Δt. For a generic conserved quantity ϕ, this may be written as

$$\frac{\phi^{n+1} - \phi^{n}}{\Delta t} + \nabla \cdot \boldsymbol{F}\left(\phi^{n+\theta}\right) = S\left(\phi^{n+\theta}\right) \tag{5.6}$$

where $\boldsymbol{F}$ represents the flux vector, S the source term, and $0 \leq \theta \leq 1$ defines the degree of implicitness. This formulation allows both fully explicit and semiimplicit schemes to be recovered as special cases.

In the incompressible limit, the same algorithm remains applicable, with the pressure field determined through a constraint equation derived from the continuity condition. Importantly, the conservation form of the momentum equation is preserved, ensuring correct representation of inertia and force balance even in strongly transient flows.

The authors emphasise that algorithms not written in conservative form may yield physically incorrect solutions, particularly in the presence of steep gradients or discontinuities. Conservation ensures that shocks, contact surfaces, and vortical structures propagate correctly without spurious numerical artefacts. This is a decisive advantage over formulations based solely on primitive variables (Parszewski et al., 1989; Szady et al., 1995).

5.2.2 Governing equations and constitutive description of fluid motion

A general numerical treatment of fluid flow must begin with a precise statement of the governing equations and the constitutive assumptions that close them. While the conservation laws of mass, momentum, and energy are universal, their practical implementation requires explicit modelling of stress, heat transfer, and thermodynamic behaviour. This section establishes the fundamental field equations in a form suitable for a unified finite element formulation applicable to both compressible and incompressible flows.

The continuity equation expresses the local balance of mass within a moving fluid element. In differential form, it is written as

$$\frac{D\rho}{Dt} + \rho \nabla \cdot \boldsymbol{u} = 0 \tag{5.7}$$

where ρ is the density, $\boldsymbol{u}$ the velocity vector, and $\frac{D}{Dt}$ denotes the material derivative. For compressible flow, this equation governs the evolution of density. In incompressible flow, density is assumed constant and the continuity equation reduces to the kinematic constraint

$$\nabla \cdot \boldsymbol{u} = 0 \tag{5.8}$$

which must be satisfied everywhere in the flow domain at all times.

The momentum balance follows directly from Newton's second law applied to a fluid continuum. In tensor notation, the governing equation is

$$\rho \frac{D\boldsymbol{u}}{Dt} = \nabla \cdot \boldsymbol{\sigma} + \rho \boldsymbol{f} \tag{5.9}$$

where $\boldsymbol{\sigma}$ is the Cauchy stress tensor and $\boldsymbol{f}$ represents body forces per unit mass, such as gravity.

The stress tensor is decomposed into an isotropic pressure component and a deviatoric viscous component,

$$\boldsymbol{\sigma} = -p\boldsymbol{I} + \boldsymbol{\tau} \tag{5.10}$$

with $\mathbf{I}$ denoting the identity tensor. The viscous stress tensor $\boldsymbol{\tau}$ captures resistance to deformation and depends on the rate of strain.

For a Newtonian fluid, the constitutive relation is linear and isotropic,

$$\boldsymbol{\tau} = 2\mu\varepsilon(\boldsymbol{u}) + \lambda(\nabla \cdot \boldsymbol{u})\boldsymbol{I} \tag{5.11}$$

where μ is the dynamic viscosity, λ the bulk viscosity, and $\varepsilon(\boldsymbol{u})$ the strain-rate tensor,

$$\varepsilon(\boldsymbol{u}) = \frac{1}{2}(\nabla \boldsymbol{u} + (\nabla \boldsymbol{u})^T) \tag{5.12}$$

For incompressible flow, the volumetric strain vanishes, and the bulk viscosity term plays no role, simplifying the stress description.

The energy equation accounts for the balance between kinetic energy, internal energy, mechanical work, and heat transfer. Expressed in terms of total specific energy (e),

$$e = \varepsilon + \frac{1}{2}\boldsymbol{u} \cdot \boldsymbol{u} \tag{5.13}$$

where ε is the specific internal energy, the conservation law including the generalisation with heat consideration be:

$$\rho\frac{De}{Dt} = \boldsymbol{\sigma}: \nabla\boldsymbol{u} - \nabla\cdot\boldsymbol{q} + \rho r \tag{5.14}$$

Here, **q** is the heat flux vector and r represents internal heat sources per unit mass.

Heat conduction is commonly modelled using Fourier's law,

$$\boldsymbol{q} = -k\nabla T \tag{5.15}$$

where k is the thermal conductivity and T the temperature field.

To complete the system of equations for compressible flow, a thermodynamic relation between pressure, density, and temperature is required. This is typically provided by an equation of state of the form

$$p = p(\rho, T) \tag{5.16}$$

For many engineering applications involving gases, the ideal gas law is adopted,

$$p = \rho RT \tag{5.17}$$

with R denoting the specific gas constant. In liquids, density variations are usually small, and pressure is treated as an independent variable determined by the momentum and continuity equations rather than by an explicit equation of state.

A key conceptual step in developing a general numerical algorithm is recognising that the same governing equations can be written in a unified vector form for all flow regimes. Introducing the vector of primary variables

$$\boldsymbol{U} = \begin{bmatrix} \rho \\ \rho\boldsymbol{u} \\ \rho e \end{bmatrix} \tag{5.18}$$

the conservation laws may be expressed compactly as

$$\frac{\partial \boldsymbol{U}}{\partial t} + \nabla\cdot\boldsymbol{F}(\boldsymbol{U}) = \nabla\cdot\boldsymbol{G}(\boldsymbol{U}) + \boldsymbol{S} \tag{5.19}$$

where **F** contains convective fluxes, **G** viscous and diffusive fluxes, and **S** source terms.

This conservative formulation is essential for numerical accuracy, particularly in flows involving sharp gradients, high Reynolds numbers, or compressibility effects. It ensures that the discrete equations respect fundamental physical balances, independent of mesh resolution or time-step size.

By casting the governing equations in this general and conservative form, they become directly amenable to weighted residual and finite element discretisation. Both compressible and incompressible flows are treated within the same mathematical structure, with differences arising naturally from constitutive assumptions and parameter values rather than from fundamentally different numerical schemes.

This unified description provides the foundation for the algorithmic developments that follow, including stabilised formulations, characteristic-based time integration, and efficient solution strategies for large-scale fluid dynamics problems.

5.2.3 Weak formulation and temporal discretisation of the governing equations

Having established the governing conservation laws and constitutive relations, the next step in constructing a general numerical algorithm is to reformulate these equations in a form suitable for finite element discretisation. This requires two essential operations: the transformation of the strong (differential) form into a weak (integral) form, and the introduction of an appropriate time discretisation strategy that remains stable for both compressible and incompressible flow regimes.

Consider a generic conservation equation written in conservative form over a spatial domain Ω,

$$\frac{\partial \phi}{\partial t} + \nabla \cdot \boldsymbol{f}(\phi) - \nabla \cdot \boldsymbol{g}(\phi) = s \tag{5.20}$$

where ϕ denotes a conserved variable, $\boldsymbol{f}$ the convective flux, $\boldsymbol{g}$ the diffusive flux, and s a source term.

To derive the weak form, the equation is multiplied by an admissible weighting function w and integrated over the domain,

$$\int_\Omega w\frac{\partial \phi}{\partial t} d\Omega + \int_\Omega w\nabla \cdot \boldsymbol{f} d\Omega - \int_\Omega w\nabla \cdot \boldsymbol{g} d\Omega = \int_\Omega wsd\Omega \tag{5.21}$$

$$\int_\Omega w\frac{\partial \phi}{\partial t} d\Omega - \int_\Omega \nabla w \cdot \boldsymbol{f} d\Omega + \int_\Gamma w\boldsymbol{f} \cdot d\Gamma + \int_\Omega \nabla w \cdot \boldsymbol{g} d\Omega - \int_\Gamma w\boldsymbol{g}.\, \boldsymbol{n} d\Gamma = \int_\Omega wsd\Omega \tag{5.22}$$

Applying the divergence theorem to the flux terms yields

$$\int_\Omega w\frac{\partial \phi}{\partial t} d\Omega - \int_\Omega \nabla w \cdot \boldsymbol{f} d\Omega + \int_\Gamma w\boldsymbol{f} \cdot \boldsymbol{n} d\Gamma + \int_\Omega \nabla w \cdot \boldsymbol{g} d\Omega - \int_\Gamma w\boldsymbol{g} \cdot \boldsymbol{n} d\Gamma = \int_\Omega wsd\Omega \tag{5.23}$$

where Γ denotes the boundary of the domain Ω and $\boldsymbol{n}$ the outward unit normal. Boundary conditions enter the formulation naturally through the boundary integrals.

This weak form reduces the differentiability requirements on the solution variables and provides the mathematical foundation for finite element approximation.

5.2.4 Application to the Navier–Stokes equations

Applying this procedure to the momentum equation yields the weak form

$$\int_\Omega w_i \rho \frac{Du_i}{Dt} d\Omega + \int_\Omega \nabla w_i \sigma_i d\Omega = \int_\Omega w_i \rho f_i d\Omega + \int_\Gamma w_i \sigma_i \cdot \boldsymbol{n} d\Gamma \tag{5.24}$$

where index notation is used and where w_i denotes the boundary of the domain Ω and $\boldsymbol{n}$ the outward unit normal. Boundary conditions enter the formulation naturally through the boundary integrals.

This weak form reduces the differentiability requirements on the solution variables and provides the mathematical foundation for finite element approximation.

Applying this procedure to the momentum equation yields the weak form are the velocity weighting functions. The pressure contribution enters through the stress tensor and is treated consistently for both compressible and incompressible flows.

For incompressible flow, the continuity equation leads to an additional weak constraint,

$$\int_{\Omega} q\nabla\cdot\boldsymbol{u}d\Omega = 0 \tag{5.25}$$

where q is a pressure weighting function. This equation enforces mass conservation and couples pressure and velocity fields.

The finite element approximation proceeds by representing the unknown fields in terms of nodal values and interpolation functions. For example the velocity field is approximated as

$$\boldsymbol{u}(\boldsymbol{x}, t) = \sum_{a=1}^{N} N_a(\boldsymbol{x})\boldsymbol{u}_a(t) \tag{5.26}$$

where N_a are shape functions and u_a nodal velocities. Similar interpolations are introduced for pressure, density, and energy as required.

Substituting these approximations into the weak form yields a system of ordinary differential equations in time,

$$\boldsymbol{M}\dot{\boldsymbol{U}} + \boldsymbol{C}(\boldsymbol{U})\boldsymbol{U} + \boldsymbol{K}(\boldsymbol{U})\boldsymbol{U} = \boldsymbol{F} \tag{5.27}$$

where $\boldsymbol{M}$ is the mass matrix, $\boldsymbol{C}$ the convection operator, $\boldsymbol{K}$ the diffusion operator, and $\boldsymbol{K}$ the force vector.

5.2.5 Time discretisation strategy

A central requirement of a general fluid algorithm is stability across a wide range of flow regimes. This is achieved by adopting a time integration scheme that treats transient and convective terms in a consistent manner.

A commonly employed discretisation of the time derivative is

$$\dot{\boldsymbol{U}}^{n+\theta} \approx \frac{\boldsymbol{U}^{n+1} - \boldsymbol{U}^{n}}{\Delta t} \tag{5.28}$$

where Δt is the time step and $0 \leq \theta \leq 1$ controls the degree of implicitness. The fully implicit case $\theta = 1$ provides unconditional stability for linear problems, while intermediate values allow a balance between stability and computational cost.

Substituting this into the semidiscrete system gives

$$\boldsymbol{M}\frac{\boldsymbol{U}^{n+1} - \boldsymbol{U}^{n}}{\Delta t} + \boldsymbol{R}(\boldsymbol{U}^{n+\theta}) = \boldsymbol{F}^{n+\theta} \tag{5.29}$$

where $\boldsymbol{R}$ denotes the residual of the spatial operators.

5.2.6 Treatment of convective transport

Convective transport is the dominant source of numerical instability in high Reynolds number and compressible flows. The algorithm addresses this by recognising that convection propagates information along characteristic directions. Rather than relying solely on central differencing, the formulation introduces stabilisation consistent with the underlying physics.

At the continuous level, the material derivative may be written as

$$\frac{D\phi}{Dt} = \frac{\partial \phi}{\partial t} + \boldsymbol{u}\cdot\nabla\phi \tag{5.30}$$

In the discrete formulation, these considerations motivate a characteristic-based approximation in which nodal variables are advanced along approximate particle trajectories, thereby aligning the numerical update with the underlying transport physics. This strategy effectively suppresses nonphysical oscillations while avoiding the introduction of excessive artificial diffusion, preserving both stability and accuracy. A key strength of the formulation is its consistent treatment of compressibility, which is handled directly through the continuity and energy equations without resorting to an artificial pressure equation for incompressible flow. Incompressibility instead arises naturally as the limiting case of vanishing density variation. Consequently, the same discrete system provides a unified framework capable of representing incompressible viscous flows, weakly compressible low Mach number regimes, and fully compressible high speed flows, simply through the appropriate selection of material properties and time step size.

5.2.7 Stabilisation, characteristic direction, and pressure–velocity coupling

As the finite element discretisation of the Navier–Stokes equations is extended to realistic flow regimes, fundamental numerical difficulties emerge. These difficulties are not artefacts of implementation but are inherent to the mathematical character of the governing equations. In particular, dominance of convective transport, weak compressibility, and the incompressibility constraint give rise to instabilities unless special care is taken in the formulation. The material developed in this section addresses these issues through stabilisation concepts grounded in the physics of information propagation.

The Navier–Stokes equations combine elliptic, parabolic, and hyperbolic behaviour. Viscous diffusion introduces ellipticity, time dependence introduces parabolicity, and convective transport introduces hyperbolicity. When convection dominates diffusion, as quantified by a large cell Peclet number. For the standard Galerkin finite element, the formulations exhibit spurious oscillations. These oscillations are nonphysical and persist even under mesh refinement. The root cause lies in the mismatch between the direction of numerical information transport and the physical direction of flow propagation.

A second source of instability arises in incompressible or nearly incompressible flow. The pressure field acts as a constraint variable enforcing mass conservation rather than as an independent thermodynamic variable. Inappropriate interpolation of velocity and pressure fields leads to violation of the inf–sup (Ladyzhenskaya–Babuška–Brezzi) condition, resulting in pressure checkerboarding or singular system matrices.

To address convective instability, it is essential to recognise that the governing equations propagate information along characteristic directions defined by the local flow velocity. For a scalar transport equation,

$$\frac{\partial \phi}{\partial t} + \boldsymbol{u}\cdot\nabla\phi = 0 \tag{5.31}$$

the exact solution remains constant along trajectories satisfying

$$\frac{dx}{dt} = \boldsymbol{u} \tag{5.32}$$

This observation motivates numerical formulations that align discretisation with characteristic paths. Rather than evaluating gradients symmetrically in space, the formulation introduces directional bias consistent with upstream information propagation.

In a finite element context, this is achieved by modifying the weighting functions or residual equations such that test functions incorporate a component aligned with the flow direction. The resulting method suppresses nonphysical oscillations while preserving sharp gradients.

The stabilised formulation augments the classical Galerkin weak form with additional terms proportional to the residual of the governing equations. For a generic variable ϕ, the stabilised weak form may be written as

$$\int_{\Omega} w\mathfrak{R}(\phi)d\Omega + \sum_e \int_{\Omega_e} \vartheta\boldsymbol{u}\cdot\nabla w\mathfrak{R}(\phi)d\Omega = 0 \tag{5.33}$$

where $\mathfrak{R}(\phi)$ is the strong residual of the governing equation, and ϑ is a stabilisation parameter with dimensions of time.

The parameter ϑ is chosen to reflect the relative importance of convection, diffusion, and temporal effects, typically expressed as

$$\vartheta = \left[\left(\frac{2}{\Delta t}\right)^2 + \left(\frac{2\,|\boldsymbol{u}|}{h}\right)^2 + \left(\frac{4\nu}{h^2}\right)^2\right]^{\frac{-1}{2}} \tag{5.34}$$

where h is a characteristic element length and ν the kinematic viscosity. This formulation introduces dissipation only in directions where it is physically required, avoiding excessive numerical diffusion (Bogaerds et al., 2003; Chung & Kwon, 2002; Jasak & Gosman, 2003; Toreja & Rizwan-uddin, 2003; Zhang et al., 2026; Zhao et al., 2024).

5.2.8 *Pressure–velocity coupling*

In incompressible and weakly compressible flows, pressure is not governed by a transport equation but arises through coupling with the velocity field. The discrete system must therefore ensure compatibility between pressure and velocity interpolations.

The weak form of the incompressibility constraint,

$$\int_{\Omega} q\nabla\cdot\boldsymbol{u}d\Omega = 0 \tag{5.35}$$

It must be satisfied simultaneously with the momentum equations. If equal-order interpolation is used for velocity and pressure, stabilisation is required to avoid violation of the inf–sup condition. This is achieved by augmenting the continuity equation with a pressure stabilisation term,

$$\int_{\Omega} q\nabla\cdot\boldsymbol{u}d\Omega + \sum_e \int_{\Omega_e} \vartheta_p \nabla q\cdot\nabla p d\Omega = 0 \tag{5.36}$$

where ϑ_p is a pressure stabilisation parameter related to element size and flow properties.

This modification suppresses spurious pressure modes while preserving global mass conservation. A major advantage of the general algorithm is that compressibility effects are handled without altering the numerical structure. For compressible flow, density is treated as a primary variable governed by the continuity equation,

$$\frac{\partial\rho}{\partial t} + \nabla\cdot(\rho\boldsymbol{u}) = 0 \tag{5.37}$$

As the Mach number decreases, density variations diminish naturally, and the system transitions smoothly toward incompressible behaviour. No artificial pressure equation or penalty parameter is introduced. The pressure field adjusts dynamically to enforce near-zero volumetric strain.

This unified treatment is particularly important in low-Mach-number flows, where traditional compressible solvers become excessively stiff and incompressible solvers fail to capture density effects.

5.2.9 Discrete system structure

After spatial discretisation and stabilisation, the governing equations lead to a nonlinear algebraic system of the form

$$\boldsymbol{K}(\boldsymbol{u})\boldsymbol{u} = \boldsymbol{F} \tag{5.38}$$

where $\boldsymbol{u}$ collects nodal fields. The system is typically solved using Newton–Raphson iteration, with the tangent matrix assembled consistently from the stabilised weak form.

Time advancement proceeds incrementally, and convergence at each time step ensures satisfaction of both momentum balance and mass conservation.

The stabilisation techniques introduced here do not alter the governing physics. Instead, they restore numerical consistency with the physical mechanisms of information transport and constraint enforcement. Convective stabilisation aligns numerical diffusion with flow direction, while pressure stabilisation enforces compatibility between velocity and pressure fields.

These methods enable accurate simulation of:

- high Reynolds number flows
- low-Mach-number compressible flows
- incompressible viscous flows
- transient flow with strong advection within a single finite element framework.

The discretised and stabilised governing equations derived in the preceding sections form a nonlinear system of algebraic equations. This nonlinearity arises primarily from convective transport, coupling between pressure and velocity, and the dependence of material properties on the flow variables. To obtain practical numerical solutions, an efficient and robust linearisation and solution strategy is required. The present section describes the linearisation procedure, the structure of the resulting discrete system, and the overall algorithmic sequence for time-dependent fluid flow analysis.

After spatial discretisation and stabilisation, the governing equations may be written compactly as a nonlinear residual equation,

$$\boldsymbol{R}(\boldsymbol{u}) = 0 \tag{5.39}$$

The residual vector $\boldsymbol{R}(\boldsymbol{u})$ represents the imbalance of the nodal field at the discrete level.

For transient problems, this residual incorporates both temporal and spatial contributions of current stat (highlighted with superscript n), and the next step (n+1),

$$\boldsymbol{R}(\boldsymbol{u}^{n+1}) = \boldsymbol{M}\frac{\boldsymbol{u}^{n+1} - \boldsymbol{u}^{n}}{\Delta t} + \boldsymbol{N}(\boldsymbol{u}^{n+1}) - \boldsymbol{F}^{n+1} \tag{5.40}$$

where $\boldsymbol{M}$ is the mass matrix, $\boldsymbol{N}$ collects nonlinear spatial operators, and $\boldsymbol{F}$ represents external forcing filed and boundary contributions.

To solve the nonlinear system, a Newton–Raphson iterative procedure is employed. At iteration k, the residual is linearised about the current approximation $\boldsymbol{u}^k$,

$$\boldsymbol{R}(\boldsymbol{u}^k + \Delta\boldsymbol{u}) \approx \boldsymbol{R}(\boldsymbol{u}^k) + \boldsymbol{K}^k\Delta\boldsymbol{u} \tag{5.41}$$

where $\Delta\boldsymbol{u}$ is the increment in the unknown vector and $\boldsymbol{K}^k$ is the consistent tangent matrix,

$$\boldsymbol{K}^k = \left.\frac{\partial \boldsymbol{R}}{\partial \boldsymbol{u}}\right|_{\boldsymbol{u}^k} \tag{5.42}$$

The correction is obtained by solving the linear system

$$\boldsymbol{K}^k\Delta\boldsymbol{u} = -\boldsymbol{R}(\boldsymbol{u}^k) \tag{5.43}$$

after which the solution is updated,

$$\boldsymbol{u}^{k+1} = \boldsymbol{u}^k + \Delta\boldsymbol{u} \tag{5.44}$$

This process is repeated until convergence is achieved according to prescribed tolerance criteria on the residual norm or solution increment.

The tangent matrix **K** exhibits a characteristic block structure reflecting the coupling between variables. For incompressible flow, a simplified representation is

$$\boldsymbol{K} = \begin{bmatrix} \boldsymbol{K}_{uu} & \boldsymbol{K}_{up} \\ \boldsymbol{K}_{pu} & \boldsymbol{K}_{pp} \end{bmatrix} \tag{5.45}$$

where $\boldsymbol{K}_{uu}$ arises from momentum equations, $\boldsymbol{K}_{up}$ and $\boldsymbol{K}_{pu}$ represent pressure–velocity coupling, and $\boldsymbol{K}_{pp}$ includes stabilisation contributions.

In compressible flow, additional blocks corresponding to density and energy variables appear, but the overall structure remains compatible with the same solution strategy. This unified structure is a key feature of the general algorithm, allowing compressible and incompressible problems to be treated within a single computational framework

Time-dependent problems are advanced incrementally using the discretised form of the governing equations. At each time step t^{n+1}, the nonlinear system is solved to convergence before proceeding to the next step. The algorithm may be summarised as:

1. Initialise $\boldsymbol{u}^{n+1,0} = \boldsymbol{u}^n$.
2. Assemble residual $\boldsymbol{R}(\boldsymbol{u}^{n+1,k})$.
3. Assemble tangent matrix $\boldsymbol{K}^k$.
4. Solve for increment $\Delta\boldsymbol{u}^k$.
5. Update the solution.
6. Check convergence; repeat if necessary.

Implicit time integration ensures stability even for stiff problems arising from high Reynolds numbers or low Mach numbers.

Boundary conditions enter the formulation through both the residual vector and the tangent matrix. Essential (Dirichlet) boundary conditions are imposed by constraining nodal degrees of freedom, while natural (Neumann) boundary conditions appear in boundary integrals of the weak form.

For compressible flow, characteristic-based boundary conditions are often employed to ensure the correct propagation of information across boundaries. These conditions distinguish between incoming and outgoing characteristic waves and prescribe data accordingly, enhancing stability and physical consistency (Mishra et al., 2006; Mnasri, 2013; Sun et al., 2023; Xu et al., 2023; Zhang, Ming, et al., 2025).

5.2.10 Convergence and robustness considerations

The convergence behaviour of the Newton–Raphson scheme depends on the quality of the initial guess, time-step size, and degree of nonlinearity. For strongly convective or transient problems, under-relaxation may be applied to improve robustness.

$$\boldsymbol{u}^{k+1} = \boldsymbol{u}^k + \alpha\Delta\boldsymbol{u},\ 0 < \alpha \leq 1 \tag{5.46}$$

No, consider the semidiscrete momentum equation written at time level $n + 1$,

$$\rho^{n+1}\frac{\boldsymbol{u}^{n+1} - \boldsymbol{u}^{n}}{\Delta t} + \nabla\cdot(\rho\boldsymbol{u} \otimes \boldsymbol{u})^{n+1} - \nabla.\,\tau^{n+1} + \nabla p^{n+1} = \rho^{n+1}\boldsymbol{f}^{n+1} \tag{5.47}$$

In the fractional step interpretation, this equation is split into a **predictor** stage in which an intermediate velocity field $\boldsymbol{u}^*$ is computed without enforcing mass conservation explicitly. The predicted velocity satisfies

$$\rho^{n}\frac{\boldsymbol{u}^{*} - \boldsymbol{u}^{n}}{\Delta t} = -\,\nabla\cdot(\rho\boldsymbol{u} \otimes \boldsymbol{u})^{n} - \nabla.\,\tau^{n} - \nabla p^{n} + \rho^{n}\boldsymbol{f}^{n} \tag{5.48}$$

This step represents pure momentum transport influenced by inertia, viscosity, body forces, and the previously known pressure field. The resulting velocity does not, in general, satisfy the continuity equation.

To enforce mass conservation at the new time level, the velocity field must be corrected using an updated pressure. The corrected velocity is expressed as

$$\boldsymbol{u}^{n+1} = \boldsymbol{u}^{*} - \frac{\Delta t}{\rho^{n}}\nabla\left(p^{n+1} - p^{n}\right) \tag{5.49}$$

Substituting this expression into the continuity equation yields a Poisson-type equation for the pressure increment,

$$\nabla\cdot\left(\frac{\Delta t}{\rho^{n}}\nabla\left(p^{n+1} - p^{n}\right)\right) = \nabla\cdot\boldsymbol{u}^{*} \tag{5.50}$$

This equation enforces the divergence-free condition in incompressible flow or the appropriate density evolution in compressible flow. Importantly, pressure emerges here as a constraint variable whose role is to restore mass conservation rather than to transport physical quantities.

Although the fractional step interpretation suggests a sequential solution procedure, it is mathematically equivalent to the fully coupled implicit formulation described earlier when linearisation is performed consistently. The splitting should therefore be understood as a conceptual aid rather than as a requirement for algorithm implementation.

In the finite element context, the pressure correction equation arises naturally from the linearised continuity and momentum equations and does not require explicit operator splitting. This ensures that conservation properties are preserved and that the algorithm remains robust for large time steps.

For compressible flow, the pressure correction equation is modified to account for density variations. The continuity equation at time level $(n+1)$,

$$\frac{\rho^{n+1} - \rho^{n}}{\Delta t} + \nabla\cdot(\rho\boldsymbol{u})^{n+1} = 0 \tag{5.51}$$

introduces coupling between pressure, density, and velocity through the equation of state,

$$p = p(\rho, T) \tag{5.52}$$

Linearisation of this relation yields

$$\delta p = \left(\frac{\partial p}{\partial \rho}\right)_T \delta\rho + \left(\frac{\partial p}{\partial T}\right)_\rho \delta T \tag{5.53}$$

which couples the pressure correction equation to the energy balance. In the low-Mach-number limit, density variations become small, and the incompressible pressure correction equation is recovered smoothly.

5.2.11 Physical interpretation of the pressure field

This formulation highlights an essential conceptual point: pressure does not evolve independently but adjusts instantaneously to enforce compatibility between velocity and density fields. In incompressible flow, pressure has no time derivative and therefore no transport equation. Its numerical treatment must reflect this constraint nature to avoid spurious oscillations or artificial compressibility.

The pressure correction framework ensures that pressure acts globally, redistributing momentum to satisfy mass conservation throughout the domain.

It is important to emphasise that for the time step during solution, the following consideration is used, such that, the fractional step interpretation clarifies why implicit time integration is crucial for stability. The pressure correction equation contains elliptic operators that impose global constraints, while the momentum prediction step contains convective terms that can be strongly nonlinear.

By solving the coupled system implicitly, the algorithm avoids restrictive time-step limitations associated with explicit pressure treatment and remains stable even for large Courant numbers,

$$Co = \frac{|\boldsymbol{u}|\ \Delta t}{h} \gg 1 \tag{5.54}$$

5.2.12 Algorithmic consistency with the continuous governing equations

A fundamental requirement of any numerical method is consistency with the underlying partial differential equations. The present formulation is consistent in the sense that, as the spatial and temporal discretisation parameters tend to zero, the discrete equations converge formally to the continuous conservation laws of mass, momentum, and energy.

Let h denote a characteristic element size and Δt the time-step length. Consistency requires that

$$\lim_{h\to 0, \Delta t\to 0} \boldsymbol{R}_h(\boldsymbol{u}) = \boldsymbol{R}(\boldsymbol{u}) \tag{5.55}$$

where $\boldsymbol{R}_h$ is the discrete residual operator and $\boldsymbol{R}$ the continuous residual. The stabilisation terms introduced earlier are constructed so that they vanish in this limit, ensuring that no artificial physics is introduced as resolution is increased.

An important test of the general algorithm is its behaviour in the incompressible limit. As the Mach number tends to zero,

$$Ma = \frac{|\boldsymbol{u}|}{a} \to 0 \tag{5.56}$$

density variations diminish, and the flow becomes effectively incompressible. In this regime, the continuity equation reduces to the divergence-free constraint,

$$\nabla \cdot \boldsymbol{u} = 0 \tag{5.57}$$

and pressure ceases to be a thermodynamic variable, acting instead as a Lagrange multiplier enforcing mass conservation.

The present formulation accommodates this limit naturally. The pressure correction equation derived earlier enforces the constraint without introducing artificial compressibility or penalty parameters. Moreover, the stabilisation terms remain well behaved, avoiding the numerical stiffness that often plagues compressible formulations at low Mach numbers.

In the opposite limit, where compressibility effects are significant, the governing equations exhibit a strongly hyperbolic character. Pressure, density, and velocity are tightly coupled through wave propagation mechanisms, and information travels at finite characteristic speeds.

The conservative form of the discretised equations ensures correct shock capturing and wave propagation. In particular, the characteristic-based stabilisation aligns numerical dissipation with physical wave directions, preventing nonphysical oscillations near steep gradients.

For compressible flow, the algorithm respects the entropy condition, provided that the stabilisation parameters are chosen appropriately and that the time integration remains sufficiently implicit to control nonlinear instabilities.

Another critical property of the algorithm is discrete conservation. When appropriate boundary conditions are applied, the finite element formulation conserves global mass, linear momentum, and energy to within solver tolerance.

For example, integrating the discrete continuity equation over the domain yields:

$$\frac{d}{dt}\int_{\Omega} \rho d\Omega = -\int_{\Gamma} \rho \boldsymbol{u} \cdot \boldsymbol{n} d\Gamma \tag{5.58}$$

demonstrating that total mass changes only through boundary fluxes. Similar statements hold for momentum and energy, confirming that the numerical method respects fundamental physical balances.

Moreover, the generality of the algorithm extends naturally to boundary condition specification. Essential boundary conditions on velocity, temperature, or density are enforced through standard finite element constraints, while natural boundary conditions appear through boundary integrals in the weak form.

For compressible flow, characteristic-based boundary conditions provide a physically consistent means of prescribing inflow and outflow data. These conditions distinguish between incoming and outgoing waves, ensuring that only physically admissible information enters the computational domain.

In incompressible flow, pressure boundary conditions are applied indirectly through normal stress specification or reference pressure constraints, preserving the constraint nature of the pressure field.

5.2.13 Extension to energy coupling, Low-Mach behaviour, and unified variable treatment

The conservation of energy introduces an additional field variable that couples strongly with density and pressure. In conservative form, the total energy equation may be written as

$$\frac{\partial(\rho e)}{\partial t} + \nabla\cdot[(\rho e + p)\boldsymbol{u}] = \nabla\cdot(\boldsymbol{\tau}\boldsymbol{u}) - \nabla\cdot\boldsymbol{q} + \rho V \tag{5.59}$$

where e is the total specific energy, and V a volumetric heat source.

Within the finite element formulation, this equation is treated in the same manner as the momentum and continuity equations. After multiplication by a weighting function and integration over the domain, a weak form is obtained that naturally incorporates boundary heat fluxes and internal dissipation effects. The energy equation is highly nonlinear due to the dependence of both pressure and internal energy on temperature and density. Linearisation within the Newton framework yields increments of the form

$$\delta(\rho e) = e\delta\rho + \rho\delta e \tag{5.60}$$

with the internal energy increment expressed as:

$$\delta e = \left(\frac{\partial e}{\partial T}\right)_{\rho}\delta T + \left(\frac{\partial e}{\partial \rho}\right)_{T}\delta\rho \tag{5.61}$$

These relations ensure that thermodynamic consistency is maintained at the discrete level. The resulting tangent matrix contains additional coupling blocks linking temperature, density, pressure, and velocity.

A particularly demanding regime for numerical simulation is the low-Mach-number limit, where fluid velocities are small compared with the speed of sound,

$$Ma = \frac{|\boldsymbol{u}|}{a} \ll 1 \tag{5.62}$$

In this regime, pressure variations required to enforce mass conservation are large compared with thermodynamic pressure fluctuations. Conventional compressible formulations become excessively stiff because acoustic wave speeds dominate the numerical time-step restriction.

The present algorithm avoids this difficulty by treating pressure implicitly and coupling it directly to the continuity equation. As a result, acoustic effects are naturally damped as the Mach number decreases, and the formulation transitions smoothly toward incompressible behaviour without introducing artificial compressibility or separate pressure equations.

To clarify the role of pressure in low-Mach flow, it is useful to conceptually decompose the pressure field into a spatially uniform thermodynamic component and a dynamic component,

$$p = p_0(t) + p'(\boldsymbol{x}, t) \tag{5.63}$$

The uniform component p_0 governs density through the equation of state, while the dynamic component p' enforces local momentum balance and incompressibility. The numerical formulation does not explicitly require this decomposition, but its structure naturally reflects this physical separation.

A defining strength of the algorithm is its use of a unified variable set across all flow regimes. The primary unknown vector may be written as

$$\boldsymbol{U} = \begin{bmatrix} \rho \\ \boldsymbol{u} \\ T \end{bmatrix} \tag{5.64}$$

or alternatively in fully conservative form,

$$\boldsymbol{U} = \begin{bmatrix} \rho \\ \rho \boldsymbol{u} \\ \rho e \end{bmatrix} \tag{5.65}$$

Both representations lead to equivalent discrete systems when consistent linearisation is employed. The choice between them is guided by numerical convenience and physical interpretation rather than by fundamental algorithmic constraints.

The inclusion of energy transport necessitates careful treatment of thermal boundary conditions. Prescribed temperature boundaries impose essential conditions on T, while prescribed heat flux boundaries enter through natural boundary integrals,

$$\int_{\Gamma_q} w_e \boldsymbol{q} \cdot \boldsymbol{n} d\Gamma \tag{5.66}$$

These conditions are handled consistently with the momentum and continuity equations, ensuring that energy exchange across boundaries is represented accurately.

The discrete energy equation ensures conservation of total energy when external work and heat fluxes are absent. Furthermore, the formulation respects the second law of thermodynamics in the sense that viscous dissipation and heat conduction produce nonnegative entropy generation,

$$\dot{S}_{seg} \geq 0 \tag{5.67}$$

provided that physical material models are used. This property is essential for long-term stability and physical realism.

With the inclusion of the energy equation, the full algorithm consists of a tightly coupled system governing mass, momentum, and energy. The Newton-based solution strategy treats all variables simultaneously, ensuring that thermodynamic and mechanical effects remain consistent at each iteration and time step.

Despite this coupling, the structure of the system remains amenable to efficient solution through block preconditioning and iterative solvers.

5.2.14 Numerical examples, algorithm validation, and concluding synthesis

Numerical examples serve a dual role. First, they verify that the algorithm reproduces known physical behaviour in canonical flow configurations. Second, they illustrate that the same formulation can be applied without structural modification to both compressible and incompressible problems.

The emphasis is not on achieving optimal accuracy for a specific benchmark, but on confirming that the algorithm converges reliably, respects conservation laws, and produces physically meaningful solutions over a wide range of parameters.

As such, typical test problems considered include:

- transient incompressible viscous flow
- weakly compressible flow at low Mach number
- compressible flow with significant density variation

In each case, the same governing equations, stabilisation strategy, and solution algorithm are employed. Differences in flow behaviour arise solely from physical parameters such as viscosity, compressibility, and boundary conditions.

For transient simulations, the algorithm advances the solution incrementally in time. At each time step, the nonlinear system is solved to convergence using the Newton–Raphson method described earlier.

Convergence is typically assessed using the norm of the residual vector,

$$\|\boldsymbol{R}(\boldsymbol{U}^{n+1})\| < \varepsilon \tag{5.68}$$

where ε is a prescribed tolerance. Experience shows that quadratic convergence is achieved once the solution enters the asymptotic Newton regime, provided that stabilisation parameters are chosen consistently and the time step is not excessively large.

The numerical examples demonstrate several key attributes of the proposed formulation. First, spurious oscillations are effectively eliminated: even in strongly convection-dominated regimes, the stabilised scheme suppresses nonphysical overshoots while avoiding undue numerical diffusion. Second, the method exhibits smooth and reliable behaviour in the low Mach number limit, where pressure naturally adjusts to enforce near incompressibility without inducing numerical stiffness or artificial pressure waves. Third, the pressure velocity coupling is robust, preventing checkerboard pressure patterns and preserving a well-conditioned discrete system, even when equal-order interpolation is employed. Finally, the transient response is consistent with physical expectations, as the temporal evolution of velocity, pressure, and density fields remains realistic and free from artificial damping of inertial phenomena.

An important aspect of validation is verification of global conservation properties. For example, total mass within a closed domain remains constant to within solver tolerance,

$$\int_{\Omega} \rho^{n+1} d\Omega = \int_{\Omega} \rho^{n} d\Omega \tag{5.69}$$

Similarly, in the absence of external work or heat transfer, the total energy remains conserved,

$$\int_{\Omega}^{\int} \rho e d\Omega = \text{constant} \tag{5.70}$$

These results confirm that the conservative formulation and consistent linearisation preserve fundamental physical balances at the discrete level.

5.2.15 Final assessment, generality of the formulation, and outlook

A central achievement of the formulation is the elimination of artificial distinctions between compressible and incompressible flow solvers. Rather than introducing separate governing equations, penalty terms, or artificial compressibility parameters, the algorithm retains the full conservation laws and allows the physics of the problem to dictate the effective behaviour.

In the incompressible limit, density variations vanish, and pressure naturally assumes the role of a constraint variable enforcing

$$\nabla \cdot \boldsymbol{u} = 0 \tag{5.71}$$

In the compressible regime, pressure, density, and temperature evolve dynamically according to the equation of state and the energy balance. Both limits emerge from the same discrete equations, demonstrating that incompressibility is not a separate physical model but a limiting case of compressible flow.

The stabilisation techniques, on the other hand, employed in the formulation are not numerical artefacts but reflect the physical mechanisms of information transport and constraint enforcement. Characteristic-based stabilisation aligns numerical dissipation with convective directions, while pressure stabilisation ensures compatibility between velocity and pressure fields.

Crucially, these stabilisation terms vanish under mesh refinement,

$$\lim_{h \to 0} S_h = 0 \tag{5.72}$$

guaranteeing consistency with the continuous governing equations. As a result, the method avoids the common trade-off between stability and accuracy that plagues many ad hoc numerical schemes.

The formulation presented here is designed to be robust, reliable, and broadly applicable across a wide range of flow conditions. It achieves stable and consistent convergence for both steady and unsteady problems and maintains numerical stability in regimes that are traditionally challenging, such as convection-dominated flows and low Mach number conditions. A key contributor to this behaviour is the use of implicit time integration, in which pressure, velocity, density, and energy are treated within a single coupled nonlinear system. This unified treatment avoids restrictive stability constraints commonly associated with explicit or segregated strategies and makes the approach well-suited to long-time simulations and strongly coupled multiphysics problems. The formulation is intentionally general and does not rely on restrictive assumptions regarding geometry, boundary conditions, or specific flow regimes, which allows it to serve as a flexible foundation for extensions such as turbulence modelling, free surface and multiphase flows, fluid–structure interaction, and thermally driven or buoyancy affected flows, all without modifying the core algorithmic structure. Beyond its practical performance, the formulation also carries conceptual significance by illustrating that many numerical difficulties in fluid flow simulation originate not from the underlying physics but from inconsistent numerical representations of conservation laws and constraints. By enforcing strict conservation, treating pressure correctly as a constraint, and aligning numerical stabilisation with physical transport mechanisms, the method achieves both clarity and generality. While this general approach is not intended to replace highly specialised solvers tailored to narrowly defined problems, it offers a high level of reliability and coherence for complex systems involving strong coupling, regime transitions, and interacting physical processes.

5.2.16 The finite element method for fluid flow: from shape functions to the Galerkin formulation

The FEM provides a systematic and mathematically rigorous framework for solving the governing equations of fluid dynamics in complex geometries. Unlike finite difference methods, which rely on structured grids, FEM is based on a variational formulation that naturally accommodates irregular domains, unstructured meshes, and mixed boundary conditions. Its strength lies in the direct connection between the continuous partial differential equations governing fluid motion and their discrete approximation through weighted residual principles.

The development of the finite element formulation proceeds through several fundamental steps: interpolation of field variables using shape functions, evaluation of their derivatives, construction of strain rate and flux terms, derivation of the weak (Galerkin) form, and assembly of the resulting system matrices. Each of these steps has a clear physical and mathematical interpretation, which is developed in detail below.

For clarity, we consider the incompressible Navier–Stokes equations in a spatial domain Ω with boundary Γ. From Mass conservation, we obtain:

$$\nabla\cdot\boldsymbol{u} = 0 \tag{5.73}$$

And for momentum conservation

$$\rho\left(\frac{\partial \boldsymbol{u}}{\partial t} + \boldsymbol{u}\cdot\nabla\boldsymbol{u}\right) = -\nabla p + \nabla\cdot\boldsymbol{\tau} + \rho\boldsymbol{f} \tag{5.74}$$

For a Newtonian fluid,

$$\boldsymbol{\tau} = 2\mu\varepsilon(\boldsymbol{u}) \tag{5.75}$$

$$\varepsilon(\boldsymbol{u}) = \frac{1}{2}\left(\nabla\boldsymbol{u} + \nabla\boldsymbol{u}^T\right) \tag{5.76}$$

Now, the domain Ω is partitioned into a finite number of nonoverlapping elements Ω^e. Within each element, the unknown field variables are approximated by interpolation (shape) functions.

For velocity, the finite element approximation is

$$\boldsymbol{u}(\boldsymbol{x}, t) = \sum_{i=1}^{n_e} N_i(\boldsymbol{x})\boldsymbol{u}_i(t) \tag{5.77}$$

where N_i are the shape functions associated with node i, n_e is the number of nodes per element, and $\boldsymbol{u}_i$ are the nodal velocity values.

Similarly, pressure is approximated as

$$p(\boldsymbol{x}, t) = \sum_{j=1}^{m_e} M_j(\boldsymbol{x})p_j(t) \tag{5.78}$$

where M_j are pressure shape functions. The choice of N_i and M_j is crucial for stability and accuracy.

The governing equations involve spatial derivatives of velocity and pressure. These derivatives are obtained directly from the shape functions. The velocity gradient is expressed as

$$\nabla\boldsymbol{u} = \sum_{i=1}^{n_e} \nabla N_i \otimes \boldsymbol{u}_i \tag{5.79}$$

The strain-rate tensor then becomes

$$\varepsilon(\boldsymbol{u}) = \sum_{i=1}^{n_e} \boldsymbol{B}_i\boldsymbol{u}_i \tag{5.80}$$

where the strain–displacement matrix $\boldsymbol{B}_i$ is constructed from the first derivatives of the shape functions. For a two-dimensional flow,

$$\boldsymbol{B}_i = \begin{bmatrix} \frac{\partial N_i}{\partial x} & 0 \\ 0 & \frac{\partial N_i}{\partial y} \\ \frac{\partial N_i}{\partial y} & \frac{\partial N_i}{\partial x} \end{bmatrix} \tag{5.81}$$

This matrix is central to the construction of viscous terms in the finite element formulation.

The FEM is based on enforcing the governing equations in an integral sense. Let $\boldsymbol{w}$ be weighting functions for. The residual of the momentum equation is multiplied by $\boldsymbol{w}$ and integrated over the domain:

$$\int_\Omega \boldsymbol{w}\cdot\left[\rho\left(\frac{\partial \boldsymbol{u}}{\partial t} + \boldsymbol{u}\cdot\nabla\boldsymbol{u}\right) + \nabla p - \nabla\cdot\boldsymbol{\tau} - \rho \boldsymbol{f}\right] d\Omega = 0 \tag{5.82}$$

In the Galerkin method, the weighting functions are chosen from the same space as the trial functions:

$$\boldsymbol{w} = N_i\boldsymbol{e} \tag{5.83}$$

And

$$q = M_j \tag{5.84}$$

In deriving the weak form of the momentum equation, the viscous and pressure terms are integrated by parts to lower the order of the derivatives. This process uses the divergence theorem to transform volume integrals into boundary integrals.

$$\int_\Omega \boldsymbol{w}\cdot\nabla p d\Omega = -\int_\Omega (\nabla\cdot\boldsymbol{w}) p d\Omega + \int_\Gamma p\boldsymbol{w}\cdot\boldsymbol{n} d\Gamma \tag{5.85}$$

$$\int_\Omega \boldsymbol{w}\cdot\nabla\cdot\boldsymbol{\tau} d\Omega = -\int_\Omega \varepsilon(\boldsymbol{w}) : \boldsymbol{\tau} d\Omega + \int_\Gamma \boldsymbol{w}\cdot\boldsymbol{\tau}\boldsymbol{n} d\Gamma \tag{5.86}$$

The weak form of the momentum equation becomes

$$\int_\Omega \boldsymbol{w}\cdot\rho\frac{\partial \boldsymbol{u}}{\partial t} d\Omega + \int_\Omega \boldsymbol{w}\cdot\rho(\boldsymbol{u}\cdot\nabla\boldsymbol{u}) d\Omega - \int_\Omega (\nabla\cdot\boldsymbol{w}) p d\Omega + \int_\Omega \varepsilon(\boldsymbol{w}) : \boldsymbol{\tau} d\Omega = \int_\Omega \boldsymbol{w}\cdot\rho\boldsymbol{f} d\Omega \tag{5.87}$$

For the weak form of the continuity equation, the incompressibility constraint is enforced weakly as

$$\int_\Omega q\nabla\cdot\boldsymbol{u} d\Omega = 0 \tag{5.88}$$

This equation couples pressure and velocity and introduces a saddle-point structure into the discrete system.

In the discretised form of the problem, substituting the finite element approximations into the weak formulation produces the element matrices, which assemble into the global stiffness matrix. As such, the viscous contribution leads to the stiffness matrix

$$\boldsymbol{K}^e = \int_{\Omega^e} \boldsymbol{B}^T\boldsymbol{D}\boldsymbol{B} d\Omega \tag{5.89}$$

where $\boldsymbol{D}$ is the constitutive matrix for a Newtonian fluid,

$$\boldsymbol{D} = 2\mu\begin{bmatrix} 1 & 0 & 0 \\ 0 & 1 & 0 \\ 0 & 0 & \frac{1}{2} \end{bmatrix} \tag{5.90}$$

This matrix originates directly from the first derivatives of the shape functions and represents viscous resistance to deformation.

The mass matrix is

$$\boldsymbol{M}^e = \int_{\Omega^e} \rho\boldsymbol{N}^T\boldsymbol{N} d\Omega \tag{5.91}$$

and the pressure–velocity coupling matrix is

$$\boldsymbol{G}^e = \int_{\Omega^e} \boldsymbol{B}_p^T\boldsymbol{N} d\Omega \tag{5.92}$$

where $\boldsymbol{B}_p$ contains derivatives of pressure shape functions.

Element matrices are assembled into global matrices according to mesh connectivity. The semidiscrete system takes the form

$$\begin{bmatrix} \boldsymbol{M} & \boldsymbol{G}^T \\ \boldsymbol{G} & 0 \end{bmatrix}\begin{bmatrix} \dot{\boldsymbol{u}} \\ \boldsymbol{p} \end{bmatrix} + \begin{bmatrix} \boldsymbol{C}(\boldsymbol{u}) & 0 \\ 0 & 0 \end{bmatrix}\begin{bmatrix} \boldsymbol{u} \\ \boldsymbol{p} \end{bmatrix} = \begin{bmatrix} \boldsymbol{f} \\ 0 \end{bmatrix} \tag{5.93}$$

Here, $\boldsymbol{C}(\boldsymbol{u})$ represents the nonlinear convection operator.

5.3 Finite volume method

The FVM is a numerical technique developed for the solution of conservation laws governing continuum phenomena such as fluid flow, heat transfer, and mass transport. Its conceptual foundations originate from control volume analyses used in classical fluid mechanics and thermodynamics, where physical laws are expressed in integral form over finite regions of space. The method matured during the 1970s alongside the rapid growth of CFD, driven by the need to simulate complex engineering systems for which analytical solutions were no longer feasible. The primary motivation behind the development of FVM was the requirement to preserve the physical meaning of conservation laws at the discrete level, ensuring that mass, momentum, and energy are conserved not only over the entire computational domain but also locally within every individual control volume.

In the FVM, the governing partial differential equations are first written in integral form and then applied to a set of nonoverlapping control volumes that discretise the domain. This procedure transforms the continuous conservation laws into algebraic balance equations involving transient storage, surface fluxes, and volumetric source terms. A defining characteristic of FVM is that the exchange of information between neighbouring control volumes takes place through fluxes evaluated at their shared faces. Each internal face contributes equally and oppositely to the adjacent volumes, which guarantees strict local and global conservation. The discrete unknowns therefore represent volume-averaged physical quantities, giving the method a direct and intuitive physical interpretation that closely reflects experimental and theoretical descriptions of transport processes.

The FVM is often contrasted with the FEM, which was originally developed for structural and solid mechanics and is based on a variational or weak formulation of the governing equations. In FEM, the solution is approximated using shape functions defined over elements, and the interaction between elements is mediated through nodal degrees of freedom. Communication within the discretised domain therefore occurs primarily through nodes, and conservation is satisfied in an averaged or global sense unless special formulations are employed. In contrast, FVM relies on face-based fluxes as the primary mechanism of communication between neighbouring regions, making conservation an inherent property of the discretisation rather than an imposed constraint.

These fundamental differences in formulation and information transfer have important practical consequences. The face-centred flux structure of FVM makes it particularly well-suited to advection-dominated flows, compressible and incompressible fluid dynamics, and multiphysics problems where accurate representation of transport and balance is essential. Its ability to operate on general polyhedral meshes without predefined shape functions further enhances its flexibility for complex geometries commonly encountered in industrial applications. For these reasons, the FVM has become the dominant numerical framework in CFD, complementing rather than replacing the FEM, which remains highly effective for problems rooted in variational principles and solid mechanics (Jiangfeng et al., 2011; Mishra et al., 2006; Sun et al., 2026; Wang et al., 2015; Xu et al., 2021, 2023; Zhang et al., 2026).

5.3.1 General algorithm for compressible and incompressible flows

The FVM provides a unified and physically consistent framework for the numerical solution of conservation laws governing fluid motion. Its defining feature is the direct enforcement of conservation principles over finite control volumes, making it equally applicable to compressible and incompressible flows. The method does not distinguish between these two regimes at the level of the discretisation philosophy; rather, the differences arise naturally from the constitutive relations and constraints imposed on the governing equations.

At the core of the finite volume formulation lies the generic transport equation for a scalar variable (\phi), which may represent mass, momentum, energy, or any passive scalar. In its most general form, the conservation law can be written as

$$\frac{\partial(\rho\phi)}{\partial t} + \nabla\cdot(\rho\boldsymbol{u}\phi) = \nabla\cdot(\Gamma\nabla\phi) + S_\phi \tag{5.94}$$

where Γ is a diffusion coefficient, and S_ϕ denotes volumetric sources or sinks. This equation is valid for both compressible and incompressible flows. In incompressible flow, density is constant, and the continuity equation imposes $\nabla\cdot\boldsymbol{u} = 0$, while in compressible flow density, becomes an additional unknown coupled to pressure and temperature through an equation of state.

The FVM begins by integrating the governing equation over a finite control volume V_C bounded by a closed surface ∂V_C. Applying Gauss' divergence theorem to the convective and diffusive terms yields

$$\frac{d}{dt}\int_{V_C} \rho\phi dV + \sum_{f\in\partial V_C}\int_{S_f} \rho\phi\boldsymbol{u}\cdot\boldsymbol{n}_f dS = \sum_{f\in\partial V_C}\int_{S_f} \Gamma\nabla\phi\cdot\boldsymbol{n}_f dS + \int_{V_C} S_\phi dV \tag{5.95}$$

Here, the summation is taken over all faces f of the control volume, S_f denotes the face area, and $\boldsymbol{n}_f$ is the outward unit normal vector. This integral form clearly shows the balance between transient storage, net convective flux, net diffusive flux, and internal sources.

A key advantage of this formulation is that it preserves conservation exactly at the discrete level, provided that the same flux evaluated at a face is shared by the two neighbouring control volumes with opposite signs.

In the finite volume approach, the spatial discretisation is performed first, leading to a set of semidiscretised equations that are continuous in time but discrete in space. The volume integrals are approximated using cell-averaged values, while surface integrals are expressed in terms of fluxes evaluated at face centroids.

The transient term is approximated as

$$\int_{V_C} \rho\phi dV \approx \rho_C\phi_C V_C \tag{5.96}$$

where ϕ_C and ρ_C denote the values at the centroid of the control volume. This approximation introduces an error of second order in space when the variable varies smoothly.

The convective flux through a face f is written as

$$F_f^{CONV} = (\rho\boldsymbol{u})_f\cdot\boldsymbol{S}_f\phi_f \tag{5.97}$$

with $\boldsymbol{S}_f = \boldsymbol{n}_f S_f$. The face value ϕ_f must be reconstructed from neighbouring cell values using an interpolation scheme. The choice of interpolation directly affects accuracy and stability, particularly in convection-dominated flows.

The diffusive flux through the same face is approximated by

$$F_f^{diff} = \Gamma_f (\nabla \phi)_f \cdot \boldsymbol{S}_f \tag{5.98}$$

For orthogonal meshes, the gradient can be approximated using a simple two-point difference aligned with the line connecting cell centroids. For general unstructured meshes, additional corrections are required to maintain accuracy.

Combining all contributions, the semidiscretised equation for a control volume (C) can be written as

$$\frac{d}{dt}(\rho_C \phi_C V_C) + \sum_f F_f^{CONV} = \sum_f F_f^{diff} + S_{\phi,C} V_C \tag{5.99}$$

This equation has a clear physical interpretation: the rate of change of the conserved quantity inside the control volume is balanced by the net fluxes across its faces and the internal sources. The same structure applies to all transported variables, which allows a unified algorithmic treatment of mass, momentum, and energy equations (Javanmard et al., 2019; Lee et al., 2002; Merabtene et al., 2026; Seyyedi et al., 2023; Xu et al., 2023, 2024).

5.3.2 Applicability to compressible and incompressible flows

The generality of the finite volume algorithm becomes evident when considering compressible and incompressible flows within the same framework. In incompressible flow, density is constant and can be taken outside the time derivative and flux terms. Pressure appears as a Lagrange multiplier enforcing mass conservation, and special pressure velocity coupling algorithms are required.

In compressible flow, density varies in space and time, and the continuity equation itself becomes a transport equation of the same form. Pressure is linked to density and temperature through an equation of state, and acoustic effects impose stricter requirements on temporal discretisation.

Despite these differences, the spatial discretisation procedure remains unchanged. Only the definitions of fluxes and source terms differ, which highlights the robustness and flexibility of the finite volume methodology (Adeeb et al., 2018; García Nieto et al., 2010; Maryška et al., 2008; Nieuwboer et al., 2023; Sun et al., 2023; Wang et al., 2019; Xu, 2026).

5.3.3 Spatial accuracy and flux evaluation in the finite volume method

A fundamental requirement of any numerical discretisation scheme is a clear understanding of its spatial accuracy. In the FVM, accuracy is governed by two closely related approximations: the assumed spatial variation of the transported variable inside each control volume and the numerical evaluation of volume and surface integrals.

Consider a scalar field $\phi(\boldsymbol{x})$ defined within a control volume whose centroid is located at $\boldsymbol{x}_C$. The exact variation of ϕ in space can be expressed through a Taylor series expansion about the centroid,

$$\phi(\boldsymbol{x}) = \phi_C + \nabla\phi_C \cdot (\boldsymbol{x} - \boldsymbol{x}_C) + \frac{1}{2}(\boldsymbol{x} - \boldsymbol{x}_C) \otimes (\boldsymbol{x} - \boldsymbol{x}_C) : \nabla\phi_C + O(|\boldsymbol{x} - \boldsymbol{x}_C|^3) \tag{5.100}$$

Here, $\phi_C = \phi(\boldsymbol{x}_C)$, and the higher-order tensor products account for curvature in the scalar field. In practical finite volume implementations, this infinite series is truncated. The most common assumption is that ϕ varies linearly within the control volume,

$$\phi(\boldsymbol{x}) \approx \phi_C + \nabla\phi_C \cdot (\boldsymbol{x} - \boldsymbol{x}_C) \tag{5.101}$$

Comparing this linear approximation with the full Taylor expansion shows that the neglected terms are of order $|\boldsymbol{x} - \boldsymbol{x}_C|^2$. Consequently, the spatial representation of ϕ inside the cell is **second-order accurate**, provided the gradient $\nabla\phi_C$ itself is evaluated with at least second-order accuracy.

This assumption forms the mathematical foundation for many widely used finite volume schemes and explains why mesh smoothness and gradient reconstruction quality play a central role in the overall accuracy of the solution (Chung et al., 2004; Liu et al., 2023; Tu et al., 2023; Yang et al., 2022; Zhang, Xu, et al., 2025).

5.3.4 Mean value approximation over control volumes

In the FVM, volume integrals are evaluated using cell-averaged quantities. The exact average of ϕ over a control volume V_C is defined as

$$\overline{\phi_C} = \frac{1}{V_C}\int_{V_C} \phi(\boldsymbol{x})dV \tag{5.102}$$

Substituting the Taylor expansion of ϕ into this expression gives

$$\overline{\phi_C} = \phi_C + \frac{1}{V_C}\int_{V_C} \nabla\phi_C \cdot (\boldsymbol{x} - \boldsymbol{x}_C)dV + O(|\boldsymbol{x} - \boldsymbol{x}_C|^2) \tag{5.103}$$

Because $\boldsymbol{x}_C$ is defined as the centroid of the control volume, the first order moment vanishes,

$$\int_{V_C} (\boldsymbol{x} - \boldsymbol{x}_C)dV = 0 \tag{5.104}$$

As a result, the cell average reduces to

$$\overline{\phi_C} = \phi_C + O(\Delta x^2) \tag{5.105}$$

demonstrating that the mean value approximation used in finite volume formulations is also second-order accurate in space. This property ensures consistency between point values at cell centres and volume-averaged quantities, which is essential for conservative discretisations.

5.3.5 Convective flux evaluation at cell faces

The convective transport of a scalar quantity occurs through the faces of the control volume. The exact convective flux across a face f is given by:

$$\int_{S_f} \rho\phi \boldsymbol{u}\cdot\boldsymbol{n}_f dS \tag{5.106}$$

To evaluate this integral numerically, the scalar field is expanded around the face centroid $\boldsymbol{x}_f$,

$$\phi(\boldsymbol{x}) = \phi_f + \nabla\phi_f\cdot(\boldsymbol{x} - \boldsymbol{x}_f) + O(|\boldsymbol{x} - \boldsymbol{x}_f|^2) \tag{5.107}$$

Substituting this expression into the surface integral yields

$$\int_{S_f} \rho\phi \boldsymbol{u}\cdot\boldsymbol{n}_f dS = (\rho\boldsymbol{u})_f\cdot\boldsymbol{S}_f\phi_f + O(\Delta x^2) \tag{5.108}$$

where

$$\boldsymbol{S}_f = \int_{S_f} \boldsymbol{n}_f dS \tag{5.109}$$

$\boldsymbol{S}_f$ is the outward oriented face area vector. The first-order correction term vanishes because the face centroid is used as the integration point. This confirms that evaluating convective fluxes using face-centred values is second-order accurate, provided that the face value ϕ_f is reconstructed consistently.

Diffusive transport involves gradients of the transported variable. The exact diffusive flux through face f is

$$\int_{S_f} \Gamma\nabla\phi\cdot\boldsymbol{n}_f dS \tag{5.110}$$

Using a Taylor expansion of the gradient about the face centroid and truncating higher order terms gives

$$\int_{S_f} \Gamma\nabla\phi\cdot\boldsymbol{n}_f dS = \Gamma_f(\nabla\phi)_f\cdot\boldsymbol{S}_f + O(\Delta x^2) \tag{5.111}$$

Thus the diffusive flux is also second-order accurate when evaluated at the face centroid. However, unlike the convective term, the accuracy and robustness of this approximation depend strongly on mesh orthogonality. On nonorthogonal meshes, additional correction terms are often required to preserve accuracy and stability.

As such, the analysis above shows that second-order accuracy in the FVM arises from two conditions:

1. Linear reconstruction of the solution within control volumes.
2. Midpoint evaluation of surface and volume integrals.

Higher order accuracy can be achieved by either improving the assumed spatial variation of ϕ or by using higher order quadrature rules for surface integrals. For example, fourth-order accurate surface integration can be obtained using

Simpson-type rules that require values at multiple integration points on each face,

$$\int_{S_f} \phi dS \approx \frac{1}{6}(\phi_{ip1} + 4\phi_{ip2} + \phi_{ip3})S_f \tag{5.112}$$

To maintain formal consistency, the reconstruction of ϕ inside the cell must then also be of higher order, involving second derivatives and accurate gradient evaluations.

5.3.6 Transient semidiscretized finite volume equations and fundamental properties

When unsteady phenomena are of interest, the FVM must account for temporal variation in addition to spatial transport. This is achieved by retaining the transient term in the general conservation equation and performing integration over both the control volume and a finite time interval. The resulting formulation bridges the continuous governing equations and their fully discrete numerical counterparts.

Starting from the general transport equation for a scalar variable ϕ,

$$\frac{\partial(\rho\phi)}{\partial t} + \nabla\cdot(\rho\boldsymbol{u}\phi) = \nabla\cdot(\Gamma\nabla\phi) + S_\phi \tag{5.113}$$

integration over a control volume V_C and a time interval $[t, t + \Delta t]$ yields

$$\int_t^{t+\Delta t}\int_{V_C}\frac{\partial(\rho\phi)}{\partial t}dVdt + \int_t^{t+\Delta t}\int_{S_f}\rho\phi\boldsymbol{u}\cdot\boldsymbol{n}_f dSdt - \int_t^{t+\Delta t}\int_{S_f}\Gamma\nabla\phi \cdot\boldsymbol{n}_f dSdt = \int_t^{t+\Delta t}\int_{V_C} S_\phi dVdt \tag{5.114}$$

This expression represents the exact balance of transient storage, convective transport, diffusive transport, and volumetric sources over a finite control volume and time interval.

For fixed computational grids, where the control volume geometry does not change with time, the transient term can be simplified by interchanging the order of differentiation and integration,

$$\int_t^{t+\Delta t}\int_{V_C}\frac{\partial(\rho\phi)}{\partial t}dVdt = \int_t^{t+\Delta t}\frac{d}{dt}\left(\int_{V_C}\rho\phi dV\right)dt \tag{5.115}$$

Defining the cell-averaged quantity as

$$(\rho\phi)|_{Cell} = \frac{1}{V_C}\int_{V_C}\rho\phi dV \tag{5.116}$$

the transient contribution becomes

$$\int_t^{t+\Delta t}\frac{d}{dt}((\rho\phi)|_{Cell}V_C)dt \tag{5.117}$$

Using midpoint integration in time, this term is approximated with second-order accuracy, consistent with the spatial discretisation. At this stage, the governing equation is said to be semidiscretized, meaning that space has been discretised while time remains continuous.

Applying midpoint approximations to the convective, diffusive, and source terms leads to the semidiscretized finite volume equation

$$\frac{d}{dt}((\rho\phi)|_{Cell}V_C) + \sum_f (\rho \boldsymbol{u})_f \cdot \boldsymbol{S}_f \phi_f - \sum_f \Gamma_f (\nabla\phi)_f \cdot \boldsymbol{S}_f = S_{\phi,C} V_C \tag{5.118}$$

This equation holds for each control volume in the mesh and forms the foundation for both steady and unsteady finite volume solvers. Further progress towards a fully discrete scheme requires assumptions regarding the temporal variation of ϕ, which leads to explicit or implicit time integration strategies.

5.3.7 Fundamental properties of the discretised equations

Conservation is the defining strength of the finite volume approach. Because fluxes are evaluated at shared faces between neighbouring control volumes, the flux leaving one cell enters the adjacent cell with equal magnitude and opposite sign. Mathematically, for a face shared by two control volumes C and F,

$$F_{C\to F} = -F_{F\to C} \tag{5.119}$$

This guarantees exact conservation of mass, momentum, and energy at the discrete level, independent of mesh structure or numerical scheme. Such a property is essential for physically meaningful simulations, particularly in long-time integrations and strongly coupled multiphysics problems.

Moreover, accuracy describes how closely the numerical solution approximates the exact solution of the governing equations. In the FVM, accuracy is assessed through the truncation error introduced during discretization. For the standard cell centred formulation with midpoint integration, the leading truncation error terms scale as $O(\Delta x^2)$, which indicatesg the second-order spatial accuracy. This implies that halving the grid spacing reduces the discretisation error by approximately a factor of four, provided the solution is sufficiently smooth. Accuracy is therefore strongly influenced by mesh quality and gradient reconstruction.

Furthermore, because the solution procedure is iterative, achieving convergence is essential for obtaining a physically meaningful simulation. The nonlinear discretised conservation equations require repeated updates starting from an initial guess, and the iteration continues until the change between successive solutions falls below a specified tolerance ε.

Convergence may refer to two related but distinct concepts:

1. Iterative convergence, where residuals decrease within a fixed grid.
2. Grid convergence, where the solution becomes independent of further mesh refinement.

Both forms are essential for ensuring numerical credibility.

Consistency refers to the requirement that a discretisation scheme faithfully reproduces the governing differential equations as the mesh is refined. In practical terms, as the spatial grid spacing and time step approach zero, the discrete equations should converge towards the exact continuous form. This behaviour is ensured when the truncation error (the difference between the exact differential operator and its discrete approximation) tends to zero in the limit. Formally, consistency requires that the truncation error vanishes as

$$\Delta x \to 0, \Delta t \to 0 \tag{5.120}$$

Consistency ensures that numerical solutions converge towards the exact solution of the governing equations, provided stability is maintained.

Also, stability concerns the boundedness of numerical errors during the solution process. For transient problems, a stable scheme prevents errors from growing uncontrollably as time advances. Explicit schemes impose strict limits on the time step, while implicit schemes improve stability at the cost of solving coupled algebraic systems. In steady problems, stability is closely linked to the properties of the resulting coefficient matrix, such as diagonal dominance.

5.3.8 Numerical behaviour, and practical implementation of finite volume discretisations

5.3.8.1 Conservation as a fundamental discrete principle

One of the most critical requirements of any numerical method intended for fluid flow simulation is the preservation of conservation laws at the discrete level. In the FVM, conservation is not an approximate property but an inherent characteristic of the formulation. Because the governing equations are integrated over control volumes, all fluxes are explicitly accounted for through the bounding surfaces of each cell.

For any interior face shared by two neighbouring control volumes C and F, the flux evaluated at that face is common to both cells. If the flux leaving cell C through face f is denoted by F_f, then the flux entering the neighbouring cell F is exactly $-F_f$. This antisymmetry ensures that, when summing the discrete equations over the entire computational domain, all internal fluxes cancel identically, leaving only boundary contributions. As a result, global conservation of mass, momentum, and energy is satisfied to machine precision.

This property is particularly important in long transient simulations and in problems involving strong coupling between physical fields, where even small conservation errors can accumulate and lead to nonphysical solutions.

5.3.8.2 Convergence of numerical solutions

Because the discretised governing equations are generally nonlinear, they must be solved using iterative techniques. Starting from an initial guess, the solution is updated repeatedly until a convergence criterion is satisfied. In most practical solvers, convergence is declared when the change in the solution or the residual norm between successive iterations falls below a prescribed tolerance ε.

Two distinct notions of convergence are commonly encountered. The first is iterative convergence, which refers to the convergence of the numerical algorithm on a fixed grid. The second is grid convergence, which refers to the convergence of the numerical solution as the grid is refined. A solution that is converged iteratively on a coarse grid may still be far from the grid-independent solution.

Ensuring grid convergence is essential for validating numerical results, particularly in research and design applications. This typically requires systematic mesh refinement studies and careful assessment of solution sensitivity to grid resolution.

5.3.8.3 *Consistency of the discretisation*

Consistency is a fundamental mathematical property linking the discrete equations to the original continuous equations. A discretisation scheme is said to be consistent if the discretised equations reduce to the exact differential equations as the grid spacing Δx and time step Δt approach zero.

Formally, consistency requires that the truncation error vanishes in the limit

$$\Delta x \to 0, \ \Delta t \to 0 \tag{5.121}$$

If a scheme is inconsistent, no amount of grid refinement will lead to the correct solution. In the FVM, consistency is ensured by the use of proper integral approximations and reconstruction procedures whose errors scale with powers of the discretization parameters.

5.3.8.4 *Stability considerations*

Stability describes how numerical errors evolve during the solution process and requires that such errors remain bounded as iterations progress or as time advances. For steady state problems, stability is closely linked to the algebraic properties of the discretised system, where diagonal dominance of the coefficient matrix, often referred to as the Scarborough criterion, provides a sufficient condition for stability and convergence of linear systems. In transient simulations, stability is governed primarily by the time integration scheme, with explicit methods relying solely on known quantities from the current time level and therefore imposing severe restrictions on the allowable time step for stability, despite their low computational cost per step. Implicit methods, by contrast, treat future time level variables as unknowns within a coupled system of equations, leading to a higher per-step cost but permitting significantly larger time steps and making them more suitable for stiff problems and low speed flows. Additional numerical robustness can be achieved through techniques such as underrelaxation or false transient formulations, which introduce controlled damping to suppress numerical oscillations. Beyond stability, practical numerical methods must also achieve computational economy by balancing accuracy, robustness, and cost, a balance that is well embodied by the FVM due to its sparse discretised structure arising from local control volume connectivity, which enables efficient storage and solution using sparse matrix techniques even in large-scale three-dimensional simulations. Finally, accurate numerical modelling must respect the inherently directional nature of transport processes in fluids, particularly in convection dominated regimes characterised by high Péclet numbers, where information propagates primarily along the flow direction rather than isotropically as in diffusion dominated, low Péclet number regimes, since failure to align the numerical scheme with this transportiveness can lead to unphysical oscillations or excessive numerical diffusion, especially in regions with sharp gradients (Amrani et al., 2018; Chen & Zhang, 2026; Cheng et al., 2012; Chu et al., 2020; Del Coz Díaz et al., 2011; García-Jurado et al., 2013, 2015; Ge et al., 2015; Huang et al., 2015; Liu et al., 2024; Luan et al., 2011; Ma, Han, et al., 2024; Ma, Xiao, et al., 2024; Mi & Bai, 2022; Mnasri, 2013; Mnasri et al., 2010; Vesvard et al., 2025; Wang, Zheng, et al., 2015; Zhang, Ming, et al., 2025; Zhi et al., 2025; Zhou, Zhang, et al., 2025).

5.3.8.5 Boundedness of interpolation schemes

Conservation alone does not guarantee physically meaningful solutions. For many transported quantities, physical constraints dictate that the solution should remain bounded within certain limits. In the absence of sources, the value of a scalar within the domain should lie between the minimum and maximum values imposed at the boundaries.

The discrete finite volume equation for a control volume C can be written in algebraic form as

$$a_C \phi_C + \sum_{F \in NB(C)} a_F \phi_F = b_C \tag{5.122}$$

where $NB(C)$ denotes neighbouring control volumes. Boundedness requires that the solution satisfy

$$\min_F (\phi_F) \leq \phi_C \leq \max_F (\phi_F) \tag{5.123}$$

Achieving boundedness places restrictions on interpolation schemes and linearization procedures. Many high resolution schemes are designed specifically to satisfy this condition while maintaining accuracy.

5.3.8.6 Variable arrangement, solution strategy, and mesh infrastructure

The effectiveness of a finite volume formulation is strongly influenced by the arrangement of variables, the choice of solution strategy, and the underlying mesh and data structures. Variables may be stored either at mesh vertices or at the centroids of control volumes, with vertex centred arrangements offering explicit interpolation capabilities through dual meshes but at the cost of increased storage, more complex boundary condition treatment, and generally lower integration accuracy due to the mismatch between storage locations and control volume centroids. Cell-centred arrangements, which store variables at control volume centroids, are therefore more widely adopted in modern CFD solvers, as they naturally support conservative flux evaluation, straightforward implementation on arbitrary polyhedral meshes, and second-order or higher accuracy on sufficiently smooth grids, although care must be taken on highly skewed or nonorthogonal meshes where additional correction techniques are required to preserve accuracy and robustness. Once discretisation is complete, the resulting nonlinear algebraic equations are solved using either explicit or implicit strategies, with implicit methods being preferred for steady, stiff, and strongly coupled problems due to their superior stability, despite the higher per iteration cost associated with solving coupled systems. In practice, the large, sparse systems arising from finite volume discretisation are handled using iterative solvers that exploit matrix sparsity to efficiently reduce residuals to convergence. Underpinning all these aspects is comprehensive mesh support, in which each control volume and face stores detailed geometric and topological information, including centroids, volumes, area vectors, and connectivity, with consistent face normal orientation being essential for correct flux evaluation. Efficient sparse data structures are therefore indispensable for managing large-scale meshes while maintaining acceptable memory usage and computational performance.

5.4 Smoothed particle hydrodynamics

Having discussed mesh-based numerical frameworks such as the FEM and the FVM, it is natural to consider alternative formulations that avoid the inherent constraints associated with fixed spatial discretisations. While FEM and FVM have proven highly successful for a wide range of sloshing problems, their reliance on meshes can lead to additional numerical complexity when large free surface deformations, strong nonlinear motion, wave breaking, or fluid fragmentation are present. In such situations, mesh distortion, remeshing, or complex interface tracking techniques may be required, which can increase computational cost and reduce robustness. These limitations have motivated the development and adoption of particle-based approaches, in which the fluid motion is described in a purely Lagrangian manner, and the numerical discretisation moves naturally with the flow. Within this context, SPH has emerged as a particularly promising method for sloshing dynamics, as it eliminates the need for meshes altogether and offers a direct and physically intuitive representation of free surface flows. In addition to its favourable physical properties, SPH is especially well suited to modern high-performance computing architectures, since its local interaction structure and neighbour-based computations map efficiently onto massively parallel GPU platforms. This combination of numerical robustness, physical consistency, and computational scalability makes SPH an attractive complement to FEM and FVM for the simulation of complex sloshing phenomena.

5.4.1 A General algorithm for compressible and incompressible smoothed particle hydrodynamics

SPH is a fully Lagrangian, mesh-free numerical method developed to solve the governing equations of continuum mechanics by discretising the physical domain into a finite set of moving particles. Each particle represents a finite portion of the continuum and carries physical quantities such as mass, density, velocity, pressure, and energy. The method was originally introduced in the late 1970s for astrophysical applications, but its ability to naturally handle large deformations, free surfaces, moving boundaries, and multiphase interactions has led to its extensive use in fluid and solid mechanics (Kincl et al., 2025; Li, Zhang, et al., 2025; Pan, Ng, et al., 2025; Wang, 2025; Zhao, Wu, et al., 2025; Zhao, Yu, et al., 2025).

The central idea of SPH is that any field variable can be reconstructed at a given spatial location through a weighted interpolation of the values stored at neighbouring particles. This interpolation is performed using a smoothing kernel with compact support, which limits the interaction to particles located within a finite neighbourhood. As a result, the governing partial differential equations of continuum mechanics are transformed into a system of ordinary differential equations in time, evaluated along particle trajectories.

The starting point of SPH is an integral representation of a scalar field $f(\boldsymbol{x})$ over the domain Ω:

$$f(\boldsymbol{x}) = \int_{\Omega} f(\boldsymbol{x}')\delta(\boldsymbol{x} - \boldsymbol{x}')d\boldsymbol{x}' \tag{5.124}$$

where δ is the Dirac delta function. In SPH, the delta function is replaced by a smooth kernel function $W(\boldsymbol{x} - \boldsymbol{x}', h)$, where h is the smoothing length that controls the width of the kernel. This leads to the SPH approximation:

$$\langle f(\boldsymbol{x})\rangle = \int_{\Omega} f(\boldsymbol{x}')W(\boldsymbol{x} - \boldsymbol{x}', h)d\boldsymbol{x}' \tag{5.125}$$

After discretising the domain into particles, the integral is replaced by a summation over neighbouring particles:

$$f_i = \sum_{j=1}^{n} \frac{m_j}{\rho_j} f_j W(\boldsymbol{x}_i - \boldsymbol{x}_j, h) \tag{5.126}$$

where m_j and ρ_j are the mass and density of particle j, respectively, and n is the number of neighbours within the kernel support. This expression constitutes the core interpolation formula used throughout SPH.

A general SPH algorithm for both compressible and incompressible flows follows a sequence of well-defined steps. First, the computational domain is initialised by distributing particles according to the problem geometry and assigning initial values of mass, density, velocity, and other required variables. The smoothing length is chosen such that each particle interacts with a sufficient number of neighbours to ensure numerical stability and accuracy.

Second, at each time step, a neighbour search is performed to identify particles located within the support radius of the kernel. This step is critical for computational efficiency, especially in large-scale three-dimensional simulations.

Third, the governing conservation equations are discretised using SPH formulations. For compressible flows, the density is typically evaluated using a summation approach or through the continuity equation, while pressure is obtained from an equation of state. The momentum equation is then solved to compute particle accelerations, including contributions from pressure gradients, viscous effects, and body forces. For incompressible flows, additional constraints are introduced, often through a pressure Poisson equation, to enforce constant density or divergence-free velocity fields.

Fourth, time integration schemes such as Euler, leapfrog, or predictor-corrector methods are employed to update particle positions, velocities, and other state variables. The time step is selected according to stability criteria, commonly based on the Courant condition and viscous diffusion limits.

Finally, boundary conditions and free surface treatments are applied in a manner consistent with continuum mechanics. One of the strengths of SPH is that free surfaces are naturally captured by the absence of neighbouring particles, without the need for explicit interface tracking.

Through this algorithmic framework, SPH provides a unified approach for simulating both compressible and incompressible flows, offering strong physical intuition, geometric flexibility, and robustness in problems involving complex kinematics and evolving domains.

5.4.2 Discretisation of the continuum domain in smoothed particle hydrodynamics

After establishing the fundamental interpolation concept of SPH, the next essential step is the discretisation of the continuum domain. In SPH, the physical domain is represented by a finite number of Lagrangian particles, each carrying mass and moving with the local material velocity. These particles act as interpolation points rather than physical molecules, which allows the method to remain fully consistent with continuum mechanics while avoiding the complexity of molecular-scale modelling (Fang & Yin, 2025; Shobuzako et al., 2025; Villodi & Ramachandran, 2025; Xu & Yu, 2025).

In a continuum description, physical quantities are defined over infinitesimal volume elements. In SPH, each particle represents such a volume element, characterised by a finite mass m and density ρ. The differential volume element is therefore written as:

$$d\boldsymbol{x} = \frac{m}{\rho} \tag{5.127}$$

Using this relation, the kernel approximation of a scalar field can be converted into a particle summation. Starting from the kernel interpolant, and substituting the discrete particle representation, the SPH approximation becomes:

$$\langle f(\boldsymbol{x})\rangle = \int_{\Omega} f(\boldsymbol{x}')W(\boldsymbol{x}-\boldsymbol{x}', h)d\boldsymbol{x}' \tag{5.128}$$

$$f_i = \sum_{j=1}^{n} \frac{m_j}{\rho_j} f_j W(\boldsymbol{x}_i - \boldsymbol{x}_j, h) \tag{5.129}$$

The accuracy of SPH interpolations depends strongly on the number and spatial distribution of neighbouring particles. For reliable approximations, a minimum number of neighbours is required. Increasing the number of neighbours beyond these values generally improves smoothness but does not necessarily lead to significant gains in accuracy, while increasing computational effort.

Once the approximation of scalar quantities is defined, spatial derivatives required by the governing equations can be derived. Consider a vector field $\boldsymbol{f}$.

The divergence of this field can be expressed in integral form as

$$\langle \nabla \cdot \boldsymbol{f}(\boldsymbol{x}) \rangle = \int_{\Omega} \nabla \cdot \boldsymbol{f}(\boldsymbol{x}') W(\boldsymbol{x} - \boldsymbol{x}', h) d\boldsymbol{x}' \tag{5.130}$$

Using vector calculus identities and applying the divergence theorem, the divergence can be written solely in terms of kernel gradients. After discretisation, the basic SPH approximation of the divergence becomes

$$\nabla \cdot \boldsymbol{f}_i = \sum_{j=1}^{n} \frac{m_j}{\rho_j} \boldsymbol{f}_j \cdot \nabla W(\boldsymbol{x}_i - \boldsymbol{x}_j, h) \tag{5.131}$$

Although this form is mathematically consistent, it does not always provide optimal numerical behaviour in fluid flow simulations. A more accurate and commonly used formulation is obtained by employing the identity

$$\nabla \cdot \boldsymbol{f}_i = \frac{1}{\rho} [\nabla \cdot (\rho \boldsymbol{f}) - \boldsymbol{f} \cdot \nabla \rho] \tag{5.132}$$

Which leads to the SPH discretisation

$$\nabla \cdot \boldsymbol{f}_i = \frac{1}{\rho_i} \sum_{j=1}^{n} m_j (\boldsymbol{f}_j - \boldsymbol{f}_i) \cdot \nabla W(\boldsymbol{x}_i - \boldsymbol{x}_j, h) \tag{5.133}$$

This formulation ensures improved numerical symmetry and is widely adopted in practical SPH implementations.

A similar procedure is used to derive the gradient of a scalar field f. The SPH approximation of the gradient is written as

$$\nabla f_i = - \sum_{j=1}^{n} \frac{m_j}{\rho_j} f_j \nabla W(\boldsymbol{x}_i - \boldsymbol{x}_j, h) \tag{5.134}$$

To improve conservation properties and guarantee pairwise force symmetry, an alternative form based on density weighted gradients is often employed:

$$\nabla f_i = - \frac{1}{\rho_i} \sum_{j=1}^{n} m_j (f_j - f_i) \nabla W(\boldsymbol{x}_i - \boldsymbol{x}_j, h) \tag{5.135}$$

In applications such as pressure force evaluation, a fully symmetric formulation is preferred, ensuring that the forces exchanged between particle pairs are equal in magnitude and opposite in direction, thereby preserving linear and angular momentum.

Through these discretisation procedures, the governing partial differential equations of mass, momentum, and energy conservation are transformed into a coupled system of ordinary differential equations in time. This particle-based formulation forms the mathematical foundation of SPH and enables the method to accurately simulate complex continuum phenomena involving large deformations, free surfaces, and evolving geometries in both compressible and incompressible flow regimes.

The Laplacian operator is required in the modelling of viscous effects, heat conduction, and surface tension. Unlike gradients and divergences, the Laplacian cannot be directly approximated by a simple kernel gradient. Instead, it is derived from a Taylor series expansion around the reference particle (Singh et al., 2025; Yang et al., 2025; Yuan, Qiu, et al., 2025; Zhou, Han, et al., 2025; Zhu, Ke, et al., 2025).

In Cartesian coordinates, the SPH approximation of the Laplacian of a scalar field f at particle i can be written as

$$\nabla^2 f_i = 2\sum_{j=1}^{n} \frac{m_j}{\rho_j}(f_i - f_j)\frac{(\boldsymbol{x}_i - \boldsymbol{x}_j)\cdot\nabla W(\boldsymbol{x}_i - \boldsymbol{x}_j, h)}{|\boldsymbol{x}_i - \boldsymbol{x}_j|^2} \tag{5.136}$$

This formulation provides a stable and consistent approximation and is widely used in practical SPH implementations. Equivalent expressions can be derived in polar or cylindrical coordinates when required by the problem geometry.

5.4.3 Smoothed particle hydrodynamics formulation of the conservation equations

Once the spatial operators are defined, the governing equations of mass, momentum, and energy conservation can be expressed in SPH form.

The conservation of mass can be evaluated using the density summation approach,

$$\rho_i = \sum_{j=1}^{n} m_j W(\boldsymbol{x}_i - \boldsymbol{x}_j, h) \tag{5.137}$$

Or alternatively, through the Lagrangian form of the continuity equation,

$$\frac{D\rho_i}{Dt} = \sum_{j=1}^{n} m_j(\boldsymbol{u}_i - \boldsymbol{u}_j)\cdot\nabla W(\boldsymbol{x}_i - \boldsymbol{x}_j, h) \tag{5.138}$$

The momentum equation expresses Newton's second law for each particle. In SPH, it is written in a symmetric form to guarantee conservation of linear and angular momentum. For a viscous fluid, the acceleration of particle i is given by

$$\begin{aligned}\frac{D\boldsymbol{u}_i}{Dt} &= \sum_{j=1}^{n} m_j\left(\frac{P_i}{\rho_i^2} - \frac{P_j}{\rho_j^2}\right)\nabla W(\boldsymbol{x}_i - \boldsymbol{x}_j, h)\\ &+ 2\nu_i\sum_{j=1}^{n} m_j\frac{\boldsymbol{u}_i - \boldsymbol{u}_j}{\rho_j}\frac{(\boldsymbol{x}_i - \boldsymbol{x}_j)\cdot\nabla W(\boldsymbol{x}_i - \boldsymbol{x}_j, h)}{|\boldsymbol{x}_i - \boldsymbol{x}_j|^2} + \boldsymbol{F}_B\end{aligned} \tag{5.139}$$

where P represents viscous dissipation, ν is the heat flux vector, and $\boldsymbol{F}_B$ denotes volumetric heat sources. In many fluid flow applications, simplified forms of this equation are adopted, depending on whether thermal effects are significant.

The evolution of the specific internal energy eie_iei is governed by the SPH form of the energy equation,

$$\frac{De_i}{Dt} = -P\sum_{j=1}^{n} m_j(\boldsymbol{u}_i - \boldsymbol{u}_j)\cdot\nabla W(\boldsymbol{x}_i + \boldsymbol{x}_j, h) + \varepsilon_v + \nabla.\boldsymbol{q} + q_H \tag{5.140}$$

where ε_v represents viscous dissipation, $\mathbf{q}$ is the heat flux vector, and q_H denotes volumetric heat sources. In many fluid flow applications, simplified forms of this equation are adopted, depending on whether thermal effects are significant.

5.4.4 Errors in smoothed particle hydrodynamics approximations

The SPH approximation of a function and its derivatives is not exact and introduces truncation errors. These errors can be analysed by expanding the function f in a Taylor series about the reference particle position $\boldsymbol{x}$:

$$f(\boldsymbol{x}') = f(\boldsymbol{x}) + \nabla f(\boldsymbol{x})\cdot(\boldsymbol{x}' - \boldsymbol{x}) + O(|\boldsymbol{x}' - \boldsymbol{x}|^2) \tag{5.141}$$

Substituting this expansion into the kernel interpolation reveals that the SPH approximation of a function is second-order accurate, provided the kernel satisfies symmetry and normalisation conditions. However, near boundaries or free surfaces, kernel truncation leads to loss of consistency and increased error. This phenomenon is known as particle inconsistency and is a major source of numerical inaccuracy in SPH simulations (Fan, Liu, et al., 2025; Ghoneim, 2025; Ricci et al., 2025; Shao & Lo, 2003; Zhan et al., 2025b).

The choice of the smoothing kernel plays a critical role in the accuracy and stability of SPH. A valid kernel must satisfy positivity, symmetry, compact support, and normalisation, and it should converge to the Dirac delta function as the smoothing length h tends to zero.

One of the most widely used kernels is the cubic spline kernel, defined as:

$$W(r, h) = \alpha_D \begin{cases} \frac{2}{3} - r^2 + \frac{1}{2}r^3, \ 0 \le r \le 1 \\ \frac{1}{6}(2 - r)^3, \ 1 \le r \le 2 \\ 0, \ r > 2 \end{cases} \tag{5.142}$$

where $r = |\boldsymbol{x}_i - \boldsymbol{x}_j|/h$ and α_D is a normalisation constant that depends on the spatial dimension.

Other kernels, such as quartic and quintic spline kernels, offer improved smoothness and higher order accuracy at the expense of increased computational cost. The selection of the kernel and the smoothing length must therefore balance accuracy, stability, and efficiency.

Together, the conservation laws, error characteristics, and kernel properties define the mathematical core of the SPH method and strongly influence its performance in simulating compressible and incompressible continuum flows.

5.4.5 Neighbour searching and free surface modelling in smoothed particle hydrodynamics

An essential computational component of the SPH method is the identification of neighbouring particles that contribute to the kernel interpolation at each particle location. Since SPH is a fully Lagrangian technique, particles move continuously in space, and therefore the neighbour search must be performed repeatedly during the simulation. The efficiency and robustness of this procedure have a direct impact on the overall computational cost of SPH simulations.

Neighbour searching is performed through a direct or brute force approach. For a given reference particle i, the distance to every other particle j in the domain is computed and compared with the kernel support radius kh. If the condition

$$|\boldsymbol{x}_i - \boldsymbol{x}_j| \leq kh \tag{5.143}$$

is satisfied, particle j is considered a neighbour and included in the summation. Although straightforward to implement, this method has a computational complexity proportional to N^2, where N is the total number of particles, making it impractical for large-scale simulations.

To reduce computational effort, spatial decomposition techniques are commonly employed. One widely used approach is the cell-linked list method, in which the computational domain is divided into regular cells whose dimensions are comparable to the kernel support radius. Each particle is assigned to a cell based on its spatial coordinates. For a reference particle, neighbour searching is then restricted to particles located in the same cell and adjacent cells. This reduces the computational complexity to approximately linear with the number of particles.

Another efficient technique is the Verlet neighbour list. In this method, for each particle, a list of potential neighbours is constructed within a cut-off radius slightly larger than the kernel support radius, typically $kh + L$, where L is a buffer distance accounting for particle motion between updates. During the simulation, only particles within the actual support radius kh are used in the SPH summations, while the neighbour list itself is updated less frequently. This approach can significantly reduce computational cost, provided that the update interval and buffer distance are chosen carefully.

For three-dimensional simulations involving large particle counts, hierarchical data structures such as octrees, often combined with parallel computing or GPU acceleration, are commonly adopted. These methods further enhance performance and make large-scale SPH simulations feasible (Kim & Kim, 2025; Wang, Zhang, et al., 2025; Zhang & Sheng, 2025; Zhang, Lourenço, et al., 2025).

5.4.6 Identification and treatment of the free surface

One of the major strengths of SPH lies in its natural treatment of free surfaces. Since particles explicitly represent the material domain, the free surface emerges automatically as the region where particles lack neighbours on one side. Unlike mesh-based methods, no explicit interface tracking or reconstruction is required.

However, the presence of a free surface introduces additional physical effects, such as surface tension, which must be modelled appropriately. Moreover, the incomplete kernel support near the free surface leads to numerical challenges, including reduced accuracy and loss of consistency.

A widely used approach for modelling surface tension in SPH is the continuum surface force method. In this formulation, the force due to surface tension acting at a point on the interface is expressed as

$$\boldsymbol{f}_s = \sigma\kappa\boldsymbol{n} + \nabla_s \sigma \tag{5.144}$$

where σ is the surface tension coefficient, κ is the local curvature of the interface, and $\boldsymbol{n}$ is the unit normal vector to the free surface. For most practical applications involving a constant surface tension coefficient, the tangential term $\nabla_s \sigma$ vanishes, and the surface tension force reduces to a purely normal contribution.

To compute the surface normal and curvature within SPH, a colour function is often introduced. This scalar field takes the value unity inside the fluid and zero outside. The smoothed colour function at particle i is defined as

$$C_i = \sum_{j=1}^{n} \frac{m_j}{\rho_j} W(\boldsymbol{x}_i - \boldsymbol{x}_j, h) \tag{5.145}$$

The gradient of this field provides an estimate of the interface normal,

$$\boldsymbol{n}_i = \nabla C_i \tag{5.146}$$

and the unit normal is obtained by normalisation. The curvature of the free surface can then be evaluated from the divergence of the unit normal. The resulting surface tension force is added as an additional term to the momentum equation for particles identified as belonging to the free surface.

A particle is typically classified as a surface particle if the magnitude of the colour function gradient exceeds a prescribed threshold. This criterion allows the localisation of the interface while avoiding numerical instabilities associated with poorly defined normals.

Through these neighbour searching strategies and free surface treatments, SPH achieves both computational efficiency and physical realism. The combination of adaptive neighbour identification and natural interface representation makes SPH particularly well suited for problems involving complex free surface dynamics, large deformations, and evolving flow domains.

5.4.7 Surface tension modelling and free surface identification in smoothed particle hydrodynamics

When modelling free surface flows using SPH, additional physical mechanisms must be incorporated to represent surface-related forces accurately. Among these, surface tension plays a dominant role in small-scale flows, multiphase systems, and phenomena involving interfaces. SPH provides a natural framework for incorporating surface tension effects, as the free surface is inherently represented by the particle distribution itself.

5.4.7.1 Continuum surface force formulation

A widely adopted approach for modelling surface tension in SPH is the continuum surface force method (Asis et al., 2025; Feng, Xiao, et al., 2025; Moore & Teixeira-Dias, 2025; Souto-Iglesias et al., 2013; Stokes & Bayandor, 2025; Wang, Chen, et al., 2025; Yu et al., 2025). In this formulation, surface tension is treated as a body force acting in a thin interfacial region rather than as a boundary condition.

The force per unit volume acting at a point on the free surface is expressed as

$$f_s(\boldsymbol{x}_s) = \sigma\kappa\boldsymbol{n}(\boldsymbol{x}_s) + \nabla_s\sigma \tag{5.147}$$

where σ is the surface tension coefficient, κ is the local curvature of the interface, and $\boldsymbol{n}$ is the outward unit normal to the free surface. The operator ∇_s denotes the surface gradient. In most engineering applications, the surface tension coefficient is assumed constant, which eliminates the tangential term $\nabla_s\sigma$. Under this assumption, the surface tension force reduces to a purely normal contribution,

$$f_{s,i} = \sigma\kappa_i\boldsymbol{n}_i \tag{5.148}$$

acting on particle i located at the interface.

5.4.7.2 Colour function and interface geometry

To evaluate the surface normal and curvature within SPH, a colour or phase indicator function is introduced. This scalar field takes the value unity within the fluid domain and zero outside it. The smoothed colour function at particle i is defined using the standard SPH interpolation, as in Eq. (5.22),

The gradient of this field provides an estimate of the interface normal direction, as in Eq. (5.23)

The corresponding unit normal vector is obtained as

$$\boldsymbol{n}_i^u = \frac{\boldsymbol{n}_i}{|\boldsymbol{n}_i|} \tag{5.149}$$

The curvature of the free surface is related to the divergence of the unit normal vector. Within SPH, it is commonly approximated as

$$\kappa_i = -\nabla\cdot\boldsymbol{n}_i^u \tag{5.150}$$

The negative sign ensures positive curvature for convex interfaces. Combining these expressions, the surface tension force acting on a surface particle becomes

$$f_{s,i} = -\sigma\nabla^2 C_i\frac{\boldsymbol{n}_i}{|\boldsymbol{n}_i|} \tag{5.151}$$

where the Laplacian of the colour function is evaluated using an SPH approximation consistent with the discretised Laplacian operator.

5.4.7.3 Identification of free surface particles

A crucial step in the application of surface tension models is the identification of particles that belong to the free surface. In SPH, this is typically achieved by examining the magnitude of the gradient of the colour function. A particle is classified as a surface particle if the condition

$$|\boldsymbol{n}_i| > \ell \tag{5.152}$$

is satisfied, where ℓ is a prescribed threshold parameter. This criterion reflects the fact that particles near the free surface experience an imbalance in neighbour distribution, leading to a nonzero gradient of the colour function.

To avoid numerical instabilities, the surface tension force is only applied when the gradient magnitude exceeds the threshold value. The choice of ℓ controls the thickness of the interfacial layer over which surface tension is active. Excessively small values may lead to noisy curvature estimates, while overly large values may suppress legitimate surface effects.

Once the surface tension force is computed, it is incorporated as an additional term in the momentum conservation equation for surface particles. The modified SPH momentum equation can be written as

$$\frac{D\boldsymbol{u}_i}{Dt} = \sum_{j=1}^{n} m_j \left(\frac{P_i}{\rho_i^2} + \frac{P_j}{\rho_j^2} \right) \nabla W(\boldsymbol{x}_i - \boldsymbol{x}_j, h) + \wp + \boldsymbol{g} + \boldsymbol{f}_{s,i} \tag{5.153}$$

where $\wp$ is an extra term to present viscosity. This formulation allows surface tension effects to be captured in a manner that is fully consistent with the particle-based nature of SPH.

Through the use of colour functions, kernel-based gradients, and curvature approximations, SPH provides a physically meaningful and computationally efficient framework for modelling free surface flows with surface tension. These techniques enable the simulation of complex interfacial phenomena while preserving the fundamental principles of continuum mechanics.

5.4.8 Incompressible formulations, free surface detection, and interface modelling in smoothed particle hydrodynamics

While the classical SPH formulation is often applied in a weakly compressible framework, many engineering problems require the explicit enforcement of incompressibility. In addition, accurate identification of free surfaces and robust treatment of material interfaces are essential for realistic simulations of multiphase and multicomponent systems. These aspects are addressed through specialised formulations and interaction models within SPH.

In incompressible ISPH, the density is assumed to remain constant, and the incompressibility condition is enforced through a pressure correction procedure rather than an equation of state. A common approach is the fractional step method. In the first step, an intermediate velocity field is computed without enforcing incompressibility. In the second step, a pressure Poisson equation is solved to correct the velocity field and ensure zero divergence.

The incompressibility condition is expressed as:

$$\nabla \cdot \boldsymbol{u} = 0 \tag{5.154}$$

Within SPH, the source term of the pressure Poisson equation is typically constructed from the discrete continuity equation, involving relative velocities between neighbouring particles. For free surface flows, special modifications are introduced to ensure stability, as particles near the surface do not possess a complete kernel support. A particle is often classified as belonging to the free surface if its density satisfies

$$\rho_i^* < \beta\rho_0 \tag{5.155}$$

where ρ_i^* is the current density, ρ_0 is a reference density, and β is a user-defined parameter. This criterion allows surface particles to be identified and treated appropriately in the pressure correction step.

An alternative and computationally efficient method for detecting free surface particles relies on evaluating the divergence of the particle position vector. For a particle (i), the SPH approximation of the divergence of position is written as

$$\nabla\cdot\boldsymbol{r}_i = \sum_{j=1}^{n} \frac{m_j}{\rho_{j=1}}(\boldsymbol{r}_i - \boldsymbol{r}_j)\cdot\nabla W(\boldsymbol{r}_i - \boldsymbol{r}_j, h) \tag{5.156}$$

In a two-dimensional domain, this quantity takes a value close to two in the interior of the fluid, where the kernel support is complete. Near the free surface, the lack of neighbouring particles leads to a reduced value. A threshold, practically between 1.2 and 1.5, is used to classify surface particles. Although this method may fail to detect a small fraction of surface particles, the resulting error is generally acceptable, as undetected particles usually experience pressures close to zero.

Once the free surface particles are identified, boundary conditions such as zero pressure can be imposed directly, ensuring a physically consistent representation of the interface.

In simulations involving multiple materials or phases, SPH must accurately represent the interaction between particles belonging to different continua. Each SPH particle represents a finite volume with associated mass and density, and is commonly idealised as a circular or spherical element whose size is related to the kernel support radius.

At material interfaces, contact detection is performed by comparing the distance between the centres of two particles with the sum of their effective radii. Contact or penetration is identified if

$$R = \frac{(kh)_a + (kh)_b}{r_{a,b}} \geq 1 \tag{5.157}$$

where $r_{a,b}$ is the distance between particles a and b, and $(kh)_a$, $(kh)_a$ are their respective support radii.

To prevent unphysical interpenetration, a repulsive contact force is often introduced, inspired by intermolecular interaction models. A commonly used form of the repulsive force between two particles at an interface is

$$\boldsymbol{f}_{a,b} = \begin{cases} \overline{P}\,(R^{N_1} - R^{N_2})\frac{\boldsymbol{x}_{ab}}{r_{ab}^2}, & R \geq 1 \\ 0, & R < 1 \end{cases} \tag{5.158}$$

where $\boldsymbol{x}_{ab} = \boldsymbol{x}_a - \boldsymbol{x}_b$, and $\overline{P}$, N_1, and N_2 are empirical parameters controlling the magnitude and stiffness of the repulsion.

Although such repulsive forces are effective in preventing penetration, they introduce concepts originating from molecular dynamics into a continuum framework.

This mixing of scales requires careful justification and calibration, particularly when accurate stress transmission across interfaces is required.

Through incompressible formulations, geometric free surface detection, and interface interaction models, SPH is able to simulate a wide range of complex flow problems involving free surfaces and multiple materials. The challenge remains to balance numerical robustness with strict adherence to the principles of continuum mechanics (Huang, Shen, et al., 2025; Pan, Hong, et al., 2025; Sandnes et al., 2025; Vakilha, 2025; Ye et al., 2025; Yilmaz, 2025).

5.4.9 Turbulence modelling, variable smoothing length, and numerical stabilisation in smoothed particle hydrodynamics

As SPH has matured, increasing attention has been given to its application in turbulent flows and highly transient regimes. At the same time, the numerical stability and consistency of the method require additional strategies, particularly when large gradients, shocks, or rapidly evolving flow structures are present.

Turbulence modelling in SPH remains a challenging topic due to the Lagrangian nature of the method and the absence of a fixed spatial grid. In principle, direct numerical simulation (DNS) can be performed using SPH by resolving all relevant spatial and temporal scales of turbulence. In this case, the governing equations are solved directly without introducing additional turbulence models. However, DNS requires extremely fine particle resolutions and very small time steps, which restricts its applicability to low Reynolds number flows or small computational domains.

To reduce computational cost, several turbulence modelling strategies have been adapted for SPH. One of the earliest approaches involved the formulation of an SPH analogue of the alpha turbulence model. In this method, two velocity fields are introduced: the physical velocity obtained from the momentum equation and a smoothed velocity used to advect the particles. This averaging process damps small-scale fluctuations while preserving large-scale flow structures. Although physically sound, this approach is computationally expensive and rarely used in large-scale simulations.

A more widely adopted strategy is based on the large eddy simulation (LES) concept. In LES SPH formulations, large-scale motions are resolved explicitly by the particle discretisation, while the dissipative effects of unresolved small scales are modelled through additional stress terms. Subparticle scale (SPS) models have been proposed for SPH to represent these effects. Such models have been successfully applied to problems involving wave breaking, dam breaking, and free surface turbulence. Nevertheless, three-dimensional LES SPH simulations remain computationally demanding due to the large number of particles required to achieve adequate resolution.

Despite these advances, turbulence modelling in SPH is still an active area of research. Accurate prediction of turbulence statistics often requires careful calibration and, in many cases, higher spatial resolution than initially expected.

The smoothing length h plays a fundamental role in SPH, as it defines the size of the kernel support and therefore the number of neighbouring particles involved in the interpolation. In flows characterised by strong density variations, shocks, or localised compression, a fixed smoothing length may lead to poor resolution or numerical instability (Chakraborty & Shaw, 2013; Erbaş et al., 2025; Huang, Jia, et al., 2025; Liang, Sun, Lyu et al., 2025; Liang, Sun, & Zhang, 2025; Niu et al., 2025; Rublev et al., 2025; Zargaran et al., 2025).

To address this issue, adaptive or variable smoothing length formulations have been proposed. A common approach relates the smoothing length to the local particle density,

$$h_i = h_0 \left(\frac{\rho_0}{\rho_i} \right)^{\frac{1}{n_d}} \tag{5.159}$$

where h_0 and ρ_0 are reference values, and n_d is the number of spatial dimensions. This expression ensures that the number of neighbouring particles remains approximately constant as the flow evolves.

An alternative formulation is obtained by differentiating the smoothing length with respect to time,

$$\frac{Dh_i}{Dt} = -\frac{1}{n_d} \frac{h_i}{\rho_i} \frac{D\rho_i}{Dt} \tag{5.160}$$

This relation couples the evolution of the smoothing length directly to the continuity equation, allowing a smooth and consistent adjustment of the kernel support during the simulation.

While adaptive smoothing lengths improve accuracy in regions of strong gradients, they introduce additional complexity in the formulation and require careful treatment of kernel normalisation and gradient corrections.

SPH simulations often require numerical stabilisation to prevent nonphysical behaviour such as particle interpenetration, excessive noise, or loss of convergence. These issues are particularly pronounced in simulations involving shocks, high velocity gradients, or tensile stress states.

One widely used stabilisation technique is artificial viscosity, which introduces controlled dissipation into the momentum and energy equations. Artificial viscosity converts kinetic energy into internal energy, mimicking the effect of physical viscosity in regions of strong compression. Its primary role is to stabilise the numerical scheme rather than to model real viscous effects.

Another important correction is the eXtended Smoothed Particle Hydrodynamics (XSPH) formulation, in which particles are advected using a locally averaged velocity rather than their individual velocities. This procedure promotes a more regular particle distribution and reduces numerical noise while preserving conservation properties.

Despite their effectiveness, numerical corrections must be applied with caution. Excessive stabilisation can obscure physical phenomena and reduce solution accuracy. Therefore each correction technique should be justified based on the characteristics of the problem being studied.

Through turbulence modelling strategies, adaptive smoothing length formulations, and numerical stabilisation methods, SPH continues to evolve as a powerful tool for simulating complex, highly dynamic continuum flows. The challenge lies in achieving a balance between numerical robustness, physical fidelity, and computational efficiency (Aly et al., 2025; Antuono et al., 2025; Deng et al., 2013; Du et al., 2025; Fan, Zhang, et al., 2025; Ghannam et al., 2025; He et al., 2025; Jiao et al., 2025; Liu et al., 2025; Ma et al., 2026; Ramezanzadeh et al., 2025; Sigalotti & Vargas, 2025; Zhang, Fan, et al., 2025).

5.4.10 Turbulence modelling, variable smoothing length, and numerical corrections in smoothed particle hydrodynamics

The application of SPH to complex and realistic flow problems requires careful consideration of turbulence effects, adaptive spatial resolution, and numerical stabilisation. As SPH is extended from laminar to more realistic flow regimes, additional physical and numerical complexities arise. High Reynolds number flows involve strong velocity gradients, vortical structures, and multi-scale dynamics that place stringent demands on spatial resolution and numerical dissipation. Without appropriate treatment, SPH simulations of such flows may suffer from excessive noise, particle disorder, or unphysical energy growth. Consequently, the accurate representation of turbulence and small-scale dynamics has become a central topic in the development of advanced SPH formulations.

Turbulent flows are characterised by a wide range of interacting temporal and spatial scales. In principle, SPH allows DNS of turbulence by solving the conservation equations without additional modelling, resolving all relevant scales of motion. In this case, the momentum equation,

$$\frac{D\boldsymbol{u}_i}{Dt} = \sum_{j=1}^{n} m_j \left(\frac{P_i}{\rho_i^2} + \frac{P_j}{\rho_j^2} \right) \nabla W_{ij} + \boldsymbol{g} + \text{viscous terms} \tag{5.161}$$

is evaluated for all particles at sufficiently fine resolution. However, DNS with SPH becomes prohibitively expensive at moderate or high Reynolds numbers, as the number of particles required scales rapidly with the smallest turbulent length scale.

To reduce computational cost, turbulence models adapted to SPH have been proposed. One of the earliest approaches is the SPH form of the alpha turbulence model, in which two velocity fields are introduced. The physical velocity evolves from the momentum equation, while a smoothed velocity field is used to advect the particles,

$$\boldsymbol{u}_i^* = \boldsymbol{u}_i + \varepsilon \sum_{j=1}^{n} \frac{m_j}{\rho_j} (\boldsymbol{u}_j + \boldsymbol{u}_i) W_{ij} \tag{5.162}$$

where ε controls the degree of smoothing. This formulation damps small-scale fluctuations while preserving large-scale motion, but it significantly increases computational cost.

More widely used are LES concepts adapted to SPH. In LES SPH, the large turbulent structures are resolved directly by the particle discretisation, while the effects of unresolved small scales are modelled through additional SPS stress terms. These approaches have shown good qualitative agreement with experimental observations in problems such as dam breaking, wave breaking, and free surface turbulence. Nevertheless, three dimensional LES SPH remains computationally demanding and requires careful resolution control (Boregowda & Liu, 2025; Chen, Meringolo, et al., 2025; Fang et al., 2025; Fraga Filho et al., 2025; Fuentes et al., 2025; Ghasemi V., Firoozabadi & Mahdinia, 2013; Huang, Yue, Zhu, et al., 2025; Jin et al., 2025; Kiara et al., 2013a, 2013b; Lanzafame, 2013; Lian et al., 2025; Martínez-Estévez et al., 2025; Nasruddin et al., 2025; Sayehvand & Bakhtiari Haftlang, 2025; Ulrich et al., 2013; Xiong et al., 2025; Yan et al., 2025; Yuan, Gao, et al., 2025; Zhang, Wang, et al., 2025).

5.4.11 Variable smoothing length formulation

The smoothing length (h) defines the radius of influence of the kernel and directly controls the number of neighbouring particles contributing to SPH interpolations. If (h) is too small, particle deficiency leads to noisy and inconsistent results; if it is too large, excessive smoothing reduces accuracy.

In flows involving shocks, impacts, or strong density variations, a fixed smoothing length is insufficient. To address this limitation, adaptive smoothing length formulations are introduced, ensuring that the number of neighbours remains approximately constant during the simulation.

A commonly used algebraic relation links the smoothing length to the local density,

$$h_i = h_0 \left(\frac{\rho_0}{\rho_i} \right)^{\frac{1}{n_d}} \tag{5.163}$$

where h_0 and ρ_0 are reference values and n_d is the number of spatial dimensions. As density increases, the smoothing length decreases, improving local resolution.

Alternatively, a fully dynamic formulation is obtained by differentiating h with respect to time and coupling it to the continuity equation,

$$\frac{Dh_i}{Dt} = -\frac{1}{n_d} \frac{h_i}{\rho_i} \frac{D\rho_i}{Dt} \tag{5.164}$$

This expression ensures a smooth temporal evolution of the smoothing length and is particularly suitable for highly transient flows. However, variable smoothing length formulations require additional corrections to maintain kernel normalisation and consistency.

5.4.12 Numerical corrections and stabilisation

SPH simulations are prone to numerical instabilities arising from particle disorder, tensile stresses, and sharp gradients. To mitigate these issues, several stabilisation techniques are commonly employed.

One of the most important is artificial viscosity, introduced to stabilise compressive flows and shocks. The artificial viscosity term π_{ij} is added to the pressure contribution in the momentum equation,

$$\frac{D\boldsymbol{u}_i^*}{Dt} = \frac{D\boldsymbol{u}_i}{Dt} - \sum_{j=1}^{n} m_j \pi_{ij} \nabla W_{ij} \tag{5.165}$$

where

$$\pi_{ij} = \begin{cases} \dfrac{-\alpha c_{ij} \chi_{ij} + \beta \chi_{ij}^2}{\overline{\rho}_{ij}}, & (\boldsymbol{u}_{ij} \cdot \boldsymbol{x}_{ij}) < 0 \\ 0, & (\boldsymbol{u}_{ij} \cdot \boldsymbol{x}_{ij}) < 0 \end{cases} \tag{5.166}$$

Here, α and β are tuning parameters, c_{ij} is the average speed of sound, and c_{ij} depends on the relative velocity and particle separation. Artificial viscosity prevents particle interpenetration and suppresses spurious oscillations but must be carefully calibrated to avoid excessive dissipation.

Another widely used correction is the XSPH formulation, in which particle positions are updated using a smoothed velocity,

$$\boldsymbol{u}_i^* = \boldsymbol{u}_i + \eta \sum_{j=1}^{n} \frac{2m_j}{\rho_i + \rho_j} (\boldsymbol{u}_j - \boldsymbol{u}_i) W_{ij} \tag{5.167}$$

with $0 \leq \eta \leq 1$. This correction improves particle regularity while preserving linear and angular momentum.

Together, turbulence modelling strategies, adaptive smoothing length formulations, and numerical stabilisation techniques extend SPH to complex, high Reynolds number, and highly transient flows. Their appropriate selection and calibration are crucial for achieving physically meaningful and numerically stable simulations.

5.4.13 Artificial viscosity, XSPH correction, and tensile instability in smoothed particle hydrodynamics

In physical fluid systems, kinetic energy is dissipated through viscous mechanisms and converted into internal energy, particularly in regions of strong compression, shocks, or rapid deformation. In numerical simulations, however, discretisation errors can lead to spurious oscillations, particle interpenetration, and nonphysical pressure fluctuations. To address these issues, artificial viscosity is introduced as a numerical dissipation mechanism.

Artificial viscosity in SPH does not represent the true physical viscosity of the fluid. Instead, it acts as a stabilising term that damps high-frequency numerical noise and ensures correct entropy generation in compressive flows. The artificial viscosity term is added to the pressure contribution in the momentum and energy equations.

The standard form of artificial viscosity between a pair of particles (i) and (j) is written as

$$\pi_{i,j} = \begin{cases} \dfrac{-\alpha_\pi c_{ij}\chi_{ij} + \beta_\pi \chi_{ij}^2}{\overline{\rho}_{ij}}, & \boldsymbol{u}_{ij}\cdot\boldsymbol{x}_{ij} < 0 \\ 0, & \boldsymbol{u}_{ij}\cdot\boldsymbol{x}_{ij} \geq 0 \end{cases} \tag{5.168}$$

where

$$\chi_{ij} = \frac{h_{ij}\boldsymbol{u}_{ij}\cdot\boldsymbol{x}_{ij}}{|\boldsymbol{x}_{ij}|^2 + \phi^2}, \ \boldsymbol{u}_{ij} = \boldsymbol{u}_i - \boldsymbol{u}_j, \ \boldsymbol{x}_{ij} = \boldsymbol{x}_i - \boldsymbol{x}_j \tag{5.169}$$

Here, α_π and β_π are dimensionless coefficients controlling the linear and quadratic components of dissipation, c_{ij} is the average speed of sound between particles, $\overline{\rho}_{ij}$ *is the average density, and* h_{ij} is the mean smoothing length. The small parameter ϕ^2, typically chosen as $(0.001h_{ij}^2)$, prevents numerical singularities when particles approach each other.

The conditional nature of the formulation ensures that artificial viscosity is activated only when particles are approaching one another, that is, in compressive regions of the flow. This prevents unnecessary dissipation in regions dominated by expansion or shear.

When artificial viscosity is included, the momentum equation becomes

$$\frac{D\boldsymbol{u}_i^*}{Dt} = \frac{D\boldsymbol{u}_i}{Dt} - \sum_{j=1}^{n} m_j\pi_{ij}\nabla W(\boldsymbol{x}_i - \boldsymbol{x}_j, h) \tag{5.170}$$

where $\frac{D\boldsymbol{u}_i^*}{Dt}$ denotes the acceleration obtained from the inviscid pressure and body force terms. The superscript (*) indicates the corrected acceleration after introducing artificial viscosity.

Similarly, the artificial viscosity contribution to the energy equation is expressed as

$$\frac{De_i^*}{Dt} = \frac{De_i}{Dt} + \sum_{j=1}^{n} m_j\pi_{ij}(\boldsymbol{u}_i - \boldsymbol{u}_j)\cdot\nabla W(\boldsymbol{x}_i - \boldsymbol{x}_j, h) \tag{5.171}$$

This term ensures that the kinetic energy dissipated by artificial viscosity is consistently transferred into internal energy, preserving the overall energy balance.

The choice of the coefficients α_π and β_π strongly influences the solution. Excessive artificial viscosity can suppress physically meaningful flow features, whereas insufficient viscosity may lead to numerical instability. Therefore calibration against analytical solutions or experimental data is essential.

Despite the stabilising effect of artificial viscosity, SPH simulations may still exhibit particle disorder and noisy velocity fields, particularly in long-time integrations. To address this issue, a velocity regularisation technique known as the XSPH correction was introduced.

The fundamental idea of XSPH is to move particles using a locally averaged velocity rather than their individual velocities. This promotes a more regular particle distribution while preserving the Lagrangian character of the method.

The corrected velocity of particle i is defined as

$$u_i^* = u_i + \eta \sum_{j=1}^{n} \frac{2m_j}{\rho_i + \rho_j}(u_j - u_i) W(x_i - x_j) W(x_i - x_j, h) \tag{5.172}$$

where η is a dimensionless parameter.

The XSPH correction modifies only the velocity used to update particle positions. The physical velocity entering the momentum and energy equations remains unchanged. This distinction is crucial, as it allows the method to improve particle ordering without introducing artificial forces or violating conservation laws.

From a physical perspective, XSPH can be interpreted as a local smoothing of the velocity field, analogous to filtering out high-frequency numerical noise. Importantly, the formulation preserves linear and angular momentum exactly, as the velocity corrections are pairwise antisymmetric.

XSPH has proven particularly effective in free surface flows, multiphase problems, and long duration simulations, where particle disorder would otherwise accumulate and degrade accuracy.

One of the most serious numerical pathologies in SPH is tensile instability. This phenomenon manifests as the unphysical clustering or clumping of particles, often forming filament-like or string structures. Tensile instability typically arises in regions where the pressure becomes negative, which can occur when pressure is computed from an equation of state or when elastic stresses dominate.

In tensile regimes, the kernel gradient induces attractive forces between particles instead of repulsive ones, leading to particle aggregation and loss of resolution. This behaviour is entirely numerical and has no physical counterpart in continuum mechanics.

To mitigate tensile instability, an artificial stress or artificial pressure term is introduced into the momentum equation. The additional term generates a short-range repulsive force that counteracts the spurious attraction.

The corrective force term is written as

$$-\sum_{j=1}^{n} m_j R_p (f_{ij})^s \nabla W(x_i - x_j, h) \tag{5.173}$$

where

$$f_{ij} = \frac{W(x_i - x_j, h)}{W(\Delta p, h)} \tag{5.174}$$

Δp is the average particle spacing, and s is a positive exponent.

The artificial stress coefficient R_p is defined as the sum of contributions from particles i and j,

$$R_p = R_{p,i} + R_{p,j} \tag{5.175}$$

with

$$R_{p,i} = \begin{cases} \frac{\lambda \mid P_i \mid}{\rho_i^2}, & P_i < 0 \\ 0, & P_i \geq 0 \end{cases} \tag{5.176}$$

And

$$R_{p,j} = \begin{cases} \frac{\lambda \mid P_j \mid}{\rho_j^2}, & P_j < 0 \\ 0, & P_j \geq 0 \end{cases} \tag{5.177}$$

Here, λ is a tuning parameter that depends on the smoothing length, particle spacing, and problem dimension.

This formulation ensures that the corrective force is activated only in regions of negative pressure, precisely where tensile instability occurs. The repulsive force grows rapidly as particles approach each other, preventing clustering while leaving compressive regions unaffected.

Although artificial stress techniques are effective, they introduce additional empirical parameters and must be applied judiciously. Overcorrection can lead to excessive repulsion and artificial stiffening of the flow, while undercorrection may fail to suppress instability (Cai et al., 2025; Chen, Zhao, et al., 2025; Feng, Gan, et al., 2025; Feng, Yin, et al., 2025; Guang et al., 2025; Huang & Xie, 2025; Li, Li, et al., 2025; Li, Wang, et al., 2025; Mahdevari & Bakhtiari Haftlang, 2025; Marrone et al., 2025; Paneer et al., 2025; Qin et al., 2025; Tsuji et al., 2025; Wang, Haidn, et al., 2025; Ying et al., 2025; Yokoyama et al., 2025; Zhan et al., 2025a, 2025c; Zhu, Zheng, et al., 2025).

5.4.14 Time integration, boundary conditions, and overall smoothed particle hydrodynamics algorithmic structure

After spatial discretisation, the governing equations of mass, momentum, and energy conservation are reduced to a coupled system of ordinary differential equations of the form

$$\frac{DU_i}{Dt} = \boldsymbol{F}_i(\boldsymbol{U}, \boldsymbol{x}, t) \tag{5.178}$$

where $\boldsymbol{U}_i$ represents the vector of particle properties, typically including position $\boldsymbol{x}_i$, velocity $\boldsymbol{u}_i$, density ρ_i, and internal energy e_i. The right-hand side $\boldsymbol{F}_i$ contains pressure forces, viscous terms, body forces, and possible stabilisation contributions.

Most SPH implementations rely on explicit time integration schemes due to their simplicity and low memory requirements. The forward Euler method represents the simplest choice,

$$\boldsymbol{U}_i^{n+1} = \boldsymbol{U}_i^n + \Delta t \boldsymbol{F}_i^n \tag{5.179}$$

where n denotes the current time level and Δt is the time step. Although easy to implement, this method is only first-order accurate in time and conditionally stable, requiring very small time steps.

A more commonly used scheme is the leapfrog method, which is second-order accurate and symplectic in nature. Velocities are updated at half-time steps,

$$\boldsymbol{u}_i^{n+\frac{1}{2}} = \boldsymbol{u}_i^{n-\frac{1}{2}} + \Delta t\boldsymbol{a}_i^n \tag{5.180}$$

followed by the position update,

$$\boldsymbol{x}_i^{n+1} = \boldsymbol{x}_i^n + \Delta t\boldsymbol{u}_i^{n+\frac{1}{2}} \tag{5.181}$$

The leapfrog scheme exhibits good long-term energy behaviour and is therefore widely adopted in SPH simulations involving large deformations and long integration times.

Another frequently used approach is the predictor-corrector method. In this case, a provisional state is first computed,

$$\boldsymbol{U}_i^* = \boldsymbol{U}_i^n + \Delta t\boldsymbol{F}_i^n \tag{5.182}$$

followed by a corrected update,

$$\boldsymbol{U}_i^{n+1} = \boldsymbol{U}_i^n + \frac{\Delta t}{2}(\boldsymbol{F}_i^n + \boldsymbol{F}_i^*) \tag{5.183}$$

This method achieves second-order temporal accuracy and offers significantly improved numerical stability when compared with the first-order Euler scheme, albeit at the expense of increased computational cost per time step. Nevertheless, explicit SPH formulations remain conditionally stable, with the maximum admissible time step primarily governed by the Courant–Friedrichs–Lewy (CFL) stability criterion.

$$\Delta t \leq C_{CFL}\frac{h_i}{c_i + |\boldsymbol{u}_i|} \tag{5.184}$$

where c_i is the local speed of sound and C_{CFL} is a safety factor.

Additional constraints arise from viscous diffusion,

$$\Delta t \leq C_v\frac{h_i^2}{\nu_i} \tag{5.185}$$

and from body forces such as gravity,

$$\Delta t \leq C_g\sqrt{\frac{h_i}{|\boldsymbol{g}|}} \tag{5.186}$$

The final time step is chosen as the minimum of all relevant stability limits, ensuring numerical robustness.

5.4.15 Boundary cndition treatment in smoothed particle hydrodynamics

One of the simplest approaches is to represent solid boundaries using fixed boundary particles. These particles remain stationary and exert forces on fluid particles through the same SPH interaction laws. Pressure and viscous forces between fluid and boundary particles are computed using the standard SPH formulations,

$$f_{ib} = m_b \left(\frac{P_i}{\rho_i^2} + \frac{P_b}{\rho_b^2} \right) \nabla W_{ib} \tag{5.187}$$

where the subscript b denotes boundary particles. To prevent fluid penetration, boundary particles are often assigned artificial pressures or repulsive forces.

While this boundary treatment is conceptually simple and easy to implement, it may suffer from reduced accuracy in the vicinity of solid walls because the kernel support is truncated near boundaries. This truncation leads to a loss of consistency in the SPH interpolation and can introduce spurious pressure gradients or velocity errors. As a result, boundary particles must be placed carefully and often in multiple layers to maintain an approximately uniform neighbour distribution. To overcome these limitations, ghost or mirror particle methods are commonly employed. In these approaches, fictitious particles are generated outside the physical domain by reflecting fluid particles across the boundary surface. The physical properties of these ghost particles are prescribed in such a way that the required boundary conditions, such as no penetration or no slip, are satisfied, while simultaneously restoring kernel symmetry and improving numerical accuracy near the wall.

For a no slip wall, the velocity of a ghost particle is defined as

$$\boldsymbol{u}_g = 2\boldsymbol{u}_w - \boldsymbol{u}_i \tag{5.188}$$

where $\boldsymbol{u}_w$ is the wall velocity and $\boldsymbol{u}_i$ is the velocity of the corresponding fluid particle. Pressure and density are typically mirrored directly,

$$P_g = P_i \;\&\; \rho_g = \rho_i \tag{5.189}$$

This method restores kernel symmetry near boundaries and significantly improves accuracy, but it increases algorithmic complexity and computational cost.

An alternative approach is based on short range repulsive forces between fluid particles and solid boundaries. The force is typically formulated as

$$f_{iw} = K \left(\frac{h - d_{iw}}{d_{iw}} \right)^2 \boldsymbol{n}_w,\ d_{iw} < h \tag{5.190}$$

where d_{iw} is the distance from the particle to the wall, $\boldsymbol{n}_w$ is the wall normal, and K and n are empirical parameters. This method is robust and easy to implement, but it lacks strict consistency with continuum mechanics.

5.4.16 Solid boundary representation in smoothed particle hydrodynamics

SPH does not possess a natural notion of boundaries. The computational domain is defined purely by the presence or absence of particles. Consequently, solid boundaries must be introduced through specialised particle treatments that reproduce the correct physical constraints on velocity, pressure, and stress.

One widely used approach is the boundary particle method, in which solid walls are discretised by layers of particles that remain fixed or move according to prescribed kinematics. These boundary particles participate in the SPH summations in the same manner as fluid particles, ensuring force transmission across the interface.

For a fluid particle (i) interacting with a boundary particle (b), the pressure force contribution is written as

$$\boldsymbol{f}_{ib}^{(p)} = m_b\left(\frac{P_i}{\rho_i^2} + \frac{P_b}{\rho_b^2}\right)\nabla W\left(\boldsymbol{x}_i - \boldsymbol{x}_b, h\right) \tag{5.191}$$

where P_b and ρ_b are artificial or prescribed values associated with the boundary particle. To enforce impermeability, boundary particles are often assigned pressures that counteract penetration and maintain a near zero normal velocity at the wall.

Viscous interaction between fluid and boundary particles can be included through additional terms in the momentum equation, ensuring approximate no slip conditions,

$$\boldsymbol{f}_{ib}^{(v)} = 2\nu m_b \frac{\boldsymbol{u}_i - \boldsymbol{u}_b}{\rho_b}\frac{(\boldsymbol{x}_i - \boldsymbol{x}_b)\cdot\nabla W_{ib}}{|\boldsymbol{x}_i - \boldsymbol{x}_b|^2} \tag{5.192}$$

Although conceptually simple, the boundary particle method requires careful placement of multiple layers of particles to prevent kernel truncation and to maintain numerical accuracy near walls.

In many applications, boundaries are not fixed but move dynamically under external forces or fluid loading. Examples include floating bodies, elastic structures, and impact driven solids. SPH naturally accommodates such problems by treating solid bodies as collections of particles governed by their own equations of motion.

For a rigid solid composed of particles k, the translational and rotational motion of the body is governed by:

$$M\frac{d\boldsymbol{u}_m}{dt} \sum_k \boldsymbol{F}_k \tag{5.193}$$

$$\boldsymbol{I}\frac{d\Omega}{dt} = \sum_k \left(\boldsymbol{r}_k \times \boldsymbol{F}_k\right) \tag{5.194}$$

where M is the total mass of the solid, $\boldsymbol{u}_m$ its centre of mass velocity, Ω its angular velocity, and $\boldsymbol{F}_k$ the hydrodynamic force acting on particle k. The force $\boldsymbol{F}_k$ is obtained directly from SPH interactions with neighbouring fluid particles.

The velocity of each solid particle is then updated as

$$\boldsymbol{u}_k = \boldsymbol{u}_m + \Omega \times \left(\boldsymbol{x}_k - \boldsymbol{x}_{cm}\right) \tag{5.195}$$

This formulationn is ensuring rigid body motion. This formulation guarantees full momentum exchange between the fluid and the solid and preserves Newton's third law at the particle level.

For deformable solids, SPH particles are endowed with elastic or viscoelastic constitutive relations. The momentum equation is augmented by stress divergence terms,

$$\frac{D\boldsymbol{u}_i}{Dt} = \sum_{j=1}^{n} m_j \left(\frac{\boldsymbol{\sigma}_i}{\rho_i^2} + \frac{\boldsymbol{\sigma}_j}{\rho_j^2} \right) \cdot \nabla W_{ij} + \boldsymbol{g} \tag{5.196}$$

where $\boldsymbol{\sigma}$ is the Cauchy stress tensor computed from strain measures and material laws. This unified particle description allows seamless coupling between fluids and solids within the same numerical framework.

Accurate boundary modelling in SPH requires the proper enforcement of fundamental kinematic constraints, most notably impermeability and no-slip conditions at solid walls. Unlike mesh-based methods, SPH does not impose these constraints explicitly through boundary equations. Instead, they arise implicitly from the interaction forces between fluid particles and boundary representations. For an impermeable wall, the essential requirement is that the normal component of the relative velocity between a fluid particle and a boundary particle vanishes,

$$(\boldsymbol{u}_i - \boldsymbol{u}_b) \cdot \boldsymbol{n}_b = 0 \tag{5.197}$$

where $\boldsymbol{n}_b$ is the outward normal of the boundary. In practice, this condition is enforced approximately through repulsive pressure forces or by introducing specialised boundary constructions. One of the most accurate approaches is the mirror particle method, in which fictitious ghost particles are created outside the physical domain to restore kernel symmetry. For a flat wall, a ghost particle g corresponding to a fluid particle i is placed at the mirrored position,

$$\boldsymbol{x}_g = \boldsymbol{x}_i - 2[(\boldsymbol{x}_i - \boldsymbol{x}_w) \cdot \boldsymbol{n}_w] \boldsymbol{n}_w \tag{5.198}$$

with velocity and pressure defined to satisfy boundary conditions,

$$\boldsymbol{u}_g = 2\boldsymbol{u}_w - \boldsymbol{u}_i \;\&\; P_g = P_i \tag{5.199}$$

This formulation significantly improves pressure and velocity accuracy near solid walls by compensating for kernel truncation. However, it also increases algorithmic complexity and computational cost, particularly for complex or moving boundaries.

5.4.17 Algorithmic consistency and conservation properties

A key requirement of any numerical method for continuum mechanics is the preservation of fundamental conservation laws. SPH achieves exact conservation of mass by construction, as particle masses are constant. Conservation of linear and angular momentum is ensured when interaction forces are antisymmetric,

$$\boldsymbol{f}_{ij} = -\boldsymbol{f}_{ji} \tag{5.200}$$

Symmetric formulations of pressure and viscous forces are therefore essential. For example, the pressure force between particles (i) and (j) must be written as

$$f_{ij}^{(p)} = m_i m_j \left(\frac{P_i}{\rho_i^2} + \frac{P_j}{\rho_j^2} \right) \nabla W_{ij} \tag{5.201}$$

which guarantees pairwise force balance.

Energy conservation is more subtle. While inviscid SPH formulations can conserve total energy exactly, the inclusion of artificial viscosity, boundary dissipation, and stabilisation terms introduces controlled energy loss. The challenge is to ensure that such dissipation is physically justified and does not dominate the solution.

Algorithmic consistency also requires careful handling of variable smoothing lengths, boundary truncation, and time integration. When smoothing lengths vary spatially, additional correction terms are required to preserve kernel normalisation,

$$\sum_j \frac{m_j}{\rho_j} W_{ij}(h_i) \approx 1 \tag{5.202}$$

Failure to satisfy this condition leads to zeroth order inconsistency and spurious pressure gradients, particularly near boundaries.

5.4.18 Accuracy, consistency, and practical limitations of smoothed particle hydrodynamics

A numerical discretisation is said to be *consistent* if it is capable of reproducing the exact solution of the governing continuum equations as the discretisation parameters, such as spatial resolution and time step, tend to zero. In the context of SPH, consistency is fundamentally linked to the properties of the smoothing kernel and to the quality of the particle distribution. The kernel approximation must satisfy a set of moment conditions to ensure that constant and linear fields are exactly reproduced. However, in practical SPH simulations, the continuous kernel integrals are replaced by discrete particle summations, and exact consistency is only achieved when particles are uniformly distributed and the kernel support is complete. Irregular particle spacing, boundary truncation, and evolving particle disorder inevitably degrade consistency, leading to interpolation errors, spurious forces, and reduced accuracy. These limitations highlight the importance of kernel selection, particle regularisation, and corrective formulations in practical SPH applications.

Consider the kernel approximation of a scalar function $f(\boldsymbol{x}')$,

$$\langle f(\boldsymbol{x}) \rangle = \int_\Omega f(\boldsymbol{x}') W(\boldsymbol{x} - \boldsymbol{x}', h) d\boldsymbol{x}' \tag{5.203}$$

Expanding $f(\boldsymbol{x}')$ in a Taylor series about $\boldsymbol{x}$,

$$f(\boldsymbol{x}') = f(\boldsymbol{x}) + \nabla f(\boldsymbol{x}) \cdot (\boldsymbol{x}' - \boldsymbol{x}) + \frac{1}{2} (\boldsymbol{x}' - \boldsymbol{x})^T \boldsymbol{H}_f (\boldsymbol{x}' - \boldsymbol{x}) + O(|\boldsymbol{x}' - \boldsymbol{x}|^3) \tag{5.204}$$

where $\boldsymbol{H}_f$ is the Hessian matrix of second derivatives, and substituting into the kernel approximation yields

$$\langle f(\boldsymbol{x})\rangle = f(\boldsymbol{x})\int W d\boldsymbol{x}' + \nabla f(\boldsymbol{x})\cdot\int(\boldsymbol{x}'-\boldsymbol{x})W d\boldsymbol{x}' + O(h^2) \tag{5.205}$$

For zeroth and first order consistency, the kernel must satisfy the moment conditions

$$\int W(\boldsymbol{r}, h) d\boldsymbol{r} = 1 \tag{5.206}$$

$$\int \boldsymbol{r} W(\boldsymbol{r}, h) d\boldsymbol{r} = 0 \tag{5.207}$$

In theory, these conditions are satisfied by properly constructed kernels. However, in practice, the integrals are replaced by particle summations,

$$\sum_j \frac{m_j}{\rho_j} W_{ij} \neq 1 \tag{5.208}$$

$$\sum_j \frac{m_j}{\rho_j}(\boldsymbol{x}_j - \boldsymbol{x}_i) W_{ij} \neq 0 \tag{5.209}$$

This loss of consistency becomes particularly pronounced near solid boundaries or in regions where the particle distribution is irregular or distorted. Under such conditions, kernel support is incomplete and the discrete particle summations no longer satisfy the required moment conditions. This phenomenon, commonly referred to as particle inconsistency, represents one of the primary sources of numerical error in SPH. It manifests through inaccurate interpolation, spurious pressure forces, and unphysical particle motion, ultimately degrading the overall quality and reliability of the numerical solution. These deficiencies are especially critical when evaluating spatial derivatives, which play a central role in the governing equations.

The accuracy of gradient operators is therefore of fundamental importance in fluid dynamics applications, as pressure gradients directly determine particle acceleration and control the evolution of the flow field. In SPH, pressure gradients are approximated using kernel-based discrete operators, and any inconsistency or irregularity in the particle distribution can be strongly amplified in the momentum equation. Although the standard SPH pressure gradient formulation is designed to preserve conservation properties, it does not guarantee the exact reproduction of constant or linear pressure fields in the presence of particle disorder. Consequently, careful treatment of gradient accuracy is essential for achieving stable, physically meaningful, and reliable SPH simulations. The standard SPH pressure gradient formulation,

$$\nabla P_i = \rho_i \sum_{j=1}^{n} m_j \left(\frac{P_i}{\rho_i^2} + \frac{P_j}{\rho_j^2}\right)\nabla W_{ij} \tag{5.210}$$

is symmetric and conserves momentum, but it does not guarantee exact reproduction of constant or linear pressure fields when particles are irregularly distributed.

To analyse this issue, consider a constant pressure field ($P_i = P_j = P_0$). Substituting into the SPH gradient expression yields

$$\nabla P_i = P_0 \sum_{j=1}^{n} m_j \left(\frac{1}{\rho_i^2} + \frac{1}{\rho_j^2} \right) \nabla W_{ij} \tag{5.211}$$

which does not vanish exactly unless the particle distribution is perfectly uniform and kernel support is complete. As a result, spurious pressure forces may arise, particularly near free surfaces and solid boundaries.

Several correction techniques have been proposed to address this issue. One approach introduces a kernel gradient correction matrix $\boldsymbol{L}_i$,

$$\nabla f_i = \sum_j \frac{m_j}{\rho_j} (f_j - f_i) \boldsymbol{L}_i \nabla W_{ij} \tag{5.212}$$

where $\boldsymbol{L}_i$ is computed such that first order consistency is restored locally. While effective, this approach increases computational cost and algorithmic complexity.

Boundary truncation is an inherent challenge in SPH. Near solid walls or free surfaces, particles lack neighbours on one side, leading to incomplete kernel support. This violates the kernel moment conditions and degrades approximation accuracy.

The density summation near a boundary illustrates this issue clearly,

$$\rho_i = \sum_{j=1}^{n} m_j W_{ij} < \rho_0 \text{near boundaries} \tag{5.213}$$

This artificial density deficit can lead to incorrect pressure evaluation when an equation of state is used,

$$P_i = c^2 (\rho_i - \rho_0) \tag{5.214}$$

resulting in nonphysical pressure gradients that push particles away from boundaries.

Boundary correction techniques, such as ghost particles, renormalised kernels, or density correction factors, are therefore essential to maintain accuracy. For example a simple renormalisation introduces a correction factor α_i,

$$\alpha_i = \sum_{j=1}^{n} \frac{m_j}{\rho_j} W_{ij} \tag{5.215}$$

ensuring that the corrected kernel satisfies

$$\sum_j \frac{m_j}{\rho_j} \frac{W_{ij}}{\alpha_i} = 1 \tag{5.216}$$

5.4.19 Convergence behaviour of smoothed particle hydrodynamics

In classical mesh-based numerical methods, convergence is typically assessed by systematically refining the computational mesh and examining the corresponding reduction in numerical error with respect to an analytical or reference solution. In

SPH, however, convergence behaviour is inherently more subtle, as the method does not rely on a fixed mesh. Instead, convergence depends simultaneously on two closely coupled discretisation parameters: the particle spacing Δp and smoothing length h.

For SPH to be formally consistent and convergent, the following limiting conditions must be satisfied:

$$\begin{cases} h \to 0 \\ \Delta p \to 0 \\ \frac{\Delta p}{h} \to 0 \end{cases} \tag{5.217}$$

In practice, simulations often maintain a fixed ratio $h = o\Delta p$, where o is a characteristic constant. Under this condition, SPH typically exhibits first order convergence in space for many practical problems, although higher order convergence may be achieved with corrected formulations and highly regular particle distributions.

Higher-order convergence can, in principle, be achieved through the use of kernel correction techniques, moving least-squares formulations, or highly regular particle distributions. Nevertheless, in realistic simulations, formal convergence rates are often obscured by secondary error sources. Particle disorder, kernel truncation near boundaries, and stabilisation mechanisms such as artificial viscosity or numerical diffusion frequently dominate the overall error budget. These effects can mask the underlying convergence behaviour, making straightforward refinement studies difficult to interpret. Consequently, convergence analysis in SPH requires careful design, including controlled particle initialisation, consistent boundary treatment, and systematic sensitivity studies. Without such precautions, apparent convergence or lack thereof may reflect numerical artefacts rather than the intrinsic accuracy of the method itself.

5.4.20 Computational cost and practical limitations

The computational cost of SPH scales with the number of particles N and the average number of neighbours N_n,

$$Cost \approx O(NN_n) \tag{5.218}$$

In three dimensions, maintaining sufficient neighbour counts for accuracy typically requires large N_n, increasing computational expense. Although neighbour searching algorithms reduce complexity, large-scale SPH simulations remain computationally demanding compared with mesh-based methods for problems involving simple geometries and small deformations.

Time step restrictions further limit efficiency. The explicit time integration commonly used in SPH must satisfy stringent stability conditions,

$$\Delta t \leq \min\left(C_{CFL}\frac{h}{c + |\boldsymbol{u}|}, C_v\frac{h^2}{v}\right) \tag{5.219}$$

which can result in very small time steps, particularly for incompressible or low Mach number flows.

Given these limitations, SPH results must be interpreted with care. Numerical artefacts such as pressure noise, particle clustering, or excessive dissipation may arise if stabilisation parameters are poorly chosen or if resolution is insufficient.

Validation against analytical solutions, benchmark problems, and experimental data is therefore essential. Typical validation cases include hydrostatic pressure tests, dam break problems, and oscillating free surface flows, where physical behaviour is well understood.

5.4.21 Extensions, applications, and comparative assessment of smoothed particle hydrodynamics

The original formulation of SPH was developed primarily to model inviscid, compressible flows, with early applications focused on astrophysical phenomena characterised by large density variations and highly dynamic motion. Over time, however, the flexibility of the particle-based framework has enabled SPH to be progressively extended to a broad range of physical processes relevant to engineering and continuum mechanics. These extensions have been achieved by enriching the governing equations, introducing additional interaction models, and adapting numerical strategies to address specific physical constraints. As a result, SPH has evolved into a versatile methodology capable of handling fluids, solids, and their interactions under complex conditions.

Weakly compressible and incompressible SPH is one of the most widely used extensions of SPH for engineering applications. The development of formulations suitable for incompressible or nearly incompressible flows. In weakly compressible SPH, incompressibility is approximated by permitting small density variations that are controlled through an artificial equation of state. Pressure is related to density according to,

$$P_i = c^2(\rho_i - \rho_0) \tag{5.220}$$

where c is an artificial speed of sound chosen such that density variations remain below a few per cent,

$$\frac{|\rho_i - \rho_0|}{\rho_0} \ll 1 \tag{5.221}$$

This approach avoids solving a pressure Poisson equation and preserves algorithmic simplicity, but it imposes a severe time step restriction through the CFL condition,

$$\Delta t \le C_{CFL} \frac{h}{c + |\boldsymbol{u}|} \tag{5.222}$$

To overcome this limitation, incompressible SPH (ISPH) formulations were developed. In ISPH, density is kept constant, and pressure is obtained by solving a

Poisson equation derived from the incompressibility constraint,

$$\nabla \cdot \boldsymbol{u}^{n+1} = 0 \tag{5.223}$$

Using a fractional step method, an intermediate velocity $\boldsymbol{u}^*$ is computed without pressure,

$$\boldsymbol{u}^* = \boldsymbol{u}^n + \Delta t \boldsymbol{a}^{np} \tag{5.224}$$

where $\boldsymbol{a}^{np}$ includes nonpressure forces. The pressure is then obtained from

$$\nabla \cdot \left(\frac{1}{\rho} \nabla p^{n+1} \right) = \frac{1}{\Delta t} \nabla \cdot \boldsymbol{u}^* \tag{5.225}$$

The corrected velocity is

$$\boldsymbol{u}^{n+1} = \boldsymbol{u}^* - \frac{\Delta t}{\rho} \nabla p^{n+1} \tag{5.226}$$

Although incompressible SPH formulations are generally more computationally demanding due to the need to solve a pressure Poisson equation at each time step, they permit significantly larger time steps and provide smoother, more accurate pressure fields than weakly compressible approaches. These advantages make ISPH particularly attractive for low Mach number flows, where density variations are physically negligible and pressure accuracy plays a dominant role in the flow dynamics.

5.4.22 Multiphase and multicomponent smoothed particle hydrodynamics

One of the most powerful and widely exploited extensions of SPH is its application to multiphase and multicomponent flows. In such problems, multiple fluids or materials coexist within the same computational domain, often exhibiting sharply contrasting physical properties, such as density, viscosity, compressibility, and equation of state. In SPH, this complexity is handled naturally by assigning each particle a phase or material identifier, allowing the governing equations to be solved with phase-dependent material properties. As a result, particles belonging to different phases may obey distinct constitutive relations, enabling the accurate representation of interfaces, interfacial forces, and complex interactions between fluids without the need for explicit interface tracking or reconstruction.

$$\begin{cases} \rho_i = \rho^{(k)} \\ \nu_i = \nu^{(k)} \\ c_i = c^{(k)} \end{cases} \tag{5.227}$$

where the superscript (k) denotes the phase. Interfacial forces are introduced through surface tension models, commonly using colour functions or phase field approaches.

For a two-phase system, the colour function C is defined as:

$$C_i = \begin{cases} 1, \textit{fluid A} \\ 0, \textit{fluid B} \end{cases} \tag{5.228}$$

The interface normal and curvature are obtained from

$$\begin{cases} \boldsymbol{n}_i = \nabla C_i \\ \kappa_i = -\nabla \cdot \left(\frac{\boldsymbol{n}_i}{|\boldsymbol{n}_i|} \right) \end{cases} \tag{5.229}$$

and the surface tension force is

$$f_{s,i} = \sigma \kappa_i \boldsymbol{n}_i \tag{5.230}$$

SPH naturally accommodates large interface deformations, breakup, and coalescence without requiring explicit interface reconstruction or tracking algorithms. This intrinsic capability arises from its fully Lagrangian, particle-based description, in which interfaces are simply represented by the spatial distribution of particles belonging to different phases. As a result, complex topological changes are captured automatically as the particles evolve in time. These characteristics have made SPH particularly attractive for a wide range of applications involving highly deformable free surfaces and multiphase interactions, including sloshing dynamics, wave breaking, emulsions, foams, and droplet formation and fragmentation.

Beyond fluid flows, SPH has also been successfully extended to solid mechanics and fluid–structure interaction problems. In solid mechanics applications, particles represent material points of a deformable body, and their motion is governed by the divergence of the stress tensor rather than pressure gradients, enabling a unified particle-based framework for modelling both fluids and solids and their mutual interaction.

The momentum equation for a deformable solid is written as

$$\frac{D\boldsymbol{u}_i}{Dt} = \sum_{j=1}^{n} m_j \left(\frac{\boldsymbol{\sigma}_i}{\rho_i^2} + \frac{\boldsymbol{\sigma}_j}{\rho_j^2} \right) \cdot \nabla W_{ij} + \boldsymbol{b} \tag{5.231}$$

where $\boldsymbol{\sigma}$ is the Cauchy stress tensor and $\boldsymbol{b}$ represents body forces.

The stress tensor is obtained from constitutive relations. For a linear elastic material,

$$\boldsymbol{\sigma} = \lambda' (\nabla \cdot \boldsymbol{d}) \boldsymbol{I} + 2\mu' \boldsymbol{\varepsilon} \tag{5.232}$$

$$\boldsymbol{\varepsilon} = \frac{1}{2} (\nabla \boldsymbol{d} + (\nabla \boldsymbol{d})^T) \tag{5.233}$$

where $\boldsymbol{d}$ is displacement and λ', μ' are Lamé constants (Ali & Shimoda, 2022).

Fluid–structure interaction is handled by coupling fluid and solid particles through interaction forces at the interface. Momentum exchange is enforced naturally through pairwise forces, ensuring conservation,

$$f_{fs} = -f_{sf} \tag{5.234}$$

This unified particle framework avoids mesh distortion and remeshing, making SPH particularly effective for problems involving large structural deformation, impact, and fracture.

5.5 Concluding remarks

This chapter has established the principal numerical frameworks that underpin contemporary simulations of free surface and sloshing phenomena, with detailed attention given to the FEM, the FVM, and SPH. Although these methodologies arise from distinct mathematical and computational philosophies, they are united by a common objective: the faithful numerical representation of the governing conservation laws of fluid motion. Across all three approaches, discretisation, approximation, and stabilisation emerge as the central mechanisms through which continuous physics is translated into a computable form.

A key message of this chapter is that numerical accuracy and robustness are not inherent attributes of a given method in isolation. Rather, they arise from the combined effect of spatial and temporal discretisation, stabilisation strategies, variable coupling, and resolution control. Even formulations that are theoretically conservative and consistent may exhibit nonphysical oscillations, excessive numerical diffusion, or poor convergence when applied without sufficient regard to these interacting factors. This is particularly evident in sloshing problems, where strong advection, moving free surfaces, and transient nonlinear dynamics place stringent demands on numerical schemes.

While the formulations presented here provide a rigorous theoretical foundation, their effectiveness in practice depends critically on how the computational domain is discretised. Mesh quality, element shape, grid smoothness, and the distribution of resolution play a decisive role in controlling truncation error, numerical stability, and convergence behaviour. In convection-dominated regimes and flows involving large free-surface deformation, inadequate meshing can undermine even the most carefully designed numerical algorithms, leading to misleading predictions and poor physical fidelity.

For this reason, the development of reliable CFD simulations cannot be separated from sound meshing practice. The next chapter therefore moves beyond numerical formulation to focus on the practices commonly adopted in CFD mesh generation. Emphasis will be placed on mesh topology selection, quality metrics, boundary layer resolution, local refinement strategies, and typical pitfalls encountered in industrial and research applications. By doing so, the discussion aims to bridge the gap between numerical theory and dependable computational practice, enabling the methods introduced in this chapter to be applied with confidence in real engineering problems.

References

Adeeb, E., Haider, B. A., & Sohn, C. H. (2018). Influence of rounded corners on flow interference between two tandem cylinders using FVM and IB-LBM. *International Journal of Numerical Methods for Heat and Fluid Flow, 28*(7), 1648–1663. https://doi.org/10.1108/HFF-08-2017-0319.

Ali, M. A., & Shimoda, M. (2022). Toward multiphysics multiscale concurrent topology optimization for lightweight structures with high heat conductivity and high stiffness using MATLAB. *Structural and Multidisciplinary Optimization, 65*, 207. https://doi.org/10.1007/s00158-022-03291-0.

Aly, A. M., Huang, C., & Alotaibi, M. (2025). Study on heat and mass transfer in a porous cavity based on artificial intelligence δ-SPH model. *International Communications in Heat and Mass Transfer, 164*, 108907. https://doi.org/10.1016/j.icheatmasstransfer.2025.108907.

Amrani, I., Cheriet, A., & Feliachi, M. (2018). Arbitrary movement simulation using 3D NC-FVM for electromagnetic problems. *COMPEL - The international journal for computation and mathematics in electrical and electronic engineering, 37*(4), 1306–1314. https://doi.org/10.1108/compel-08-2017-0352.

Antuono, M., Marrone, S., & Colagrossi, A. (2025). Coalescing and break-up of viscous drops with surface tension through the smoothed particle hydrodynamics. *Computer Methods in Applied Mechanics and Engineering, 442*, 118014. https://doi.org/10.1016/j.cma.2025.118014.

Asis, M. A., Mutsuda, H., Nakashima, T., Baso, S., & Ardianti, A. (2025). Numerical study on the reduction effect of sloshing phenomenon and energy harvesting using a flexible piezoelectric nanogenerator. *Results in Engineering, 25*, 104543. https://doi.org/10.1016/j.rineng.2025.104543.

Bogaerds, A. C. B., Hulsen, M. A., Peters, G. W. M., & Baaijens, F. P. T. (2003). Time dependent finite element analysis of the linear stability of viscoelastic flows with interfaces. *Journal of Non-Newtonian Fluid Mechanics, 116*(1), 33–54. https://doi.org/10.1016/S0377-0257(03)00099-5.

Boregowda, P., & Liu, G.-R. (2025). A boundary integral based particle initialization algorithm for smooth particle hydrodynamics. *Computer Physics Communications, 310*, 109531. https://doi.org/10.1016/j.cpc.2025.109531.

Cai, J., Wang, J., Huang, Y., Li, Q., & Fan, Y. (2025). Analysis of interaction between aircraft tire and slush-contaminated pavement. *Cold Regions Science and Technology, 240*, 104607. https://doi.org/10.1016/j.coldregions.2025.104607.

Chakraborty, S., & Shaw, A. (2013). A pseudo-spring based fracture model for SPH simulation of impact dynamics. *International Journal of Impact Engineering, 58*, 84–95. https://doi.org/10.1016/j.ijimpeng.2013.03.006.

Chen, J. Y., Zhao, J., Feng, R. F., Ni, R. C., & Peng, C. (2025). Coupled FDM-SPH modeling of CFRP-reinforced concrete damage under combined blast and fragment impact. *Computers and Structures, 319*. https://doi.org/10.1016/j.compstruc.2025.107980.

Chen, Y., & Zhang, Z. (2026). Error estimates of augmented compact FVM for nonlinear singular parabolic problems with degenerate interface. *Journal of Computational and Applied Mathematics, 473*, 116899. https://doi.org/10.1016/j.cam.2025.116899.

Chen, Y.-kun, Meringolo, D. D., Liu, Y., & Liang, J.-ming (2025). Energy balance during Bragg wave resonance by submerged porous breakwaters through a mixture theory-based δ-LES-SPH model. *Coastal Engineering, 196*, 104652. https://doi.org/10.1016/j.coastaleng.2024.104652.

Cheng, Z. D., He, Y. L., Cui, F. Q., Xu, R. J., & Tao, Y. B. (2012). Numerical simulation of a parabolic trough solar collector with nonuniform solar flux conditions by coupling FVM and MCRT method. *Solar Energy, 86*(6), 1770–1784. https://doi.org/10.1016/j.solener.2012.02.039.

Chu, Y. M., Abu-Hamdeh, N. H., Ben-Beya, B., Hajizadeh, M. R., Li, Z., & Bach, Q. V. (2020). Nanoparticle enhanced PCM exergy loss and thermal behavior by means of FVM. *Journal of Molecular Liquids, 320*. https://doi.org/10.1016/j.molliq.2020.114457.

Chung, D. H., & Kwon, T. H. (2002). Numerical studies of fiber suspensions in an axisymmetric radial diverging flow: The effects of modeling and numerical assumptions. *Journal of Non-Newtonian Fluid Mechanics, 107*(1-3), 67–96. https://doi.org/10.1016/S0377-0257(02)00142-8.

Chung, S. W., Somekawa, H., Kinoshita, T., Kim, W. J., & Higashi, K. (2004). The non-uniform behavior during ECAE process by 3-D FVM simulation. *Scripta Materialia, 50*(7), 1079–1083. https://doi.org/10.1016/j.scriptamat.2003.11.062.

Connelly, R. K., & Kokini, J. L. (2004). The effect of shear thinning and differential viscoelasticity on mixing in a model 2D mixer as determined using FEM with particle tracking. *Journal of Non-Newtonian Fluid Mechanics, 123*(1), 1–17. https://doi.org/10.1016/j.jnnfm.2004.03.006.

Del Coz Díaz, J. J., García Nieto, P. J., Castro-Fresno, D., & Menéndez Rodríguez, P. (2011). Steady state numerical simulation of the particle collection efficiency of a new urban sustainable gravity settler using design of experiments by FVM. *Applied Mathematics and Computation, 217*(21), 8166–8178. https://doi.org/10.1016/j.amc.2011.03.044.

Deng, L., Liu, Y., Wang, W., Ge, W., & Li, J. (2013). A two-fluid smoothed particle hydrodynamics (TF-SPH) method for gas-solid fluidization. *Chemical Engineering Science, 99*, 89–101. https://doi.org/10.1016/j.ces.2013.05.047.

Du, J., Jiang, Y., Yu, Y., Liu, Z., Shen, Y., Ma, H., & Zhao, Y. (2025). Numerical investigation of solid sphere entry into inelastic non-Newtonian fluids by smoothed particle hydrodynamics. *Ocean Engineering, 340*, 122395. https://doi.org/10.1016/j.oceaneng.2025.122395.

Erbaş, M., Atcı, M., Karaca, M., & Bıyıkoğlu, A. (2025). Accurate computation of multiphase flows with high-density and pressure ratios using smoothed particle hydrodynamics. *Computers & Fluids, 301*, 106805. https://doi.org/10.1016/j.compfluid.2025.106805.

Fan, W., Liu, J., & Xiao, D. (2025). An adaptive fifth-degree kernel for suppressing stress instability in SPH for compressible flows. *Computers & Fluids, 302*, 106844. https://doi.org/10.1016/j.compfluid.2025.106844.

Fan, Y., Zhang, A., Peng, M., Yan, R., Liu, Y., Cui, Y., Zhu, G., Guo, J., Dai, Y., Chen, L., & Zou, Y. (2025). Development and verification of a two-step core neutronics simulator and its preliminary application in the PC-MGCR design. *Nuclear Engineering and Design, 443*, 114307. https://doi.org/10.1016/j.nucengdes.2025.114307.

Fang, X. L., Wang, P. P., Meng, Z. F., Ming, F. R., & Zhang, A. M. (2025). A low-dissipation and accurate Riemann SPH for multiphase flows with large density ratios and strong shocks. *Journal of Computational Physics, 539*. https://doi.org/10.1016/j.jcp.2025.114210.

Fang, H., & Yin, Z.-Y. (2025). Lagrangian hybrid element particle method (LHEPM) for incompressible fluid dynamics. *Journal of Computational Physics, 541*, 114281. https://doi.org/10.1016/j.jcp.2025.114281.

Feng, D., Gan, L., Xiong, M., Li, W., & Huang, Y. (2025). Uncertainty analysis of 3D post-failure behavior in landslide and reinforced slope based on the SPH method and the random field theory. *Engineering Geology, 350*, 108017. https://doi.org/10.1016/j.enggeo.2025.108017.

Feng, H., Yin, Z.-Y., & Qiu, Y. (2025). Two-phase two-point MPM modeling of submarine granular flows considering solid-to-fluid phase transition over frictional plane. *Ocean Engineering, 341*, 122798. https://doi.org/10.1016/j.oceaneng.2025.122798.

Feng, J., Xiao, Y., Imin, R., & Rozi, A. (2025). Kernel derivative free-based physics-informed neural network for steady incompressible flows. *Engineering Analysis with Boundary Elements, 181*, 106523. https://doi.org/10.1016/j.enganabound.2025.106523.

Fraga Filho, C. A. D., Ferreira, R. M. L., Aleixo, R., Canelas, R., Amaral, S., & Viseu, T. (2025). SPH modelling of mass detachment processes in overtopped dams. The limits of Froude similarity. *Advances in Water Resources, 202*. https://doi.org/10.1016/j.advwatres.2025.104987.

Fuentes, B., Cruchaga, M., & Núñez, J. (2025). Experimental validation of a numerical model for the analysis of erosion depth in coarse sand induced by a dam break of water. *International Journal of Multiphase Flow, 185*, 105118. https://doi.org/10.1016/j.ijmultiphaseflow.2024.105118.

García Nieto, P. J., del Coz Díaz, J. J., Castro-Fresno, D., & Ballester Muñoz, F. (2010). Numerical simulation of the performance of a snow fence with airfoil snow plates by FVM. *Journal of Computational and Applied Mathematics, 234*(4), 1200–1210. https://doi.org/10.1016/j.cam.2009.07.048.

García-Jurado, D., Mainé, J. M., Batista, M., Shaw, L., Hausotte, T., & Marcos, M. (2013). Metrological evaluation of secondary adhesion wear effects in the dry turning of UNS-A92024-T3 alloy through focus-variation microscopy (FVM). *Procedia Engineering, 63*, 804–811. https://doi.org/10.1016/j.proeng.2013.08.251.

García-Jurado, D., Vazquez-Martinez, J. M., Gámez, A. J., Batista, M., Puerta, F. J., & Marcos, M. (2015). FVM based study of the influence of secondary adhesion tool wear on surface roughness of dry turned Al-Cu aerospace alloy. *Procedia Engineering, 132*, 600–607. https://doi.org/10.1016/j.proeng.2015.12.537.

Ge, J., Zhang, D., Tian, W., Wang, K., Qiu, S., & Su, G. H. (2015). Steady and transient solutions of neutronics problems based on finite volume method (FVM) with a CFD code. *Progress in Nuclear Energy, 85*, 366–374. https://doi.org/10.1016/j.pnucene.2015.07.012.

Ghannam, A., Chehade, A., Generous, M. M., Alazzam, A., Kleinstreuer, C., Ahmadi, G., & Abu-Nada, E. (2025). A comprehensive review of particle-laden flows modeling: Single/multiphase modeling approaches, benchmarks, heat transfer, intermolecular interactions, recent advances and future directions. *Physics Reports, 1118-1120*, 1–96. https://doi.org/10.1016/j.physrep.2025.03.001.

Ghasemi V, A., Firoozabadi, B., & Mahdinia, M. (2013). 2D numerical simulation of density currents using the SPH projection method. *European Journal of Mechanics - B/Fluids, 38*, 38–46. https://doi.org/10.1016/j.euromechflu.2012.10.004.

Ghoneim, A. Y. (2025). Meshfree phase-field modeling of three-phase flow using smoothed particle hydrodynamics with differential reproducing kernels and artificial compressibility. *Journal of Computational and Applied Mathematics, 469*, 116654. https://doi.org/10.1016/j.cam.2025.116654.

Guang, H., Li, B., Chai, Z., Liu, C., Liu, Q., Lu, Q., Wei, Z., Deng, J., & Hong, J. (2025). Shape optimization of fuel rods in nuclear reactor pressure vessel via smoothed particle hydrodynamics (SPH) optimization framework. *Applied Thermal Engineering, 280*, 128250. https://doi.org/10.1016/j.applthermaleng.2025.128250.

He, F., Chen, Y., Zhang, Y., Jiang, H., & Huang, C. (2025). A multi-physics SPH method for simulating heat transfer, mass transfer and phase transition of hydrate. *International Communications in Heat and Mass Transfer, 165*, 109012. https://doi.org/10.1016/j.icheatmasstransfer.2025.109012.

Huang, C., Shen, L., Yu, W., Alkayem, N. F., Han, Y., Tian, Z., Yin, H., & Cusatis, G. (2025). High-fidelity SPH-DEM framework for mesoscopic rheological behavior of fresh fiber-reinforced concrete. *International Journal of Mechanical Sciences, 289*, 110061. https://doi.org/10.1016/j.ijmecsci.2025.110061.

Huang, H., Jia, Y., & Naceur, H. (2025). Elastoplastic bending analysis of functionally graded materials structure under variable loads by a meshless method. *Composite Structures, 372*, 119544. https://doi.org/10.1016/j.compstruct.2025.119544.

Huang, X., Chen, X., Shuai, Y., Yuan, Y., Zhang, T., Li, B., & Tan, H. (2015). Heat transfer analysis of solar-thermal dissociation of NiFe2O4 by coupling MCRTM and FVM method. *Energy Conversion and Management, 106*, 676–686. https://doi.org/10.1016/j.enconman.2015.10.013.

Huang, X., Yue, Y., Zhu, B., Chen, Y., Lu, Q., Zhang, D., Kong, X., & Fang, Q. (2025). Effects of underground explosions on soil and structures based on PD–SPH modeling. *Engineering Geology, 356*, 108296. https://doi.org/10.1016/j.enggeo.2025.108296.

Huang, Y., & Xie, B. (2025). A new velocity reconstruction algorithm for multiphase flows with moving bodies. *Journal of Computational Physics, 535*, 114048. https://doi.org/10.1016/j.jcp.2025.114048.

Jasak, H., & Gosman, A. D. (2003). Element residual error estimate for the finite volume method. *Computers & Fluids, 32*(2), 223–248. https://doi.org/10.1016/s0045-7930(02)00004-x.

Javanmard, E., Yari, E., Mehr, J. A., & Mansoorzadeh, S. (2019). Hydrodynamic characteristic curves and behavior of flow around a surface-piercing propeller using computational fluid dynamics based on FVM. *Ocean Engineering, 192*, 106445. https://doi.org/10.1016/j.oceaneng.2019.106445.

Jiangfeng, S., Zhiyong, H., Lianxia, L., Chang'an, C., & Deli, L. (2011). One-dimensional simulation of hydrogen isotopes diffusion in composite materials by FVM. *International Journal of Hydrogen Energy, 36*(9), 5702–5706. https://doi.org/10.1016/j.ijhydene.2011.02.040.

Jiao, J., Jiang, M., Xie, H., & Chen, Y. (2025). Coupled effects of incident waves forcing and internal tank sloshing on the dynamics of twin floating bodies. *Fluid Dynamics & Materials Processing, 21*(9), 2075–2100. https://doi.org/10.32604/fdmp.2025.069064.

Jin, Y., Wang, H., Liu, Y. P., Yu, Z., Guo, L., Liao, L., Xu, H., Chi, M., & Chan, S. L. (2025). Numerical study on high-speed train derailment caused by debris flow deposition: A case study in China. *Engineering Failure Analysis, 181*. https://doi.org/10.1016/j.engfailanal.2025.109960.

Kallivokas, L. F., & Bielak, J. (1993). An element for the analysis of transient exterior fluid-structure interaction problems using the FEM. *Finite Elements in Analysis and Design, 15*(1), 69–81. https://doi.org/10.1016/0168-874x(93)90071-w.

Kalumuck, K. M., Duraiswami, R., & Chahine, G. L. (1995). Bubble Dynamics Fluid-Structure Interaction Simulation by Coupling Fluid Bem and Structural Fem Codes. *Journal of Fluids and Structures, 9*(8), 861–883. https://doi.org/10.1006/jfls.1995.1049.

Kiara, A., Hendrickson, K., & Yue, D. K. P. (2013b). SPH for incompressible free-surface flows. Part II: Performance of a modified SPH method. *Computers and Fluids, 86*, 510–536. https://doi.org/10.1016/j.compfluid.2013.07.016.

Kiara, A., Hendrickson, K., & Yue, D. K. P. (2013a). SPH for incompressible free-surface flows. Part I: Error analysis of the basic assumptions. *Computers and Fluids, 86*, 611–624. https://doi.org/10.1016/j.compfluid.2013.05.023.

Kim, J. W., & Kim, E. S. (2025). Total Lagrangian smoothed particle hydrodynamics for large-strain elastoplasticity with particle resolution refinement using an anisotropic Lagrangian kernel. *Engineering Analysis with Boundary Elements, 173*. https://doi.org/10.1016/j.enganabound.2025.106173.

Kincl, O., Peshkov, I., & Boscheri, W. (2025). Semi-implicit quasi-Lagrangian Voronoi approximation for compressible viscous fluid flows. *Computers & Fluids, 289*, 106530. https://doi.org/10.1016/j.compfluid.2024.106530.

Kokkinos, F. T., & Reddy, J. N. (1995). BEM and penalty FEM models for viscous incompressible fluids. *Computers & Structures, 56*(5), 849–859. https://doi.org/10.1016/0045-7949(95)00013-7.

Lanzafame, G. (2013). Implicit integrations for SPH in semi-Lagrangian approach: Application to the accretion disc modeling in a microquasar. *Computer Physics Communications, 184*(3), 671–688. https://doi.org/10.1016/j.cpc.2012.11.001.

Lavooij, C. S. W., & Tusseling, A. S. (1991). Fluid-structure interaction in liquid-filled piping systems. *Journal of Fluids and Structures, 5*(5), 573–595. https://doi.org/10.1016/s0889-9746(05)80006-4.

Lee, B. K., Kwon, H. H., & Cho, H. Y. (2002). A study on the automated process planning system for cold forging of non-axisymmetric parts using FVM simulation. *Journal of Materials Processing Technology, 130-131*, 524–531. https://doi.org/10.1016/s0924-0136(02)00799-9.

Li, C., Wang, G., Chen, G., Gao, J., Yu, P., & Peng, X. (2025). Dynamic analysis of soil-rock mixtures based on improved 3-D DDA-SPH method. *International Journal of Rock Mechanics and Mining Sciences, 195*, 106286. https://doi.org/10.1016/j.ijrmms.2025.106286.

Li, H., Zhang, A., Zhao, L., Mei, Y., Sun, Y., & Liu, S. (2025). Study on the damage effect of hypervelocity bullets on sandstone and concrete at different inclination angles. *International Journal of Impact Engineering, 199*, 105233. https://doi.org/10.1016/j.ijimpeng.2025.105233.

Li, X., Li, B., Qie, L., Li, Y., & Wang, H. (2025). Dynamic response of a frame structure impacted by debris flow containing large boulders. *Structures, 74*, 108543. https://doi.org/10.1016/j.istruc.2025.108543.

Lian, Y., Bui, H. H., & Nguyen, G. D. (2025). A coupled THM-SPH computational framework for frozen soils and its applications for thawing-induced slope failure. *Computer Methods in Applied Mechanics and Engineering, 446*. https://doi.org/10.1016/j.cma.2025.118252.

Liang, G. Q., Sun, P. N., & Zhang, A. M. (2025). Development of a multi-resolution SPH-PD model for simulating ice sheet fragmentation under underwater explosion loads. *CMES - Computer Modeling in Engineering and Sciences, 145*(3), 3405–3431. https://doi.org/10.32604/cmes.2025.072496.

Liang, G. Q., Sun, P. N., Lyu, H. G., & Zhang, G. Y. (2025). A coupled δ+-SPH-NOSB-PD method: Towards fluid-structure interaction problems involving violent free-surface flows and isotropic structural linear elastic deformations and failures. *Applied Mathematical Modelling, 145*. https://doi.org/10.1016/j.apm.2025.116104.

Liu, S., Yang, G., Xu, Y., & Hu, D. (2024). A SPH-FVM coupling method based on triangular mesh for the simulation of two-phase flows. *Engineering Analysis with Boundary Elements, 159*, 237–258. https://doi.org/10.1016/j.enganabound.2023.12.009.

Liu, X., Harrison, S. M., Fletcher, D. F., & Cleary, P. W. (2023). Numerical simulation of buoyancy–driven flow in a human stomach geometry: Comparison of SPH and FVM models. *Applied Mathematical Modelling, 124*, 367–392. https://doi.org/10.1016/j.apm.2023.08.004.

Liu, X., Yang, Y., Yang, Q., & Xu, F. (2025). A robust low-dissipation Riemann-SPH solver with a novel hybrid boundary treatment method for FSI problems. *Journal of Fluids and Structures, 135*, 104316. https://doi.org/10.1016/j.jfluidstructs.2025.104316.

Luan, H. B., Xu, H., Chen, L., Sun, D. L., He, Y. L., & Tao, W. Q. (2011). Evaluation of the coupling scheme of FVM and LBM for fluid flows around complex geometries. *International Journal of Heat and Mass Transfer, 54*(9-10), 1975–1985. https://doi.org/10.1016/j.ijheatmasstransfer.2011.01.004.

Luo, X. L. (1998). An incremental difference formulation for viscoelastic flows and high resolution FEM solutions at high Weissenberg numbers. *Journal of Non-Newtonian Fluid Mechanics, 79*(1), 57–75. https://doi.org/10.1016/s0377-0257(98)00063-9.

Ma, Y., Han, X., Chen, Z., & Khosa, A. A. (2024). Performance analysis of a nanofluid spectral splitting concentrating PV/T system with triangular receiver based on MCRT-FVM coupled method. *Applied Thermal Engineering, 239*, 122096. https://doi.org/10.1016/j.applthermaleng.2023.122096.

Ma, Y., Xiao, X., Li, W., Desbrun, M., & Liu, X. (2024). Hybrid LBM-FVM solver for two-phase flow simulation. *Journal of Computational Physics, 506*, 112920. https://doi.org/10.1016/j.jcp.2024.112920.

Ma, Y., Asai, M., Su, B., Han, Z., Li, C., & Chen, G. (2026). A two-point three-phase resolved coupling framework for modeling boulder-laden debris flows. *Journal of Rock Mechanics and Geotechnical Engineering, 18*(2), 932–953. https://doi.org/10.1016/j.jrmge.2025.06.021.

Mahdevari, S., & Bakhtiari Haftlang, P. (2025). Coupled SPH–FEM modeling of waterjet-assisted coal cutting: Numerical simulation and experimental validation. *Geomechanics for Energy and the Environment, 43*, 100732. https://doi.org/10.1016/j.gete.2025.100732.

Marrone, S., Antuono, M., Agresta, A., & Colagrossi, A. (2025). A study on the energy consistency in SPH surface tension modelling. *Computer Methods in Applied Mechanics and Engineering, 433*, 117473. https://doi.org/10.1016/j.cma.2024.117473.

Martínez-Estévez, I., Tagliafierro, B., Domínguez, J. M., Marzeddu, A., deCastro, M., & Crespo, A. J. C. (2025). Coupling between an SPH-based fluid solver and mooring dynamics for floating offshore devices: DualSPHysics and MoorDynPlus. *Journal of Ocean Engineering and Science*. https://doi.org/10.1016/j.joes.2025.12.001.

Maryška, J., Severýn, O., Tauchman, M., & Tondr, D. (2008). Modelling of processes in fractured rock using FEM/FVM on multidimensional domains. *Journal of Computational and Applied Mathematics, 215*(2), 495–502. https://doi.org/10.1016/j.cam.2006.04.074.

Merabtene, T., Garoosi, F., & Mahdi, T.-F. (2026). Comparative study of Eulerian FVM-VOF and Lagrangian MPS approaches for single- and multiphase flow simulations. *Ocean Engineering, 343*, 123378. https://doi.org/10.1016/j.oceaneng.2025.123378.

Mi, B., & Bai, X. (2022). A novel FEM-FVM coupled method of evaluating the explosion shock wave of a thermobaric bomb on the internal and external flow characteristics of aircraft. *Aerospace Science and Technology, 126*, 107558. https://doi.org/10.1016/j.ast.2022.107558.

Mishra, S. C., Chugh, P., Kumar, P., & Mitra, K. (2006). Development and comparison of the DTM, the DOM and the FVM formulations for the short-pulse laser transport through a participating medium. *International Journal of Heat and Mass Transfer, 49*(11-12), 1820–1832. https://doi.org/10.1016/j.ijheatmasstransfer.2005.10.043.

Mnasri, T. (2013). Green's functions theory combined with the FVM–BEM method for more accuracy in the storage energy domain. *International Communications in Heat and Mass Transfer, 44*, 100–107. https://doi.org/10.1016/j.icheatmasstransfer.2013.03.013.

Mnasri, T., Younès, R. B., Mazioud, A., & Durastanti, J. F. (2010). FVM-BEM method based on the Green's function theory for the heat transfer problem in buried co-axial exchanger. *Comptes Rendus - Mecanique, 338*(4), 220–229. https://doi.org/10.1016/j.crme.2010.04.004.

Moore, K., & Teixeira-Dias, F. (2025). Hypervelocity impact on ceramic composite shields: Modelling strategies and damage propagation. *Mechanics of Materials, 210*, 105454. https://doi.org/10.1016/j.mechmat.2025.105454.

Nasruddin, N. A., Islam, N., Vernuccio, S., & Oyekan, J. (2025). Hybridised mechanistic and machine learning digital twins for modelling and optimising chemical processes in flow: A comparative analysis of parallel and series-based hybridisation. *Chemical Engineering Journal Advances, 23*, 100775. https://doi.org/10.1016/j.ceja.2025.100775.

Nieuwboer, B. J., van Rhee, C., & Keetels, G. H. (2023). Towards simulating flow induced spillage in dredge cutter heads using DEM-FVM. *Ocean Engineering, 275*, 113922. https://doi.org/10.1016/j.oceaneng.2023.113922.

Niu, P., Zhang, H., Fu, H., Fan, Z., Yang, X., & Stelmakh, O. (2025). Study on splash oil supply and lubrication characteristics of conrod small end bearing with C-shaped oil collection structure. *Tribology International, 210*, 110768. https://doi.org/10.1016/j.triboint.2025.110768.

Pan, S., Ng, K. C., Shao, S., & Ho, J. H. (2025). Stress Intensity Factor and fatigue life prediction using the total Lagrangian smoothed particle hydrodynamics method. *Engineering Fracture Mechanics, 327*. https://doi.org/10.1016/j.engfracmech.2025.111388.

Pan, Y., Hong, J., Xue, D., Zhao, X., Luo, M., Liu, C., & Li, L. (2025). A SPH method to assess the dynamic behavior for multiconnected floating photovoltaic units consisting of non-uniform supporting frames in waves. *Ocean Engineering, 323*, 120622. https://doi.org/10.1016/j.oceaneng.2025.120622.

Paneer, M., Bašić, J., Sedlar, D., & Peng, C. (2025). Multiphase Lagrangian differencing dynamics method with sharp interfaces. *Engineering Analysis with Boundary Elements, 175*, 106197. https://doi.org/10.1016/j.enganabound.2025.106197.

Parszewski, Z. A., Nan, X., & Li, D. X. (1989). FEM For thermo-fluid bearing dynamic characteristics. *Computers and Structures, 31*(1), 93–102. https://doi.org/10.1016/0045-7949(89)90171-5.

Pramila, A. (1986). On the gyroscopic terms appearing when the vibration of fluid conveying pipe is analyzed using the FEM. *Journal of Sound and Vibration, 105*(3), 515–516. https://doi.org/10.1016/0022-460x(86)90178-1.

Qin, M., Yu, J., Wang, K., Huang, F., He, L., Gu, G., & Yang, L. (2025). Research on the protective performance and structural optimization based on TiB2-based composite ceramic multilayer target plates. *Structures, 81*, 110183. https://doi.org/10.1016/j.istruc.2025.110183.

Rajagopalan, D., Armstrong, R. C., & Brown, R. A. (1993). Comparison of computational efficiency of flow simulations with multimode constitutive equations: Integral and differential models. *Journal of Non-Newtonian Fluid Mechanics, 46*(2-3), 243–273. https://doi.org/10.1016/0377-0257(93)85049-G.

Ramezanzadeh, S., Ozbulut, M., & Yildiz, M. (2025). Comprehensive numerical study on the behavior of floating structures under challenging ocean conditions using WCSPH. *Journal of Fluids and Structures, 133*, 104273. https://doi.org/10.1016/j.jfluidstructs.2025.104273.

Ricci, F., Vacondio, R., Domínguez, J. M., & Tafuni, A. (2025). Three-dimensional variable resolution for multi-scale modeling in Smoothed Particle Hydrodynamics. *Computer Physics Communications, 313*, 109609. https://doi.org/10.1016/j.cpc.2025.109609.

Rublev, G. D., Parshikov, A. N., & Dyachkov, S. A. (2025). Improving approximation accuracy in Godunov-type smoothed particle hydrodynamics methods. *Applied Mathematics and Computation, 488*, 129128. https://doi.org/10.1016/j.amc.2024.129128.

Sandnes, T. D., Eke, V. R., Kegerreis, J. A., Massey, R. J., Ruiz-Bonilla, S., Schaller, M., & Teodoro, L. F. A. (2025). REMIX SPH – Improving mixing in smoothed particle hydrodynamics simulations using a generalised, material-independent approach. *Journal of Computational Physics, 532*, 113907. https://doi.org/10.1016/j.jcp.2025.113907.

Sayehvand, H.-ollah, & Bakhtiari Haftlang, P. (2025). A novel numerical investigation of particle waterjet parameters on the cutting characteristics of rock samples. *Journal of Manufacturing Processes, 135*, 161–178. https://doi.org/10.1016/j.jmapro.2025.01.027.

Seyyedi, S. M., Hashemi-Tilehnoee, M., Palomo del Barrio, E., Dogonchi, A. S., & Sharifpur, M. (2023). Analysis of magneto-natural-convection flow in a semi-annulus enclosure filled with a micropolar-nanofluid; A computational framework using CVFEM and FVM. *Journal of Magnetism and Magnetic Materials, 568*, 170407. https://doi.org/10.1016/j.jmmm.2023.170407.

Shao, S., & Lo, E. Y. M. (2003). Incompressible SPH method for simulating Newtonian and non-Newtonian flows with a free surface. *Advances in Water Resources, 26*(7), 787–800. https://doi.org/10.1016/S0309-1708(03)00030-7.

Shimizu, T., & Ohtani, M. (1996). Numerical simulation of heat transfer and fluid flow applied to exhaust manifold. *JSAE Review, 17*(1), 17–23. https://doi.org/10.1016/0389-4304(95)00049-6.

Shobuzako, K., Yoshida, S., Kawada, Y., Nakashima, R., Fujioka, S., & Asai, M. (2025). A generalized smoothed particle hydrodynamics method based on the moving least squares method and its discretization error estimation. *Results in Applied Mathematics, 26*, 100594. https://doi.org/10.1016/j.rinam.2025.100594.

Sigalotti, L. D. G., & Vargas, C. A. (2025). Smoothed Particle Hydrodynamics (SPH) Simulations of Drop Evaporation: A Comprehensive Overview of Methods and Applications. *CMES - Computer Modeling in Engineering and Sciences, 142*(3), 2281–2337. https://doi.org/10.32604/cmes.2025.060497.

Singh, V., Peng, C., & Islam, M. R. I. (2025). Three-dimensional SPH modeling of brittle fracture under hydrodynamic loading. *Computers and Fluids, 301*. https://doi.org/10.1016/j.compfluid.2025.106763.

Souto-Iglesias, A., MacIà, F., González, L. M., & Cercos-Pita, J. L. (2013). On the consistency of MPS. *Computer Physics Communications, 184*(3), 732–745. https://doi.org/10.1016/j.cpc.2012.11.009.

Stokes, S., & Bayandor, J. (2025). Validation of a coupled element-particle approach to debris analysis during hypervelocity impact. *International Journal of Impact Engineering, 203*, 105340. https://doi.org/10.1016/j.ijimpeng.2025.105340.

Sun, M. Z., Li, H. W., & Sun, J. N. (2026). Simulation and experimental verification of multi-process aerosol dynamics evolution in the wet flue gas cooling towers under crosswind conditions using FVM-TEMOM. *Journal of Aerosol Science, 192*. https://doi.org/10.1016/j.jaerosci.2025.106740.

Sun, W. Y., Nakamura, T., Cho, Y.-H., & Mizutani, N. (2023). Numerical investigation of solitary wave attenuation by a vertical plate-type flexible breakwater using an FVM–FEM coupled model. Part 1: Linear elastic isotropic material. *Ocean Engineering, 285*, 115368. https://doi.org/10.1016/j.oceaneng.2023.115368.

Szady, M. J., Salamon, T. R., Liu, A. W., Bornside, D. E., Armstrong, R. C., & Brown, R. A. (1995). A new mixed finite element method for viscoelastic flows governed by differential constitutive equations. *Journal of Non-Newtonian Fluid Mechanics, 59*(2-3), 215–243. https://doi.org/10.1016/0377-0257(95)01370-b.

Toreja, A. J., & Rizwan-uddin (2003). Hybrid numerical methods for convection-diffusion problems in arbitrary geometries. *Computers and Fluids, 32*(6), 835–872. https://doi.org/10.1016/S0045-7930(02)00031-2.

Tsuji, K., Fujioka, S., Morikawa, D. S., & Asai, M. (2025). A reliable SPH(2) formulation for Darcy–Forchheimer–Brinkman equation using a density-based particle shifting in the ALE description. *Computational Particle Mechanics, 12*(2), 1045–1073. https://doi.org/10.1007/s40571-024-00856-2.

Tu, L., Ma, Y., Han, X., & Zhu, M. (2023). Numerical analysis of a modified combined volumetric filtered concentrating PV/T system using MCRT and FVM coupled method. *Journal of Cleaner Production, 410*, 137296. https://doi.org/10.1016/j.jclepro.2023.137296.

Ulrich, C., Leonardi, M., & Rung, T. (2013). Multi-physics SPH simulation of complex marine-engineering hydrodynamic problems. *Ocean Engineering, 64*, 109–121. https://doi.org/10.1016/j.oceaneng.2013.02.007.

Vakilha, M. (2025). A fully explicit SPH method for modeling 2-D incompressible two-phase fluid-structure interaction with modified periodic and open boundary conditions. *Computers and Mathematics with Applications, 197*, 43–70. https://doi.org/10.1016/j.camwa.2025.07.038 Available from.

Vesvard, D., Karamian-Surville, P., & Abcha, N. (2025). Coupling FVM and FEM using coupling code with Code Saturne and Code ASTER. *Applications in Engineering Science, 23*, 100237. https://doi.org/10.1016/j.apples.2025.100237.

Villodi, N., & Ramachandran, P. (2025). Adaptive Compressible Smoothed Particle Hydrodynamics. *Journal of Computational Physics, 539*, 114257. https://doi.org/10.1016/j.jcp.2025.114257.

Wang, J., Zheng, H., & Tian, Z. (2015). Numerical simulation with a TVD–FVM method for circular cylinder wake control by a fairing. *Journal of Fluids and Structures, 57*, 15–31. https://doi.org/10.1016/j.jfluidstructs.2015.05.008.

Wang, K., Zhang, Q., Xia, X., Wang, L., & Liu, X. (2015). Analysis of hydraulic fracturing in concrete dam considering fluid-structure interaction using XFEM-FVM model. *Engineering Failure Analysis, 57*, 399–412. https://doi.org/10.1016/j.engfailanal.2015.07.012.

Wang, X., Luo, S., Tang, T., Liu, X., & He, Y. (2019). A MCRT-FVM-FEM coupled simulation for optical-thermal-structural analysis of parabolic trough solar collectors. *Energy Procedia, 158*, 477–482. https://doi.org/10.1016/j.egypro.2019.01.138.

Wang, Z. (2025). Accurate implicit moving particle simulation method with angular momentum conservation for high-viscous free-surface flow. *Computational Particle Mechanics, 12*(3), 1491–1505. https://doi.org/10.1007/s40571-024-00865-1.

Wang, Z., Haidn, O. J., & Hu, X. (2025). An algorithm for the incorporation of relevant FVM boundary conditions in the Eulerian SPH framework. *Computer Physics Communications, 307*, 109429. https://doi.org/10.1016/j.cpc.2024.109429.

Wang, Z., Zhang, C., Haidn, O. J., Adams, N. A., & Hu, X. (2025). FVM realization in Eulerian SPH: A comparative study within unified codebase. *Computer Physics Communications, 314*. https://doi.org/10.1016/j.cpc.2025.109683.

Wang, Z., Chen, L., Chen, J., Yao, J., Zhang, Z., Luo, Y., & Huang, S. (2025). A novel WENO-CSPH method for simulating strong elastic–plastic impact dynamics in solids.

Computer Methods in Applied Mechanics and Engineering, 444, 118142. https://doi.org/10.1016/j.cma.2025.118142.

Wang, Z. M., & Tan, S. K. (1998). Vibration and pressure fluctuation in a flexible hydraulic power system on an aircraft. *Computers and Fluids, 27*(1), 1–9. https://doi.org/10.1016/S0045-7930(97)00033-9.

Xiong, H., Cai, J., Shi, X., Jin, Y., Bao, X., & Chen, X. (2025). Numerical investigation of ice-wave-structure interaction using coupled SPH-DEM. *Ocean Engineering, 329*, 121006. https://doi.org/10.1016/j.oceaneng.2025.121006.

Xu, X., & Yu, W. (2025). SPH simulations of non-isothermal viscoplastic free-surface flows incorporating Herschel-Bulkley-Papanastasiou model. *Computers & Fluids, 287*, 106491. https://doi.org/10.1016/j.compfluid.2024.106491.

Xu, Y. (2026). Numerical investigation of bubble-induced heat transfer under external electric field based on ISPH-FVM coupling method. *International Journal of Heat and Fluid Flow, 117*, 110132. https://doi.org/10.1016/j.ijheatfluidflow.2025.110132.

Xu, Y., Yang, G., & Hu, D. (2023). A three-dimensional ISPH-FVM coupling method for simulation of bubble rising in viscous stagnant liquid. *Ocean Engineering, 278*, 114497. https://doi.org/10.1016/j.oceaneng.2023.114497.

Xu, Y., Yang, G., & Hu, D. (2024). Comparison of surface tension models for the simulation of two-phase flow in an ISPH-FVM coupling method. *European Journal of Mechanics, B/Fluids, 105*, 57–96. https://doi.org/10.1016/j.euromechflu.2023.12.012.

Xu, Y., Yang, G., Zhu, Y., & Hu, D. (2021). A coupled SPH–FVM method for simulating incompressible interfacial flows with large density difference. *Engineering Analysis with Boundary Elements, 128*, 227–243. https://doi.org/10.1016/j.enganabound.2021.04.005.

Xu, Y., Yang, G., Liu, S., & Hu, D. (2023). Improvement of surface tension discrete model in the ISPH-FVM coupling method. *International Journal of Multiphase Flow, 160*, 104347. https://doi.org/10.1016/j.ijmultiphaseflow.2022.104347.

Yan, S.-lin, Zhang, X.-bao, & Luo, Z.-hong (2025). Adaptive mesh refinement for VOF modeling gas-liquid two-phase flow: A summary of some algorithms and applications. *Chemical Engineering Science, 306*, 121291. https://doi.org/10.1016/j.ces.2025.121291.

Yang, B., Liu, S., Zhang, R., & Yu, X. (2022). Influence of reflector installation errors on optical-thermal performance of parabolic trough collectors based on a MCRT - FVM coupled model. *Renewable Energy, 185*, 1006–1017. https://doi.org/10.1016/j.renene.2021.12.102.

Yang, X., Zhang, Z., Zhang, G., Zheng, K., & Liang, G. (2025). An acceleration-based stable approach transferring information from fluid to solid in the SPH-based FSI solver. *Applied Ocean Research, 162*, 104706. https://doi.org/10.1016/j.apor.2025.104706.

Ye, M., Zhang, C., Ren, Y., Liu, Z., Haidn, O. J., & Hu, X. (2025). Adaptive optimization of wave energy conversion in oscillatory wave surge converters via SPH simulation and deep reinforcement learning. *Renewable Energy, 246*. https://doi.org/10.1016/j.renene.2025.122887.

Yilmaz, A. (2025). A hybrid R-LES-SPH model for numerical simulation of hydrodynamic problems with violent free-surface flows. *Engineering Analysis with Boundary Elements, 179*, 106405. https://doi.org/10.1016/j.enganabound.2025.106405.

Ying, H., Xu, J., & Zhu, C. (2025). SPH modeling of soil structure interactions using a novel non-local contact method. *Computers and Geotechnics, 184*, 107272. https://doi.org/10.1016/j.compgeo.2025.107272.

Yokoyama, R., Wang, K., Miwa, S., & Okamoto, K. (2025). A fully explicit Lagrangian-Lagrangian fluid-structure interaction solver with GPU acceleration based on a physically consistent moving particle hydrodynamics. *Computers & Structures, 319*, 108012. https://doi.org/10.1016/j.compstruc.2025.108012.

Yu, S., Li, Y., Gao, Y., Sun, G., & Liu, H. (2025). Investigating the crack propagation processes of grouted / ungrouted fissured tunnel specimens using 3D printing technology and SPH simulation method. *Construction and Building Materials, 500*, 144213. https://doi.org/10.1016/j.conbuildmat.2025.144213.

Yuan, M., Gao, T., & Fu, L. (2025). A diffuse-interface smoothed particle hydrodynamics method for compressible multiphase flows. *Journal of Computational Physics, 524*, 113717. https://doi.org/10.1016/j.jcp.2024.113717.

Yuan, T. J., Qiu, L. C., Zhao, K. L., & Liu, Y. (2025). SPH modeling of impact forces from heterogeneous soil-rock mixture landslides on a rigid obstruction. *Computers and Geotechnics, 184*. https://doi.org/10.1016/j.compgeo.2025.107302.

Zargaran, A., Hartwig, F., & Janoske, U. (2025). Three dimensional modeling of filtration process in complex filter structures using a data-free physics-informed neural network coupled with smoothed particle hydrodynamics. *AI Thermal Fluids, 4*, 100017. https://doi.org/10.1016/j.aitf.2025.100017.

Zhan, Y., Luo, M., & Khayyer, A. (2025c). DualSPHysics+: An enhanced DualSPHysics with improvements in accuracy, energy conservation and resolution of the continuity equation. *Computer Physics Communications, 306*, 109389. https://doi.org/10.1016/j.cpc.2024.109389.

Zhan, Y., Luo, M., & Khayyer, A. (2025a). An enhanced numerical wave tank for wave-structure interaction based on DualSPHysics+. *Ocean Engineering, 340*, 122413. https://doi.org/10.1016/j.oceaneng.2025.122413.

Zhan, Y., Luo, M., & Khayyer, A. (2025b). An enhanced SPH-based hydroelastic FSI solver with structural dynamic hourglass control. *Journal of Fluids and Structures, 135*, 104295. https://doi.org/10.1016/j.jfluidstructs.2025.104295.

Zhang, B., Wang, Z., Wan, D., & Hu, X. (2025). Implicit-splitting physics-driven particle relaxation for enhancement of Eulerian SPH. *Engineering Analysis with Boundary Elements, 176*, 106239. https://doi.org/10.1016/j.enganabound.2025.106239.

Zhang, S., Lourenço, S. D. N., & Hu, X. (2025). Multiphase SPH for surface tension: Resolving zero-surface-energy modes and achieving high Reynolds number simulations. *Computer Methods in Applied Mechanics and Engineering, 444*. https://doi.org/10.1016/j.cma.2025.118147.

Zhang, S., Fan, Y., Zhang, C., Adams, N. A., & Hu, X. (2025). A diffusive wetting model for water entry/exit based on the weakly-compressible SPH method. *Ocean Engineering, 325*, 120818. https://doi.org/10.1016/j.oceaneng.2025.120818.

Zhang, Y., Xu, W., & Chen, Y. (2025). A GPU-based FVM–DEM coupling framework and its application on gas-liquid-solid three-phase flow. *Particuology, 104*, 139–152. https://doi.org/10.1016/j.partic.2025.06.009.

Zhang, Y. D., Ming, F. R., Wang, P. P., & Liang, D. F. (2025). An adaptive SPH-FVM method with optimized particle-mesh interpolation scheme for strongly compressible multiphase flows. *International Journal of Multiphase Flow, 189*. https://doi.org/10.1016/j.ijmultiphaseflow.2025.105256.

Zhang, Y. Y., Xu, W. J., & Liu, L. (2026). A hybrid FVM-DEM framework for simulating explosion-driven dynamics of granular materials. *Powder Technology, 472*. https://doi.org/10.1016/j.powtec.2026.122136.

Zhang, Z., & Sheng, X. (2025). Numerical study of smoothed particle Galerkin method in orthogonal cutting simulations. *Computational Particle Mechanics, 12*(2), 933–945. https://doi.org/10.1007/s40571-024-00843-7.

Zhao, C., Yu, Y., Haidn, O. J., & Hu, X. (2025). Physics-driven complex relaxation for multi-body systems of SPH method. *Computer Physics Communications, 313*, 109615. https://doi.org/10.1016/j.cpc.2025.109615.

Zhao, C., Wu, Y., Yu, Y., Haidn, O. J., & Hu, X. (2025). A unified FSI framework for modeling liquid sloshing and baffle suppression in the elastic tank. *Applied Ocean Research, 157*, 104498. https://doi.org/10.1016/j.apor.2025.104498.

Zhao, K., Wang, C., He, T., Luo, G., Qin, Y., & Fang, S. (2024). Theoretical and experimental study on lubrication and friction of slipper pair of valve distribution piston pump based on FVM-TRD coupling method. *Tribology International, 194*, 109456. https://doi.org/10.1016/j.triboint.2024.109456.

Zhi, Y., Liu, S.-shuai, Yang, B., & Yu, X.-hui (2025). Analysis of the thermal performance of parabolic solar dish collector using MCRT-FVM coupled optical-thermal model. *Energy, 322*, 135579. https://doi.org/10.1016/j.energy.2025.135579.

Zhou, Z., Han, P., Sun, Y., Sun, Z., & Xi, G. (2025). Incompressible-compressible moving particle and moving Voronoi mesh method for multiphase flow. *Journal of Computational Physics, 537*, 114062. https://doi.org/10.1016/j.jcp.2025.114062.

Zhou, Z., Zhang, D., Gao, C., Li, Z., Guo, X., Shang, C., & Wang, P. (2025). Interaction analysis of subway construction and groundwater seepage in spring area using PD-FVM coupling method. *Computational Particle Mechanics, 12*(3), 1529–1547. https://doi.org/10.1007/s40571-024-00874-0.

Zhu, G., Zheng, S., Ren, Y., & Li, Y. P. (2025). Iterative high-order weakly compressible smoothed particle hydrodynamics model for viscous fluid flows. *Computer Methods in Applied Mechanics and Engineering, 447*, 118339. https://doi.org/10.1016/j.cma.2025.118339.

Zhu, Z., Ke, Q., & Su, X. (2025). Comparative analysis of Pelton turbine internal flow fields using the smoothed particle hydrodynamics method and the volume of fluid method with experimental validation. *Renewable Energy, 255*, 123828. https://doi.org/10.1016/j.renene.2025.123828.x.

Surface mesh techniques for sloshing dynamics

6

6.1 Introduction

In modern engineering practice, particularly within mechanical and civil engineering disciplines, digital verification has become a central pillar of the product and system development cycle, fundamentally reshaping how complex engineering artefacts are conceived, evaluated, and delivered. The transition from predominantly experimental, prototype-driven workflows to simulation-led digital development has compressed development timelines dramatically, reducing processes that once spanned many years to cycles measured in months, and in some highly optimised industrial contexts, even weeks. This shift has been driven not only by advances in computational power and numerical methods, but more critically by the maturation of digital representation techniques that allow physical behaviour to be predicted with sufficient confidence before any material is cut or assembled. At the heart of this transformation lies discretisation, the process by which continuous physical domains are converted into finite numerical representations suitable for computational analysis, and it is this step that ultimately governs the fidelity, robustness, and predictive capability of any digital verification workflow.

Within the broader digital development cycle, discretisation acts as the bridge between abstract design intent, typically expressed through CAD geometry, and physics-based simulation, such as computational fluid dynamics or structural mechanics. In industrial digital production pipelines, geometry generated during conceptual and detailed design stages is passed downstream to simulation environments, where virtual testing replaces or supplements costly physical experimentation. These simulations inform design iterations, material choices, manufacturing strategies, and performance optimisation, often feeding directly into automated or semiautomated design loops. Consequently, the quality of the discretised model is no longer a secondary technical detail but a strategic factor that influences decision-making speed, development cost, and ultimately product competitiveness.

In computational fluid dynamics (CFD)-based analyses, discretisation acquires a distinctive character compared to solid mechanics. While structural simulations require volumetric discretisation of solid bodies to resolve stress, strain, and deformation fields, fluid simulations are primarily concerned with the behaviour of a continuum occupying a region of space bounded by solid surfaces. In the vast majority of engineering CFD applications, it is the interaction between the fluid and the solid boundary that dominates the physics of interest, whether through momentum exchange, heat transfer, or mass transport. As a result, the accurate definition of the solid region is achieved not through volumetric meshing of the solid itself, but through precise representation of its surface, which defines the interface across which boundary conditions are imposed.

Slosh Dynamics. DOI: https://doi.org/10.1016/B978-0-443-44330-5.00008-6

This distinction has profound implications for meshing strategy. In many CFD applications, the solid body is treated as rigid, stationary, or thermally prescribed, and its internal stress state is irrelevant to the fluid solution. What matters is the geometry of the surface exposed to the fluid and the physical conditions imposed on that surface. Consequently, the solid can often be represented purely as a surface envelope, while the fluid region is discretised volumetrically. This approach significantly reduces computational cost and model complexity, while retaining the essential physics of fluid–solid interaction. However, it also places extraordinary importance on the quality of the surface mesh, because this mesh becomes the sole numerical representation of the solid domain within the CFD model (Fosas et al., 2026; Marty, 2014; Pei & Rim, 2021).

A representative industrial example can be found in automotive and mechanical system development, such as the analysis of lubrication flow within an engine oil pan. Although the oil pan is a three-dimensional solid component with finite thickness, material heterogeneity, and manufacturing constraints, its structural behaviour is often of secondary importance when the objective is to predict oil flow patterns, splash behaviour, thermal transport, or aeration under dynamic operating conditions. In such cases, the oil pan may be modelled as a rigid envelope defined by its inner surfaces, with prescribed thermal or motion boundary conditions, while the fluid domain occupies the enclosed volume. The solid material is effectively treated as void from a meshing perspective, yet its influence on the fluid is fully retained through the surface definition. Similar modelling philosophies are widely employed in civil engineering hydraulics, Heating, Ventilation, and Air Conditioning systems, turbomachinery casings, cooling jackets, and process engineering equipment.

This surface-based representation simplifies the discretisation of complex assemblies, particularly when dealing with thin-walled components, castings, or sheet-metal structures, where volumetric meshing of the solid would be prohibitively expensive or unnecessary for the analysis objectives. However, the apparent simplicity of surface meshing can be misleading. While generating a surface mesh may appear easier than constructing a fully conformal solid volume mesh, the numerical consequences of poor surface mesh quality are often more severe in CFD, because the surface mesh directly controls the shape, quality, and resolution of the fluid volume mesh that is subsequently generated.

The surface mesh effectively acts as the template from which the entire fluid discretisation is derived. Any deficiencies at the surface level, such as highly skewed elements, abrupt size transitions, tiny isolated elements embedded within coarse regions, or inconsistent normals, propagate into the volume mesh, often becoming amplified rather than mitigated. For example a single small surface triangle surrounded by much larger elements can force the volume mesher to generate a cascade of disproportionately small cells, leading to severe local refinement, numerical stiffness, and inefficient solver performance. Similarly, poor surface skewness degrades the quality of near-wall cells, undermining gradient reconstruction and compromising the accuracy of wall shear stress, heat flux, and turbulence modelling.

From the perspective of the digital development cycle, these issues have far-reaching implications. A poorly prepared surface mesh can cause meshing failures late in the workflow, necessitating repeated geometry cleanup and remeshing that erodes the time savings promised by digital verification. Worse still, it can produce simulations that converge numerically but deliver misleading results, creating false confidence in design decisions. In industrial environments where simulation outputs directly inform design approvals, tooling investments, or safety assessments, such risks are unacceptable.

Therefore surface mesh quality must be treated as a first-class engineering concern, on equal footing with solver selection or physical modelling assumptions. Uniformity of element size, controlled growth rates, low skewness, and the avoidance of excessively small features are not merely numerical preferences but practical requirements to ensure that the fluid volume mesh is smooth, well-conditioned, and capable of resolving the relevant physical phenomena without introducing artificial numerical artefacts. The surface mesh must also be sufficiently refined in regions of expected high gradients, such as sharp curvature, flow separation zones, or thermal hotspots, while remaining economical elsewhere to maintain computational efficiency.

6.2 Critical requirements for surface mesh integrity in computational fluid dynamic

In computational fluid dynamics the preparation of a high-quality surface mesh is absolutely fundamental because the surface mesh forms the numerical and geometric scaffold upon which the entire volume mesh and subsequent solution depend, and as experienced practitioners know, deficiencies at the surface level invariably propagate into the volume, manifesting as solver instability, inaccurate physics, or even outright failure to generate a valid mesh; for this reason, surface mesh quality must be regarded as an integrated synthesis of topological validity, geometric regularity, and readiness for near-wall treatments rather than a mere set of isolated metrics, and the first and most critical requirement is that the surface representation must be globally coherent and topologically valid, meaning that surfaces must be completely water-tight, with no gaps, holes, or open edges whatsoever, because any break in the boundary allows the volume mesher to misinterpret interior versus exterior, often leading to catastrophic negative volumes or unmatched volumes that cannot be tessellated, and just as importantly, the surface must contain no intersections either between distinct surfaces or within a single surface patch, because intersecting faces destroy the unambiguous definition of the domain and confuse proximity based algorithms that drive both surface triangulation and subsequent volume building; equally pernicious are duplicate surfaces, overlapping faces, or unintended coincident geometry imported from CAD assemblies, which create nonmanifold regions where more than two faces share an edge or where T-junctions appear without proper vertex connectivity, and such conditions violate the mathematical assumptions inherent in most mesh generation engines, preventing the formation of a consistent connectivity graph and frequently resulting in unexplained failures or excessive local refinement that wastes computational resources.

Beyond pure validity, the numerical quality of the surface discretisation is equally important and must be judged with a suite of quantitative measures that collectively ensure that the mesh elements represent the underlying geometry faithfully and support accurate reconstruction of gradients and fluxes; among these, skewness is widely recognised as the most influential single metric because it quantifies the deviation of a surface element from its ideal geometric shape, typically an equilateral triangle in the case of triangular meshes or an optimal polygon for mixed element types, and high skewness values correspond to highly distorted elements that degrade numerical stability and accuracy by distorting the local coordinate basis in which derivative approximations are computed, leading to poor approximation of curvature, pressure gradients, and shear layers, effects that can be especially damaging in regions of high gradient such as wakes, separation bubbles, or shocklets; the practical engineering objective is therefore to keep skewness low across the entire surface, recognising that worst-case peaks may be tolerated in isolated benign regions but that broad distributions of skewness above acceptable thresholds significantly increase the risk of solver divergence or nonphysical oscillations.

Linked closely to skewness is the aspect ratio of surface elements, which expresses the degree of elongation of an element by comparing its longest edge to its shortest, and while very low aspect ratios are generally preferable for isotropic flow regions, high aspect ratios are often unavoidable and sometimes desirable near boundaries where flow scales are anisotropic, such as in thin shear layers or regions with high curvature, provided that the elements remain aligned with the local geometric and flow directions and that transitions to more isotropic regions occur smoothly; abrupt changes in aspect ratio, or long, thin elements oriented poorly with respect to the local flow or surface normal, create numerical stiffness and can cause the volume mesher to produce poorly shaped cells that invalidate gradient reconstruction or induce spurious diffusion.

In addition to shape metrics, size continuity is critical, because sudden changes in element size force the volume mesher to generate transition cells with poor quality, and such abrupt size variations may not only increase the total cell count unnecessarily but also introduce artificial numerical artefacts due to inconsistent spatial resolution; engineers therefore seek to enforce a controlled surface element size distribution with smooth gradation, often guided by physical length scales such as boundary layer thickness, wake width, or expected turbulent structures, and in practice a moderate growth rate between adjacent elements is targeted so that the local mesh smoothly adapts from fine resolution near walls or critical features to coarser resolution in free shear regions, recognising that the ideal surface mesh does not exist in isolation but as a precursor to a volume mesh that must honour these spatial heterogeneities without generating extreme differences in cell volumes.

Another frequently overlooked but practically important geometric measure is face warping, which describes the degree to which a surface element deviates from being planar; although triangular facets are intrinsically planar, quadrilaterals or higher order polygons can become twisted, and even triangles can be considered warped when their vertices do not lie close to the actual geometric surface they are intended to represent, particularly on curved patches; such warping complicates boundary layer extrusion and degrades surface curvature representation, leading to an inaccurate mapping of curvature driven phenomena such as separation or reattachment points.

Closely related to warping is the control of minimum and maximum edge lengths, which are essential for balancing geometric fidelity against computational performance; edges that are excessively short relative to the characteristic domain dimensions force the mesh generator to create extremely small cells, dramatically increasing the computational burden and introducing stiffness into the numerical matrices, whereas edges that are overly long fail to capture critical geometric detail, effectively smoothing out curvature and flow complexity that are physically significant, so pragmatic engineering practice involves setting sensible bounds on edge lengths based on expected physical scales, manufacturing tolerances, and solver capabilities.

Perhaps the most consequential aspect of surface mesh quality for many industrial flows is normal consistency, particularly when preparing for boundary layer resolution or prism layer generation, because the direction and orientation of surface normals determine how near-wall layers will be extruded; if normals are inconsistent, flipped, or misaligned, the boundary layer mesh can grow inward or fold back on itself, producing inverted cells or negative volumes that cannot be resolved by the solver, and such errors often masquerade in practice as turbulence model failures or near-wall convergence issues, so rigorous normal orientation cheques are indispensable.

For flows where the fidelity of near wall resolution is critical—for example high Reynolds number external aerodynamics, turbomachinery blade passages, or internal duct flows with strong curvature—additional geometric considerations come to the fore, because surface curvature must be resolved with sufficient mesh density to support the growth of structured boundary layer prisms or hexahedral layers; if curvature changes too abruptly relative to surface element size, prism layers may intersect, collapse, or taper to zero thickness, all of which lead to solver breakdown or loss of wall shear fidelity, and such failures are often traced back to unresolved CAD curvature rather than inadequate volume mesh settings, underscoring the fact that surface mesh quality and CAD geometry fidelity are inseparably linked.

Indeed, the influence of the underlying CAD geometry quality on surface meshing cannot be overstated, because the mesh is only as good as the geometric model it represents, and it is a well-established engineering truth that minute CAD artefacts such as sliver surfaces, micro features, tiny fillets, engraved text, or unnecessarily detailed decorative features are almost always counterproductive for CFD; such details introduce unwarranted complexity that forces local overrefinement, generates poor quality elements, and drastically inflates computational cost without corresponding physical benefit, so experienced CFD practitioners routinely engage in geometry clean-up or simplification workflows that remove these micro features while preserving the essential physical surfaces that govern the flow.

The cleanliness of topology is also vital for special boundary conditions such as periodic, cyclic, or symmetry planes, because mismatches in surface connectivity across corresponding boundaries lead to incompatible meshes that corrupt the application of symmetry conditions, cause discontinuities in numerical fluxes, or induce artificial modes in iterative solvers; proper stitching, imprinting, and alignment of corresponding surface patches are therefore essential when designing for symmetric or repeating domains.

Beyond these geometric and topological issues, there are a number of common failure modes that engineers encounter in practice regardless of CFD code family, and awareness of these phenomena is part of professional intuition: for example bad envelope meshes created during automated surface wrapping often contain folded, tangled, or inverted regions that are not immediately obvious visually yet destroy volume meshing, and intersection meshes resulting from imperfect surface repair commonly cause local nonmanifold clusters that must be resolved using dedicated cleaning tools before any volume discretisation can succeed; zero-thickness or near-zero-thickness regions, where two surfaces lie extremely close together due to CAD tolerance mismatch or duplicated geometry, are another notorious source of prism layer collapse and negative volumes, and highly concave regions frequently challenge layer extrusion algorithms to the point of failure unless special curvature controls or local mesh refinement strategies are applied.

Because these issues are pervasive, engineering practice has evolved pragmatic thresholds for surface quality that tend to be broadly applicable across CFD software: low skewness values that approach ideal shapes, moderate aspect ratios aligned with flow and geometry, smooth element size gradation controlled by physical feature scales, minimal warping, sensible bounds on edge lengths, and unequivocal normal consistency are all field-tested desiderata that dramatically increase the likelihood of successful volume meshing and stable solver performance, and while no single numerical threshold can guarantee success in every situation, these targets serve as invaluable guides during mesh preparation.

Taken together the criteria governing surface mesh quality reveal that quality is not a monolithic attribute but a multidimensional synthesis of validity, geometry, numerical readiness, and physical representativeness, and that the most reliable CFD simulations are built upon a surface mesh that has been meticulously crafted and validated through both automated metric cheques and expert engineering judgement; neglecting any one of these aspects almost invariably leads to downstream difficulties that are far more costly to fix after volume meshing or during solution than they would have been to address at the surface stage, and this is why high-fidelity CFD practitioners invest significant time in surface mesh inspection, validation, and iterative improvement long before turning on a solver.

6.3 Geometric wrapping and its role in computational remeshing

In modern computational fluid dynamics, the reliability of any numerical prediction is fundamentally limited by the quality of the computational mesh. While advances in solvers and physical models continue to improve, the discretised geometry on which these solvers operate remains a dominant source of numerical error. This is

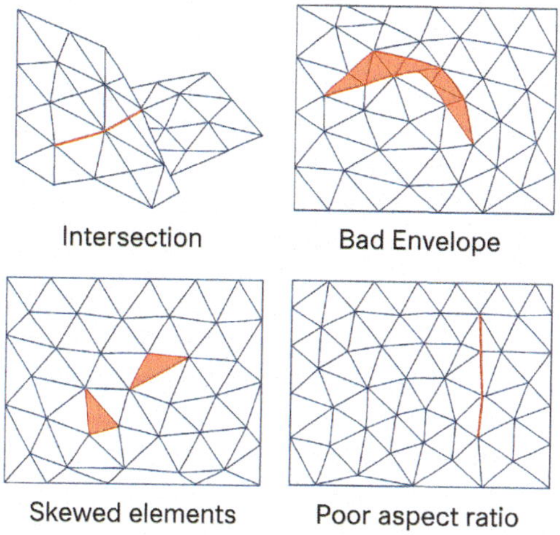

Figure 6.1 Typical surface mesh defects including gaps, overlaps, intersections, and nonmanifold regions.

Figure 6.2 Surface mesh envelopes illustrating internal and external boundary reconstruction.

particularly evident when simulations involve complex industrial geometries derived from CAD data, experimental reconstructions, or multicomponent assemblies. In such cases, geometric wrapping has become an indispensable preprocessing step.

Geometric wrapping may be understood as the process of constructing a clean, watertight, and topologically consistent surface that faithfully represents the intended physical domain, while deliberately suppressing irrelevant geometric detail. Rather than meshing the original geometry directly(as illustrated in Fig. 6.1 and Fig. 6.2), which often contains gaps, overlaps, nonmanifold edges, and excessively small features, a wrapped surface is generated as an intermediate representation. This surface acts as a mathematically well-posed boundary for subsequent volume meshing.

The need for wrapping arises from a fundamental mismatch between design geometry and numerical requirements. CAD models are created to describe shape and function, not to satisfy the strict topological and numerical constraints imposed by discretisation algorithms. Small fillets, sharp corners, tolerances, and assembly artefacts that are insignificant from a design perspective can severely degrade mesh quality, leading to skewed cells, poor orthogonality, and solver instability. Wrapping resolves this mismatch by enforcing geometric coherence and scale separation before remeshing is performed.

From a numerical standpoint, wrapping serves three essential purposes. First, it guarantees surface closure, ensuring that the computational domain is unambiguously defined. Second, it regularises geometric complexity, allowing the mesh resolution to be governed by physical length scales rather than accidental design features. Third, it provides a stable foundation for automated and repeatable mesh generation, which is crucial for large-scale parametric studies and validation exercises.

This work is situated within the broader objective of establishing reliable, high-fidelity simulation workflows for complex thermal–hydraulic systems. The focus is not merely on producing individual computational results, but on demonstrating a systematic process by which detailed simulations can be conducted, validated, and transferred across modelling fidelities.

At the core of this effort lies the integration of commercial CFD tools within a structured verification and validation framework. High-resolution simulations are employed to resolve local flow and thermal phenomena, while lower-fidelity models are used for system-level analysis. For such a high-to-low modelling strategy to be credible, the geometric and numerical consistency of the CFD simulations must be rigorously controlled. Geometric wrapping and remeshing therefore play a central enabling role.

The study demonstrates how carefully prepared computational domains, generated through wrapping and controlled remeshing, support both accurate numerical solutions and meaningful comparison with experimental data. Particular emphasis is placed on the repeatability of the workflow, the traceability of modelling assumptions, and the quantification of uncertainty arising from numerical discretisation.

Rather than treating mesh generation as a purely technical preprocessing step, it is addressed here as an integral component of the simulation methodology. The quality of the wrapped geometry directly influences solver convergence, solution robustness, and ultimately the credibility of the validation exercise. By embedding wrapping within the overall modelling strategy, the work illustrates how practical engineering simulations can be elevated from isolated analyses to components of a coherent predictive framework.

The implications of this approach extend beyond the immediate simulations presented. In validation studies, numerical uncertainty must be sufficiently controlled so that discrepancies between computation and experiment can be meaningfully interpreted. Poor geometric representation introduces uncontrolled error that cannot be separated from modelling or physical uncertainty. Wrapping mitigates this risk by enforcing a consistent geometric baseline across simulations.

Furthermore, when results from detailed CFD models are used to inform reduced-order or system-scale models, geometric consistency becomes critical. The transfer of information between modelling levels implicitly assumes that both are based on compatible representations of the physical domain. The wrapped geometry provides this compatibility, enabling reliable down-selection of models and parameters.

6.3.1 Role of wrapping in numerical consistency

A key modelling decision in this work is the explicit decoupling of geometric fidelity from numerical resolution. Wrapping enables this decoupling by allowing the surface representation to be controlled independently of the underlying CAD complexity. As a result, the mesh resolution may be chosen based on physical gradients rather than geometric artefacts.

From a numerical perspective, discretisation error represents the difference between the exact solution of the governing equations and its approximation on a computational mesh. This error is normally expected to decrease as the mesh is refined. However, poor surface quality can introduce an additional and uncontrolled source of error that does not reduce in a predictable manner with mesh refinement. Wrapping helps suppress this error by improving geometric smoothness and maintaining topological consistency.

Consequently, mesh refinement studies become meaningful, and convergence behaviour can be assessed with confidence. This is essential for verification activities and for establishing grid-independent solutions.

6.4 Advanced surface wrapping practice and scientific interpretation in STAR CCM+

In computational fluid dynamics and multiphysics simulation, geometry preparation is not a preliminary technical detail but a foundational scientific operation. The predictive reliability of numerical analysis is inseparable from the integrity of the geometric domain upon which governing equations are discretised. Regardless of how advanced the turbulence modelling, discretisation schemes, or solver algorithms may be, numerical accuracy is fundamentally constrained by the representation of physical boundaries. For this reason, surface wrapping in STAR CCM+ must be understood not merely as a meshing utility but as a geometric abstraction methodology that enables the transformation of engineering design models into mathematically consistent computational domains.

Engineering CAD models are typically constructed for manufacturing, assembly documentation, or structural detailing. These models often contain geometric complexities that are irrelevant or even detrimental to numerical simulation. Small overlaps, micro gaps, sliver faces, inconsistent stitching, intersecting solids, duplicated surfaces, and unresolved tolerance mismatches are common. In addition, assemblies may include internal mechanical details such as bolts, fasteners, fillets, or hidden cavities that do not influence the physical behaviour of interest. While such details are acceptable in manufacturing contexts, they compromise mesh generation and numerical stability.

Surface wrapping addresses this challenge through algorithmic reconstruction. Rather than manually repairing geometry or editing CAD entities, STAR CCM+ interprets the spatial envelope of the imported model and reconstructs a new triangulated surface that is watertight, manifold consistent, and optimised for simulation. This process is computationally robust, scalable to large industrial assemblies, and highly controllable. It reduces preprocessing time while enhancing solver stability and predictive reliability.

6.4.1 Initialisation of the STAR CCM+ environment and geometry import

The wrapping workflow begins with the initialisation of a STAR CCM+ simulation environment. The software operates within a structured simulation file format with the extension sim. This file contains geometry, mesh definitions, physics continua, boundary conditions, reports, scenes, and solver configurations. Opening or creating a simulation file establishes the computational workspace in which all preprocessing and analysis steps are executed.

Geometry import is a decisive stage in the workflow. STAR CCM+ supports multiple CAD formats, including initial graphics exchange specification (IGES), Parasolid, and surface triangulations such as standard tessellation language (STL). The import process is governed by tolerance settings that determine how small gaps and discontinuities are interpreted. These tolerances influence stitching operations, surface merging, and part connectivity. If the tolerance is set too small, minor CAD inconsistencies may remain unresolved, leading to wrapping failure. If the tolerance is set excessively large, distinct features may merge unintentionally, altering the intended geometry.

Therefore import settings require engineering judgement. The user must consider the smallest physically relevant feature in the simulation and ensure that tolerance parameters are aligned with that scale. For high-resolution aerodynamic simulations, small geometric details may need preservation. For large-scale external flow studies, minor manufacturing gaps may be safely eliminated.

Once imported, geometry can be organised into parts and surfaces. STAR CCM+ allows grouping and labelling of geometric entities, facilitating later selection during wrapping operations. Proper organisation at this stage streamlines the entire workflow.

6.4.2 Advanced visualisation and geometric diagnostics

One of the distinguishing features of STAR CCM+ is its advanced visualisation environment. Visualisation is not merely aesthetic but diagnostic. Before performing surface wrapping, the engineer must inspect the imported geometry carefully. Wireframe mode reveals connectivity and edge continuity. Shaded rendering exposes curvature and surface smoothness. Colouring by part allows identification of separate assemblies. Surface normal display ensures orientation consistency. Section cuts enable inspection of internal cavities. Isosurface visualisation assists in identifying complex intersections.

This diagnostic phase is critical. Surface wrapping reconstructs geometry based on its spatial interpretation. If unintended cavities or disconnected parts exist, the wrapping result may not align with the intended domain. Visualisation enables proactive correction or parameter adjustment before executing computationally expensive operations.

In industrial contexts, geometry inspection may reveal thousands of small surface defects. Manual repair would be time-consuming and inefficient. Instead, wrapping parameters can be tuned to eliminate these inconsistencies automatically during reconstruction.

6.4.3 Conceptual foundations of surface wrapping

Surface wrapping in STAR CCM+ is fundamentally grounded in a volumetric sampling strategy rather than direct manipulation of CAD surfaces. Instead of attempting to repair or stitch the original faces explicitly, the algorithm conceptually interrogates the spatial domain occupied by the geometry. It overlays a background discretisation field across the entire model and evaluates which regions of space are considered solid and which are void. Through this spatial interrogation, the software reconstructs a new surface that encloses the detected volume. The outcome is a triangulated boundary that approximates the external or internal envelope of the geometry in a mathematically consistent manner. Importantly, this reconstructed surface is not a direct replica of the original CAD faces. Rather, it is a controlled geometric abstraction designed specifically to serve as a stable and mesh-ready boundary for numerical analysis. The fidelity of this reconstructed envelope is primarily governed by the base size parameter. The base size defines the characteristic spatial resolution of the background sampling field. In practical terms, it determines how finely the geometry is interrogated during reconstruction. When a smaller base size is selected, the sampling resolution increases, allowing the wrapper to capture finer geometric variations. Curved regions are represented with greater precision, sharp edges are preserved more accurately, and small but potentially significant features remain intact. This higher fidelity representation improves geometric accuracy, particularly in regions where curvature strongly influences flow behaviour or heat transfer characteristics. However, increasing geometric fidelity is not without computational consequences. A smaller base size results in a substantial increase in the number of surface triangles generated during wrapping. This increase propagates downstream into the volume mesh, raising memory requirements and computational time. The wrapping process itself becomes more resource-intensive, and the subsequent simulation may demand additional solver iterations due to increased cell count. Therefore while a fine base size enhances geometric representation, it must be selected judiciously to avoid unnecessary computational burden. In contrast, choosing a larger base size simplifies the reconstructed geometry. Small scale details and minor surface irregularities are smoothed out, and the overall triangulation becomes coarser. This simplification accelerates the wrapping process and reduces memory consumption. In many engineering applications, such smoothing is not detrimental and may even be beneficial if the removed details are physically irrelevant. For example minor manufacturing imperfections or decorative features that do not influence flow physics can be safely abstracted away. The key consideration is whether a particular geometric feature contributes meaningfully to the physical phenomena under investigation. The selection of base size therefore represents a balance between geometric fidelity and computational efficiency. It requires engineering judgement rather than arbitrary refinement. The engineer must evaluate which features govern the dominant physical processes in the simulation. Regions where strong gradients, boundary layer development, or structural interaction occur may require higher fidelity, whereas smooth and uniform areas can tolerate coarser representation. The objective is to ensure that the reconstructed surface faithfully captures physically significant geometry without preserving unnecessary detail. One of the major advantages of this volumetric abstraction approach is its inherent robustness. Because the wrapper does not rely on direct editing of the original CAD topology, it is less sensitive to imperfections such as nonmanifold edges, overlapping surfaces, or small open gaps. Even when the imported geometry contains inconsistencies that would normally prevent successful meshing, the volumetric sampling process can reconstruct a continuous and watertight envelope, provided that wrapping parameters are chosen appropriately. This robustness greatly reduces the need for manual CAD repair and enables efficient preprocessing of complex industrial assemblies.

6.4.4 Control of surface resolution and adaptive refinement

Beyond the global base size, STAR CCM+ offers additional refinement mechanisms. Minimum and maximum surface sizes can be specified to constrain local triangulation. Curvature based refinement allows automatic reduction of element size in regions where surface curvature exceeds a defined threshold. Proximity refinement improves resolution in narrow gaps where flow gradients may be steep.

These adaptive controls prevent uniform overrefinement across the entire geometry. Instead, computational effort is concentrated where it is physically meaningful. In aerospace applications, leading edges, intake openings, and control surfaces often require higher resolution than smooth fuselage sections. In electronic cooling systems, heat transfer surfaces demand finer representation than outer casings.

The ability to combine global and local refinement ensures efficient use of computational resources. It also aligns the geometric representation with the physics of interest.

6.4.5 Wrapping volume of interest and domain extraction

A defining feature of STAR CCM+ wrapping is the Wrapping Volume of Interest option. This setting determines which spatial domain the wrapper preserves during reconstruction. The selection fundamentally distinguishes between internal and external flow simulations.

When the Largest Internal option is selected, the wrapper identifies all enclosed cavities within the geometry and preserves the largest connected internal region. This configuration is typically used for simulations of internal flow, such as cooling channels, intake manifolds, or duct networks. Smaller cavities that may originate from manufacturing gaps are discarded. This automatic ranking simplifies domain extraction without manual editing.

The External option, by contrast, extracts the outer aerodynamic envelope. All internal cavities are sealed, and irrelevant internal details are removed. This configuration is essential for external flow simulations involving vehicles, aircraft, buildings, or marine hulls. The wrapper reconstructs a clean outer shell, eliminating interior complexity.

The Seed Point option provides precise manual control. By placing a point within the desired region, the engineer instructs STAR CCM+ to preserve only the connected volume containing that point. This method is valuable when multiple cavities of similar size exist or when selective extraction of a specific chamber is required.

The Nth Largest option extends flexibility by allowing selection of cavities according to ranked volume. In complex assemblies such as multi pass heat exchangers or compartmentalised fuel tanks, this functionality enables extraction of secondary regions without modifying geometry.

These volume of interest options transform wrapping from a simple surface repair operation into a powerful domain extraction tool.

6.4.6 Surface controls for localized precision

In high-fidelity simulations, uniform geometric resolution is inefficient. Consider the case of auxiliary intake doors on an aircraft fuselage. These doors open automatically under certain flight conditions to admit additional airflow. The aerodynamic behaviour around these components influences pressure distribution, mass flow rate, and boundary layer characteristics. Small geometric variations may significantly affect flow separation.

To capture such sensitivity, high-resolution wrapping is required in the vicinity of the intake doors. However, refining the entire fuselage at the same level would dramatically increase computational cost. STAR CCM+ addresses this challenge through Surface Control definitions. Surface Control allows the engineer to locally override the global base size for selected surfaces. Smaller target sizes can be applied only to critical components while maintaining coarser resolution elsewhere.

This targeted refinement strategy reflects modern computational efficiency principles. It preserves physical accuracy where necessary while minimising unnecessary computational burden.

6.4.7 Comparison of pre wrapped and post wrapped geometry

The contrast between geometry before and after surface wrapping illustrates clearly why this operation is scientifically and computationally indispensable. Imported CAD assemblies, even when produced with high manufacturing precision, are rarely suitable for direct numerical simulation. Prior to wrapping, the geometry often contains a range of imperfections that are benign in design environments but problematic in computational analysis. Open edges may exist where surfaces fail to stitch correctly. Small gaps can appear between adjacent components due to tolerance mismatches. Overlapping faces may occupy the same spatial location without proper Boolean resolution. Inconsistent surface normal orientation may arise from complex assembly operations. These irregularities may be subtle and visually insignificant, yet they carry serious numerical consequences. During volume meshing, such defects frequently lead to failure in region detection or the generation of invalid cells. Open boundaries can prevent the identification of enclosed fluid domains. Small holes may cause unintended leakage paths that distort the computational region. Overlapping surfaces may confuse the mesher, resulting in highly skewed or inverted elements. Inconsistent normal directions can disrupt boundary layer extrusion and compromise wall treatment algorithms. The cumulative effect is reduced mesh quality, unstable discretization, and increased risk of solver divergence. In extreme cases, the simulation may not proceed at all due to topological inconsistency. Surface wrapping fundamentally transforms this condition. After wrapping, the geometry is reconstructed as a single, continuous triangulated surface that is fully watertight and topologically coherent. The wrapper closes small gaps automatically, merges intersecting surfaces, removes redundant overlaps, and eliminates micro-scale cavities that have no physical relevance. Surface normals are aligned consistently across the entire domain, ensuring uniform outward or inward orientation as required. The reconstructed surface no longer reflects the fragmented structure of the original CAD assembly but instead represents a clean computational boundary. This transformation has direct implications for downstream meshing. Polyhedral and trimmed cell volume meshers rely on a stable surface definition to generate high-quality elements. A watertight wrapped surface allows the mesher to detect fluid regions reliably and construct uniform volume cells with controlled skewness and aspect ratio. Because the wrapper simplifies unnecessary geometric detail while preserving essential shape characteristics, the resulting volume mesh is both computationally efficient and physically representative. The risk of generating highly distorted or nonphysical cells is significantly reduced. Beyond meshing convenience, the improvement in geometric integrity directly enhances solver stability. Numerical methods for fluid flow, heat transfer, or multiphysics interaction assume that governing equations are applied within a closed domain bounded by well-defined surfaces. If the domain contains unintended openings or overlapping regions, conservation laws may be violated locally. Mass imbalance can occur if leakage paths exist. Pressure oscillations may develop near artificial discontinuities. Artificial recirculation zones can appear where geometry is inconsistent. Such numerical artifacts are not reflections of physical behaviour but consequences of geometric imperfection. By reconstructing a watertight and consistent boundary, high-quality wrapping eliminates these artificial sources of instability. Boundary conditions can be applied confidently, and conservation of mass, momentum, and energy can be maintained throughout the domain. The solver benefits from improved numerical conditioning because surface smoothness and continuity support stable discretisation of gradients and fluxes. Convergence behaviour becomes more predictable, and iterative residuals typically decrease more steadily compared to simulations based on poorly prepared geometry. In addition, the removal of nonphysical cavities contributes to improved computational efficiency. Small enclosed voids that are irrelevant to the analysis may otherwise generate unnecessary mesh regions, increasing cell count and solver cost. Wrapping filters these features automatically, ensuring that only physically meaningful volumes are preserved. This selective abstraction enhances both numerical stability and computational economy. The comparison between prewrapped and postwrapped geometry therefore highlights a critical principle in computational engineering. Simulation accuracy and stability do not depend solely on sophisticated physics models or high-order discretisation schemes. They depend equally on the integrity of the geometric foundation upon which those models operate. Surface wrapping serves as the mechanism that transforms manufacturing geometry into a mathematically valid computational domain. Through closure of gaps, alignment of normals, elimination of redundant features, and enforcement of topological consistency, wrapping establishes the conditions necessary for reliable numerical prediction.

6.4.8 Computational performance and industrial scalability

In modern engineering environments, computational efficiency is not a secondary consideration but a defining constraint. Simulation-driven design workflows often operate under strict time limitations, particularly in industries such as automotive and aerospace, where development cycles are highly compressed and multiple design iterations must be evaluated in parallel. Within this context, the computational performance of the surface wrapping algorithm becomes critically important. STAR CCM+ has been developed with a strong emphasis on parallel execution, allowing the wrapping process to exploit multicore and multiprocessor architectures effectively. As a result, even very large assemblies containing millions of surface facets can be reconstructed within practical time frames. The wrapping procedure distributes spatial sampling and surface reconstruction tasks across available processors. This parallelisation significantly accelerates the volumetric interrogation of complex geometry. Memory allocation is likewise managed efficiently, scaling with processor availability so that large industrial models can be handled without excessive hardware demands. Such scalability enables engineers to process full vehicle models, complete aircraft assemblies, or detailed electronic systems without decomposing them into smaller fragments for manual treatment. The ability to handle large-scale geometry directly within a unified computational framework enhances both workflow simplicity and model integrity. In contrast, traditional CAD repair workflows often rely on manual editing, stitching, trimming, and surface reconstruction operations performed within dedicated geometry tools. These processes are labour-intensive, time-consuming, and sensitive to user experience. In addition, manual repair may introduce inconsistencies when applied across multiple design variants. Less advanced wrapping utilities may struggle with intersecting solids, micro gaps, or nonmanifold conditions, requiring repeated intervention. STAR CCM+ addresses these challenges through robust automated reconstruction algorithms capable of resolving geometric imperfections with minimal manual input. The reduction in preprocessing time is substantial, particularly when repeated simulations are required. This efficiency has direct industrial implications. In automotive aerodynamics, for example, iterative optimisation of body shape may require evaluation of dozens or hundreds of design variations. If geometry preparation consumes excessive time, it becomes a bottleneck in the development process. By minimising preprocessing effort, STAR CCM+ enables engineers to dedicate more resources to physics analysis, performance evaluation, and design interpretation. The same principle applies in aerospace development, where configuration changes to nacelles, winglets, or intake systems must be evaluated rapidly. Efficient wrapping supports agile simulation workflows aligned with modern digital engineering practices. Beyond speed, robustness contributes significantly to industrial scalability. The wrapping algorithm in STAR CCM+ demonstrates consistent performance across diverse geometric complexities. It can handle assemblies with overlapping components, small inconsistencies, and mixed CAD formats without extensive manual intervention. This robustness enhances reproducibility. When standardised wrapping parameters are defined within an organisation, they can be applied consistently across multiple projects or product generations. Such parameter standardisation ensures that geometric abstraction remains uniform throughout a development programme, reducing variability introduced by subjective preprocessing decisions. Reproducibility is particularly important in regulated industries where traceability and validation are required. Consistent wrapping settings allow engineers to compare simulation results across design iterations with confidence that geometric preparation has been performed under identical conditions. This consistency strengthens the reliability of performance comparisons and supports data-driven decision-making. Moreover, industrial scalability extends beyond single simulations to enterprise-level workflows. STAR CCM+ integrates wrapping within a broader automation framework, allowing batch processing and scripting. Multiple geometries can be wrapped automatically using predefined parameter sets. This capability supports high-throughput simulation campaigns, design of experiments studies, and optimisation routines. In such scenarios, automated and parallelised wrapping becomes an enabling technology rather than a simple preprocessing step.

6.4.9 *Scientific implications of wrapping quality*

The quality of the wrapped surface has a direct and measurable influence on the reliability of numerical simulations. In computational fluid dynamics, the governing equations are discretised over a mesh whose outer boundary is defined by the surface representation. Any geometric inaccuracy introduced during wrapping propagates into the volume mesh and ultimately affects the predicted flow field. For this reason, surface wrapping is not merely a preprocessing convenience but a scientifically consequential operation. Regions characterised by strong curvature demand particular attention. Leading edges, sharp corners, intake lips, turbine blades, and control surfaces exhibit rapid geometric variation over short spatial scales. In such areas, the boundary layer develops under the influence of precise local curvature. If the wrapped surface under resolves these features, the discretised geometry may become artificially flattened or distorted. Even small deviations can alter local pressure gradients, modify shear stress distribution, and shift flow separation points. Since aerodynamic forces and moments are sensitive to these phenomena, insufficient surface resolution can result in inaccurate drag, lift, or torque predictions. At the same time, excessive smoothing during wrapping can artificially simplify geometry in ways that compromise physical realism. If curvature refinement is insufficient, subtle geometric transitions may be lost, leading to unrealistic streamline behaviour. For example the smoothing of a wing root fillet or intake edge can modify vortex formation mechanisms. Such geometric abstraction may appear visually minor, yet its influence on integrated aerodynamic coefficients can be significant. The scientific implication is clear: geometric fidelity must be preserved wherever physical gradients are sensitive to boundary shape. However, the opposite extreme is equally problematic. Overrefinement of the wrapped surface increases the total number of surface elements and consequently inflates the volume mesh size. While finer resolution may appear desirable, it does not automatically guarantee improved predictive accuracy. Once the surface representation exceeds the level of geometric detail required to capture relevant flow physics, further refinement primarily increases computational cost. Solver memory requirements grow, iteration times lengthen, and convergence behaviour may even deteriorate due to excessive mesh density without proportional gain in solution quality. Effective wrapping therefore requires discrimination between physically meaningful features and geometrically incidental details. This balance highlights the necessity of guiding wrapping decisions with physical understanding rather than arbitrary parameter selection. The engineer must consider which geometric scales influence the dominant transport mechanisms in the simulation. In external aerodynamics, leading edges and separation zones warrant higher fidelity, whereas smooth fuselage sections may tolerate coarser representation. In internal flows, narrow passages and abrupt expansions require accurate definition, while large uniform ducts may not. Thoughtful parameter selection ensures that computational effort is directed toward resolving phenomena that affect solution accuracy.

6.4.10 Application domains and broader relevance

Surface wrapping plays a foundational role in modern computational engineering and has become an indispensable stage in simulation-driven design workflows across multiple industrial sectors. Its relevance extends far beyond simple geometry repair. It represents a systematic transformation of engineering assemblies into numerically robust computational domains. In automotive aerodynamics, surface wrapping enables rapid and reliable evaluation of vehicle exterior performance. Production-level vehicle CAD models typically contain intricate internal components, mounting brackets, sealing elements, and manufacturing tolerances that are irrelevant to external airflow analysis. The wrapping process reconstructs a clean aerodynamic envelope by eliminating interior complexity while preserving the true outer shape of the vehicle. This abstraction allows engineers to investigate drag, lift, underbody flow behaviour, wake structures, and thermal management performance with improved stability and reduced preprocessing time. As development cycles in the automotive industry continue to shorten, automated and reliable surface reconstruction has become essential for iterative aerodynamic optimisation. In aerospace engineering, the importance of wrapping is even more pronounced due to the geometric sophistication of aircraft assemblies. Fuselage sections, engine nacelles, wing body junctions, control surfaces, auxiliary intake doors, and high lift devices involve complex curvature transitions and tightly integrated components. Raw CAD models frequently contain intersecting surfaces, small mechanical details, and internal compartments that compromise direct mesh generation. Surface wrapping reconstructs a watertight external shell that accurately captures aerodynamic curvature while removing structural details that do not influence flow physics. This ensures stable boundary layer resolution and consistent pressure distribution prediction. In propulsion systems, wrapping can also extract internal flow passages for intake and exhaust analysis by isolating the relevant cavities. The ability to selectively preserve external or internal domains allows aerospace engineers to transition efficiently between aerodynamic and internal flow simulations without extensive geometry modification. In marine hydrodynamics, the quality of hull representation directly influences wave formation, resistance prediction, and flow separation behaviour. Ship hull geometries often contain discontinuities, design seams, and structural supports that are not intended to represent fluid boundaries. Surface wrapping reconstructs a smooth, continuous outer hull surface that eliminates minor imperfections while maintaining global geometric fidelity. This clean representation is critical for the accurate prediction of hydrodynamic resistance and wake characteristics. Even small geometric discontinuities can alter pressure distribution and induce artificial turbulence. Wrapping therefore ensures that the simulated hydrodynamic performance reflects the intended design rather than CAD artifacts. In electronics cooling applications, surface wrapping provides a different but equally critical function. Electronic assemblies often include numerous screws, connectors, circuit board details, and micro-scale structural features that have negligible impact on global thermal performance. Attempting to mesh these details dramatically increases computational cost without improving predictive accuracy. Surface wrapping removes irrelevant mechanical elements while preserving heat-generating components and heat transfer surfaces. This selective abstraction enables efficient thermal analysis of enclosures, heat sinks, and airflow channels. By maintaining only thermally meaningful surfaces, the engineer can focus computational resources on temperature gradients and convective pathways that govern device reliability. The relevance of wrapping extends further into multiphysics and structural fluid interaction simulations. In coupled analyses, consistent geometric interfaces between fluid and solid domains are mandatory. Any discontinuity or mismatch at the interface can introduce numerical instability and compromise coupling accuracy. Surface wrapping ensures that the fluid domain boundary is continuous and watertight, providing a clean interface for load transfer, heat exchange, or momentum coupling between domains. In such simulations, wrapping is not merely a convenience but a prerequisite for stable and physically meaningful interaction modelling. Despite its automation capabilities, successful wrapping is not a purely mechanical process. It requires informed parameter selection guided by engineering judgement. The starting point of any wrapping procedure is the definition of an appropriate base size. This parameter should be aligned with the smallest geometrical feature that has physical relevance to the simulation objective. Selecting a base size that is too coarse may smooth out critical curvature or eliminate small but important flow passages. Conversely, selecting an excessively small base size may preserve insignificant manufacturing details and unnecessarily increase computational cost. Visualisation immediately after wrapping is an essential verification step. The engineer should inspect the reconstructed surface to ensure that important geometric features are retained and that unintended cavities have not been sealed incorrectly. If critical features appear overly smoothed or distorted, local surface controls may be applied to refine specific regions without modifying the entire domain. Surface control strategies allow targeted enhancement of geometric fidelity in areas of high physical sensitivity, such as leading edges, intake openings, cooling channel entrances, or sharp curvature transitions. On the other hand, if the wrapped geometry retains excessive detail in regions that do not influence the physics of interest, the base size may be increased or curvature refinement parameters adjusted. The goal is to achieve a balanced abstraction that captures essential geometry while eliminating computationally irrelevant features. This iterative refinement process is central to efficient simulation practice. It prevents both underrepresentation and overrefinement.

6.5 Wrapping of fuselage in STAR CCM+

Surface wrapping in STAR CCM+ can be understood as an advanced geometric reconstruction technique that systematically converts complex and imperfect CAD assemblies into numerically stable, watertight computational surfaces. In practical engineering workflows, the geometry created for manufacturing is rarely directly suitable for simulation. It often contains gaps, overlaps, unresolved intersections, duplicated faces, and tolerance inconsistencies. The surface wrapping methodology resolves these issues by performing volumetric sampling of the geometry, applying adaptive refinement strategies, and reconstructing a continuous triangulated surface that preserves essential physical features while eliminating noncritical details.

This process effectively creates a bridge between design intent and numerical analysis. By defining an appropriate base size, selecting a suitable volume of interest, and applying localised surface controls when necessary, the engineer can tailor the abstraction level of the geometry according to the governing physics of the problem. The wrapper therefore does not merely repair geometry; it filters and interprets it through a computational lens. As a result, numerical stability is improved, boundary condition application becomes more reliable, and the quality of the subsequent volume mesh is significantly enhanced.

The scientific relevance of surface wrapping extends beyond preprocessing convenience. High-quality wrapped surfaces directly influence mesh orthogonality, skewness, and smoothness, which in turn affect solver convergence and predictive accuracy. In demanding engineering systems such as aircraft auxiliary intake doors, internal cooling channels, or vehicle external aerodynamics, selective wrapping strategies allow local refinement in flow-sensitive regions while maintaining a coarser representation elsewhere. This balance ensures accurate physical resolution without unnecessary computational expense.

Although many commercial and open source tools provide surface wrapping capabilities, STAR CCM+ is widely recognised as one of the most established solutions in industrial practice. Its wrapping algorithm is computationally efficient, capable of parallel execution, and designed to handle large-scale assemblies with high robustness. The level of controllability is substantial, enabling detailed adjustment of base size, minimum and maximum surface resolution, feature capture thresholds, and volume of interest selection. At the same time, the workflow remains structured and relatively straightforward to configure, which reduces setup time compared with several traditional geometry repair tools.

To begin the workflow, the STAR-CCM+ software must first be launched and the required case opened as a .sim model, ensuring that the simulation environment is properly loaded and ready for further preprocessing and setup operations (see Fig. 6.3).

Before importing any external geometry into STAR-CCM+, the user must navigate through the software's file management interface to access the appropriate import tools (as shown in Fig. 6.4). This step is essential because STAR-CCM+ supports multiple data formats—such as surface meshes, volume meshes, and CAD models—and selecting the correct import pathway ensures that the geometry is brought into the simulation environment with the intended structure and fidelity. To begin this process, the user opens the File menu from the main toolbar, which reveals a set of organised import options. These options allow the user to choose whether they want to bring in a surface mesh, a volume mesh, or a full CAD model, depending on the requirements of the preprocessing workflow. The highlighted selection in the menu demonstrates the specific command used to import a surface mesh, guiding the user towards the correct action for preparing the geometry for subsequent cleaning, wrapping, or meshing operations.

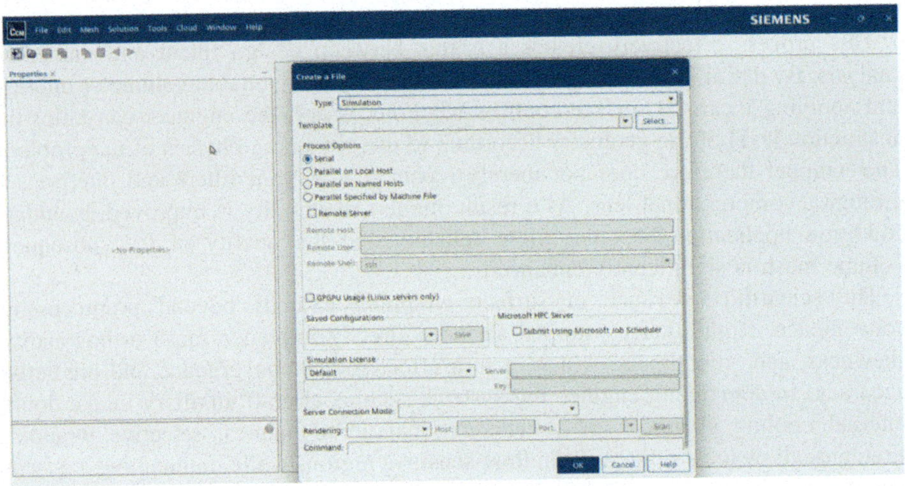

Figure 6.3 Launching the STAR CCM+ simulation environment.

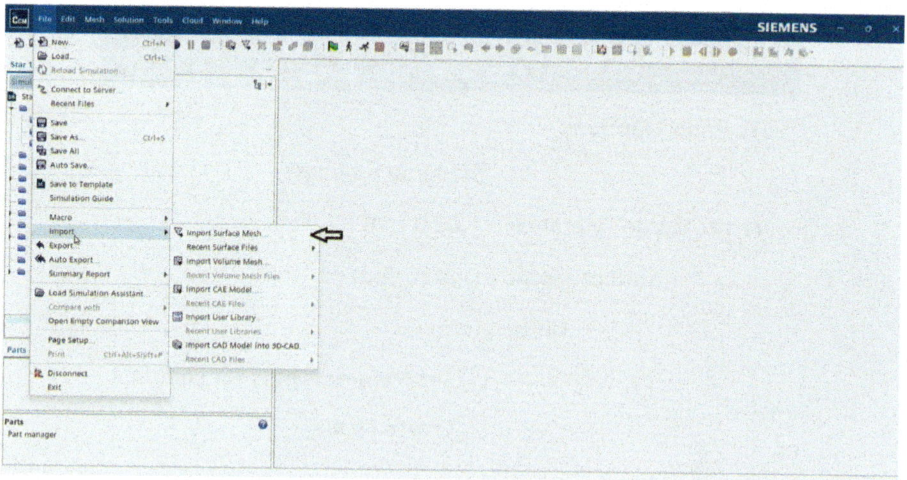

Figure 6.4 File menu interface showing geometry import options in STAR CCM+.

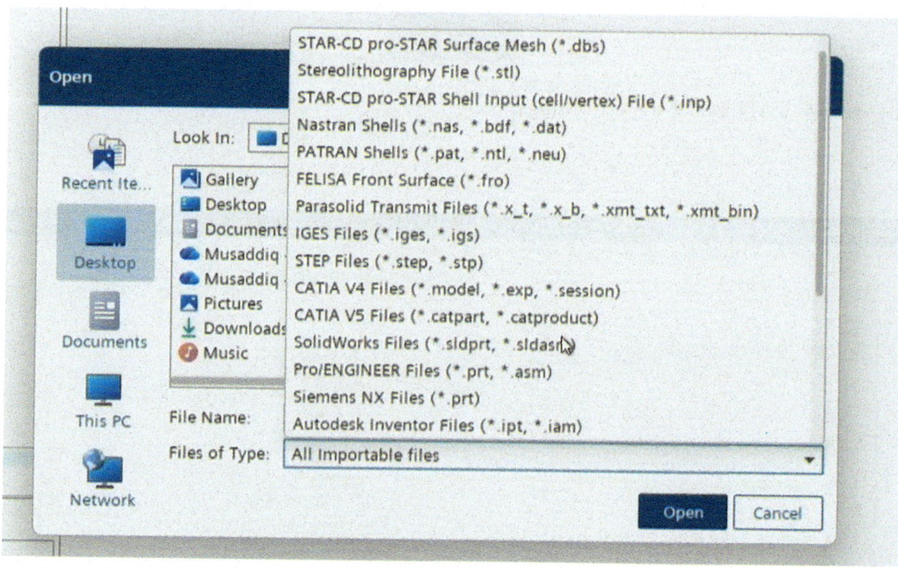

Figure 6.5 Supported model import formats and geometry selection dialogue.

Once the user selects the option to import a surface mesh, STAR-CCM+ opens a file-selection dialogue that allows the engineer to browse and choose from a wide range of supported geometry formats, as shown in Fig. 6.5.

After importing the geometry, you can adjust settings such as element size and tolerance, both of which play a crucial role in ensuring a clean and accurate CAD model import, as illustrated in Fig. 6.6.

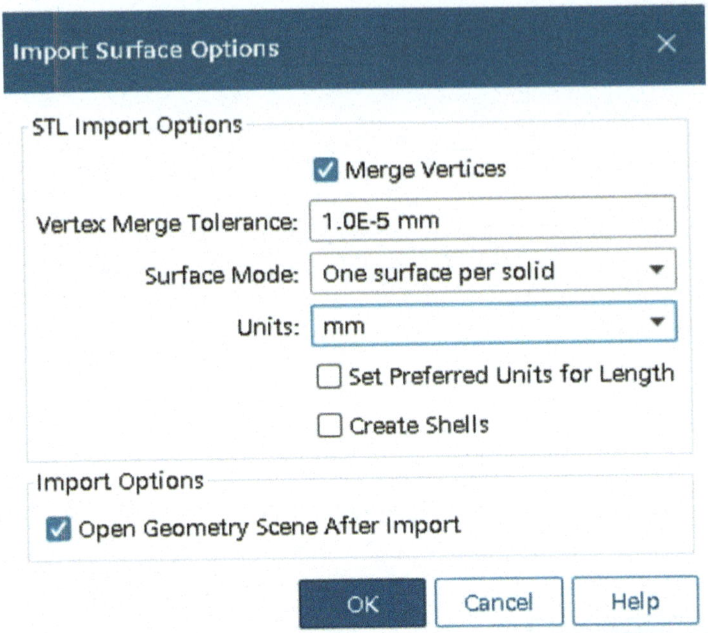

Figure 6.6 CAD model import settings.

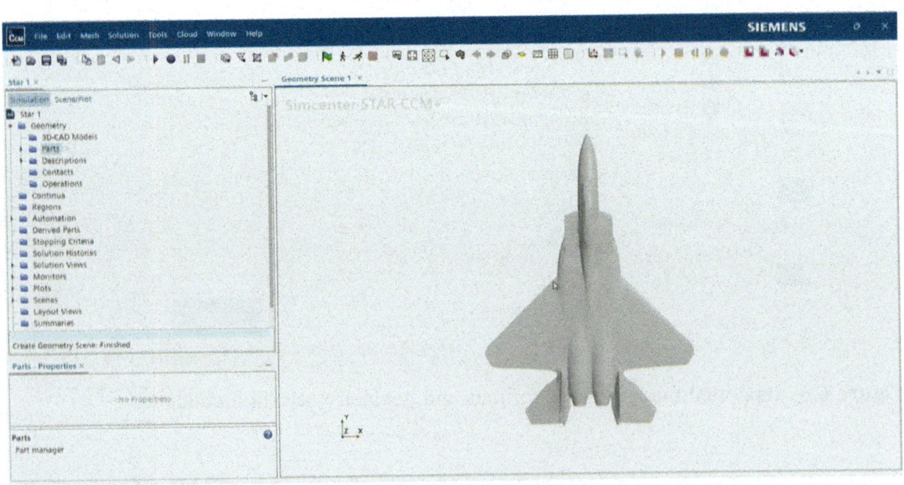

Figure 6.7 Isosurface visualisation within STAR CCM+ for geometric inspection.

STAR-CCM+ offers one of the most advanced visualisation environments available, particularly when it comes to rendering complex surfaces and applying high-quality colouration for detailed inspection and analysis as illustrated in Fig. 6.7.

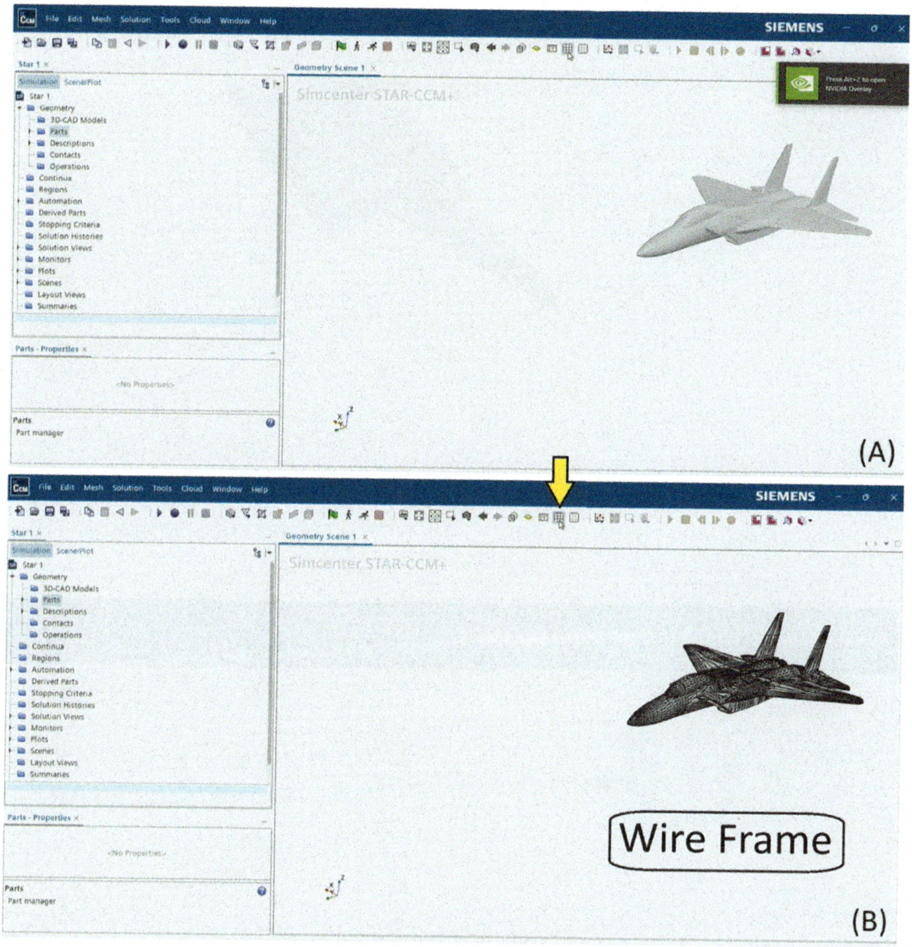

Figure 6.8 Visualisation modes: (A) isosurface rendering and (B) wireframe display of the geometry.

To display the mesh, simply select the mesh-view icon, which switches the geometry from a shaded representation to a wireframe view for detailed inspection (as shown in Fig. 6.8).

As shown in Fig. 6.9, at this stage of the workflow, the user navigates to the surface-preparation tools within STAR-CCM+ to initiate the wrapping process. This menu provides access to several geometry-cleaning and preprocessing operations, and selecting the surface wrapper option is the key step that launches the automated wrapping procedure.

After selecting the surface-wrapping operation, the next step is to choose the specific part on which the wrapping process will be applied (as shown in Fig. 6.10).

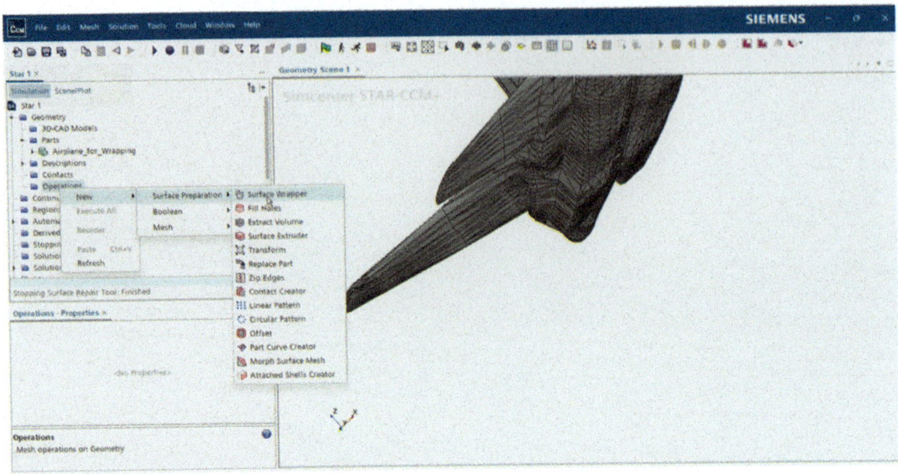

Figure 6.9 Accessing the surface wrapper operation within the surface preparation tools.

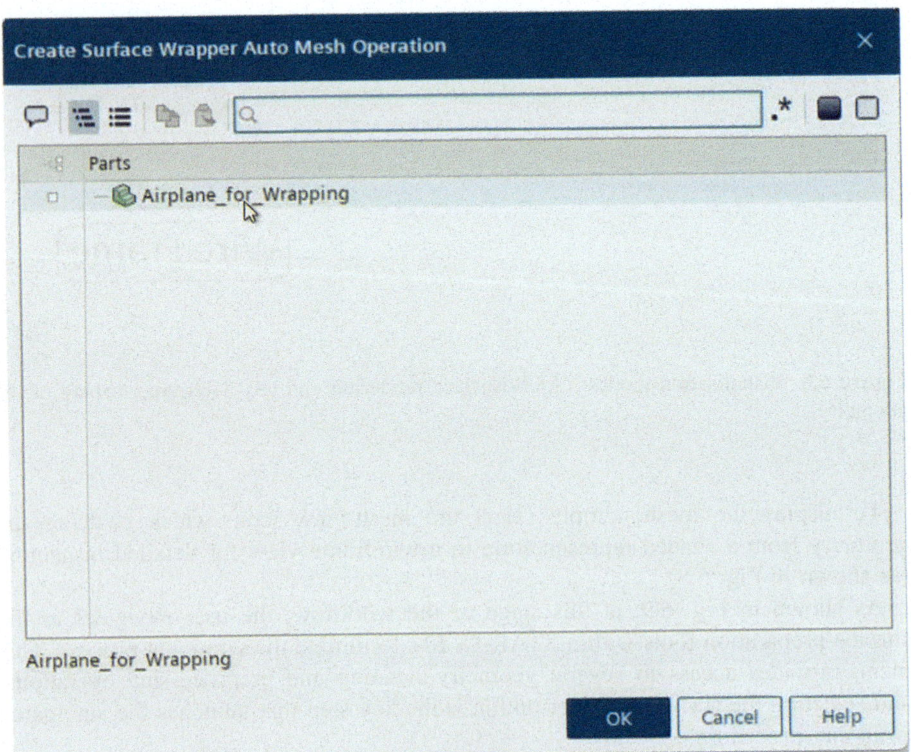

Figure 6.10 Part selection interface prior to initiating surface wrapping.

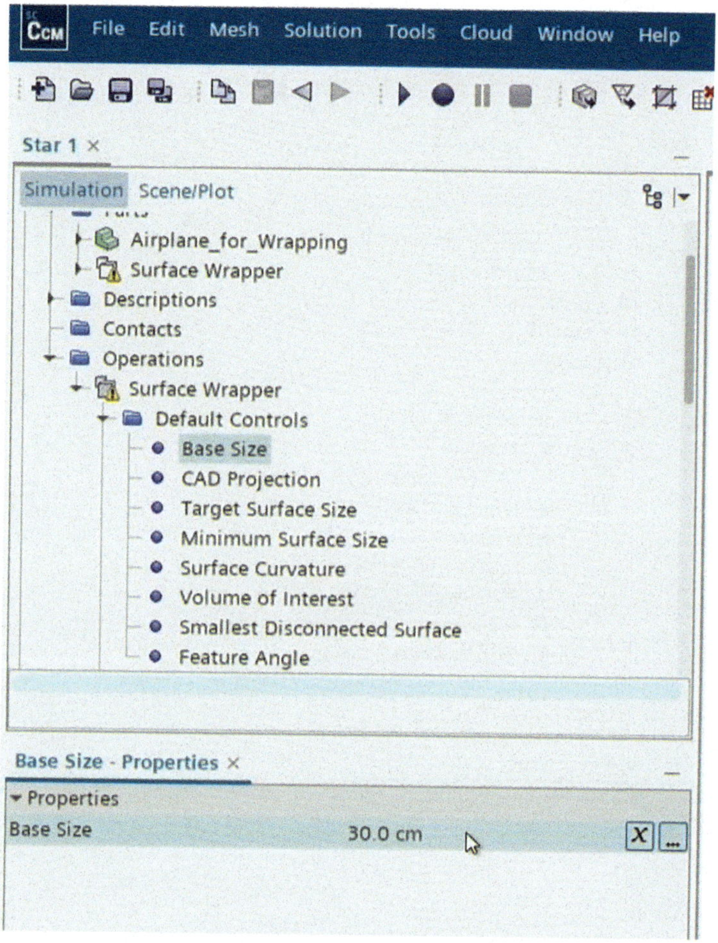

Figure 6.11 Base mesh size definition for surface wrapping control.

After this step, the surface-wrapping settings must be configured (as shown in Fig. 6.11), beginning with the selection of the base mesh size. This base size serves as the reference from which the maximum and minimum element sizes are determined, allowing the user to measure, adjust, and apply consistent sizing parameters. In many cases, the base size can also be used as a universal mesh-size control for the entire wrapping process.

Fig. 6.12 illustrates the available options for defining the Wrapping Volume of Interest in STAR CCM+. The Volume of Interest determines how the surface wrapper interprets the geometry and, more specifically, which region of the model is treated as the computational domain. In practical terms, this setting defines what the algorithm preserves and what it removes during the reconstruction of a watertight surface. The choice is particularly important for complex assemblies that contain

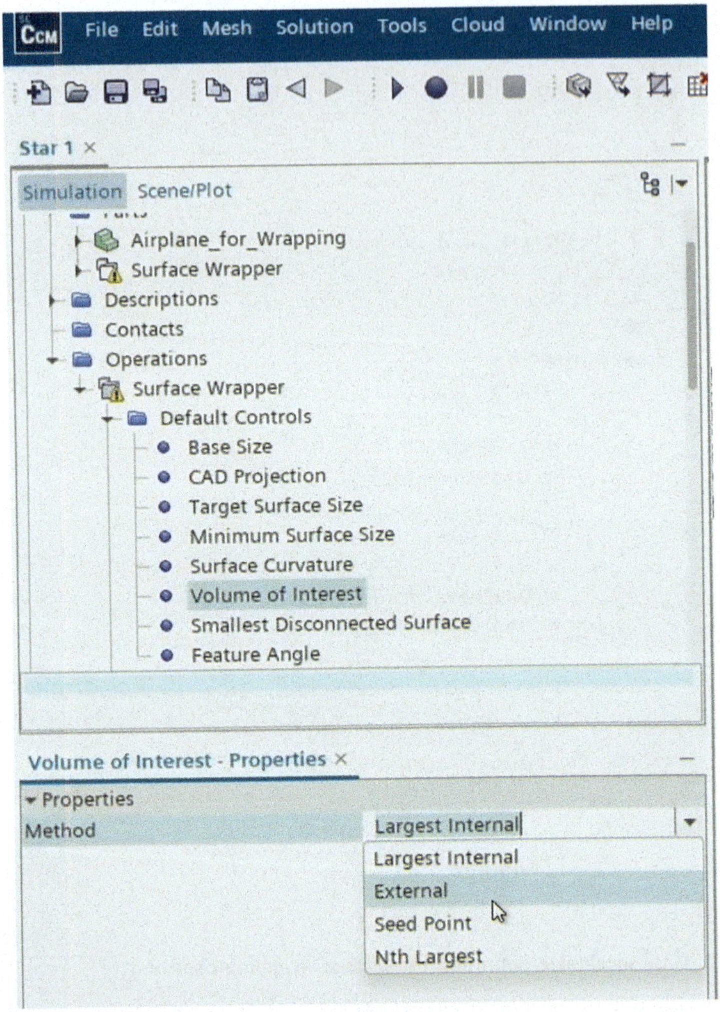

Figure 6.12 Wrapping Volume of Interest options: Largest Internal, External, Seed Point, and Nth Largest.

internal cavities, thin clearances, nested parts, or multiple disconnected volumes. The Volume of Interest acts as a filtering mechanism. It instructs the wrapper to identify the physically relevant domain while disregarding unnecessary geometric details. Each option produces a different interpretation of the model and therefore leads to a different final surface representation.

6.5.1 Largest Internal

The Largest Internal option directs STAR CCM+ to automatically detect all enclosed cavities within the geometry and rank them according to volume. The largest cavity is then selected as the primary region of interest. This configuration is typically used for internal flow simulations, such as flow inside ducts, manifolds, pipes, or cooling passages, where the fluid domain exists within a surrounding solid structure. Smaller voids, isolated pockets, manufacturing gaps, or irrelevant internal spaces are ignored unless specifically preserved. This automatic filtering makes the option highly practical for industrial CAD models that contain numerous small imperfections. It reduces unnecessary mesh generation in regions that do not influence the physical problem, thereby improving efficiency and stability.

6.5.2 External

The External option instructs the wrapper to retain only the outermost surface of the geometry. All internal cavities, embedded components, and interior details are removed. The algorithm effectively seals openings and reconstructs a continuous outer shell. This option is most appropriate for external flow simulations, such as aerodynamic analysis of vehicles, aircraft, drones, or buildings. Even if the CAD model includes internal structural details, bolt holes, engine compartments, or hollow regions, these are automatically excluded. The result is a smooth, watertight external boundary suitable for computational meshing. This approach is particularly advantageous when working with complex assemblies where internal features are irrelevant to the external flow field.

6.5.3 Seed Point

The Seed Point option provides direct user control over domain selection. Instead of relying on the automatic ranking of cavities, the user specifies a point located inside the desired region. The wrapper then identifies the connected volume surrounding that point and treats it as the Volume of Interest. This method is especially useful when multiple internal cavities have comparable sizes or when the automatically detected largest volume is not the one required for simulation. By carefully positioning the seed point, the engineer can isolate a specific chamber, passage, or subvolume without performing manual geometry cleanup. This approach offers precision and flexibility in complex geometrical configurations.

6.5.4 Nth Largest

The Nth Largest option extends the automatic cavity selection mechanism. Rather than selecting only the largest internal cavity, the user can choose the second, third, or any other ranked internal void. This feature is valuable for geometries containing several meaningful internal domains, such as multichamber devices, heat exchangers, or systems with parallel flow passages. It allows selective targeting of specific cavities based on their relative size, enabling accurate domain extraction without modifying the original CAD assembly.

In this configuration, only the external aerodynamic envelope of the fuselage is required for simulation; therefore, the **External** option within the Volume of Interest settings is selected. This choice instructs the wrapper to discard all internal cavities

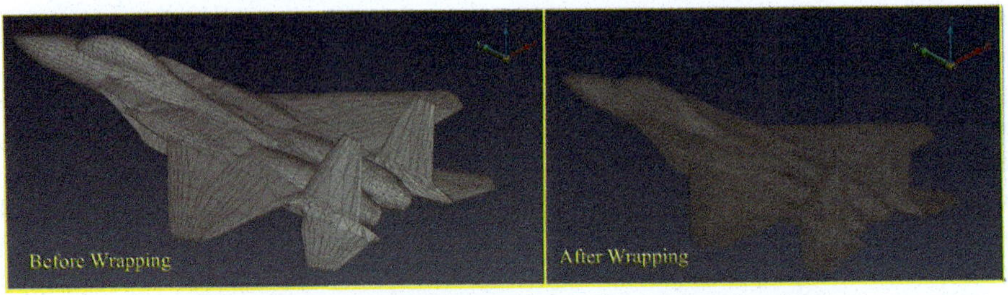

Figure 6.13 Comparison of surface mesh quality before and after wrapping.

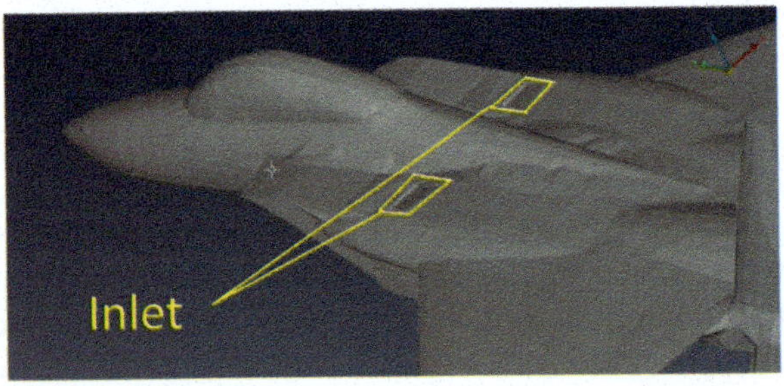

Figure 6.14 Auxiliary intake doors region requiring localised surface refinement.

and structural details, reconstructing a clean and continuous outer shell suitable for external flow analysis.

The resulting wrapped surface exhibits a high standard of mesh quality. Surface continuity is fully restored, triangulation is uniform and well distributed, and previously existing geometric defects such as gaps, overlaps, and inconsistent intersections are eliminated. This reconstructed envelope provides a watertight, topologically coherent boundary that supports stable volume meshing and reliable aerodynamic prediction, as illustrated in the corresponding Fig. 6.13.

In this case, the objective is to investigate the aerodynamic behaviour of the auxiliary intake doors, which open automatically to allow additional airflow to enter from the upper or lateral regions of the fuselage. Because these components directly influence local pressure distribution, flow separation, and transient air ingestion patterns, a refined surface resolution is required to accurately capture the relevant flow physics. However, refining the entire geometry would significantly increase computational time and resource consumption. To maintain efficiency while preserving accuracy, localised mesh refinement can be applied exclusively to the intake door region using surface control settings, as illustrated in Fig. 6.14.

Fig. 6.15 illustrates the Surface Control configuration menu within STAR CCM+. This option is accessed under the mesh operation settings and allows the user to define localised surface refinement parameters.

By selecting *New* → *Surface Control*, the engineer can assign specific mesh size criteria to chosen parts or surfaces of the geometry. This enables precise

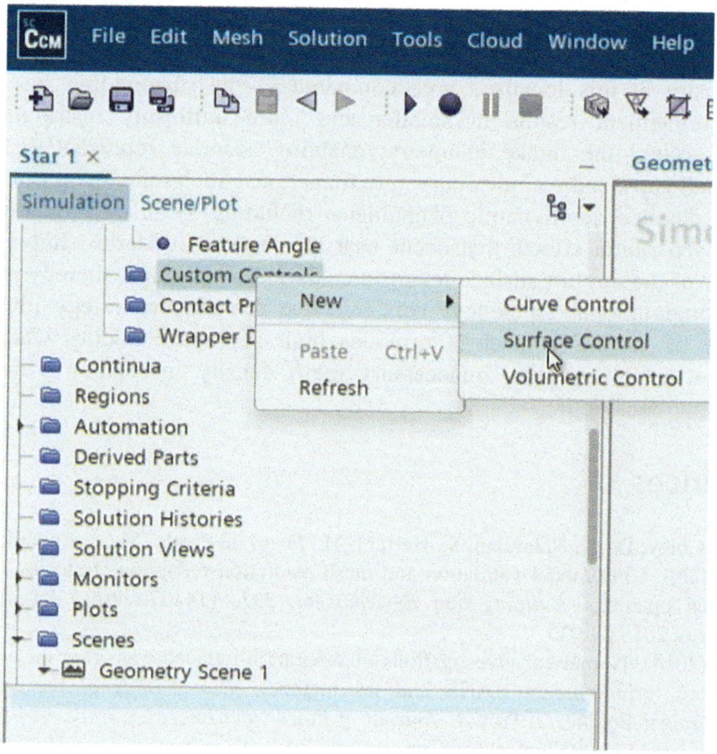

Figure 6.15 Surface Control configuration menu for localised mesh refinement.

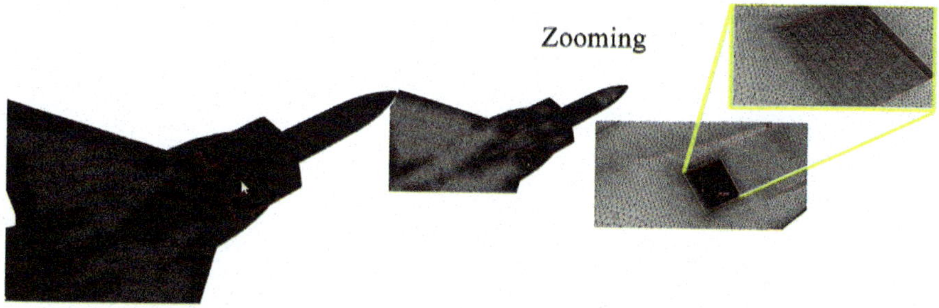

Figure 6.16 Wrapped surface with localised refinement applied to the intake door region.

adjustment of surface resolution in regions where higher geometric fidelity is required, such as aerodynamic control surfaces, intake openings, sharp edges, or flow-sensitive components.

Fig. 6.16 demonstrates the effect of applying localised surface control during the wrapping process. The global wrapped surface is first generated using the defined base size; however, additional refinement has been introduced specifically in the region of the auxiliary intake doors. The zoomed views clearly show a higher surface mesh density in this localised area compared to the surrounding fuselage. This targeted refinement results in smaller and more uniformly distributed surface elements around the intake geometry, enabling accurate representation of sharp edges, small gaps, and complex curvature transitions. Such refinement is essential for resolving detailed aerodynamic phenomena, including localised pressure gradients and flow separation effects that occur near the intake boundaries. Importantly, the remainder of the aircraft surface retains a coarser discretisation, thereby maintaining overall computational efficiency. This selective refinement strategy illustrates the advantage of surface control: it provides high geometric fidelity where physics demands it while avoiding unnecessary mesh density in regions with relatively smooth flow behaviour.

References

Fosas, D., Coley, D. A., Natarajan, S., Herrera, M., Fosas de Pando, M., & Ramallo-Gonzalez, A. (2026). CFD model validation and mesh resolution reduction strategies for complex louvred openings. *Building and Environment, 291*, 114207. https://doi.org/10.1016/j.buildenv.2018.07.033.

Marty, J. (2014). Numerical investigations of separation-induced transition on high-lift low-pressure turbine using RANS and les methods. *Proceedings of the Institution of Mechanical Engineers, Part A: Journal of Power and Energy, 228*(8), 924–952. https://doi.org/10.1177/0957650914548741.

Pei, G., & Rim, D. (2021). Quality control of computational fluid dynamics (CFD) model of ozone reaction with human surface: Effects of mesh size and turbulence model. *Building and Environment, 189*(1), 107513. https://doi.org/10.1016/j.buildenv.2020.107513.

Sloshing simulations

7

7.1 Numerical modeling framework in STAR CCM+

STAR-CCM+ is a high-fidelity computational fluid dynamics platform developed by Siemens Digital Industries Software (see Fig. 7.1). It is widely regarded as one of the most reliable industrial simulation environments for modelling fluid flow, heat transfer, multiphase dynamics, and strongly coupled multiphysics systems. The software integrates geometry preparation, mesh generation, physics definition, numerical solution, post-processing, and design exploration within a single unified framework. This integration significantly reduces workflow fragmentation and enhances numerical consistency, which is essential for reproducible engineering analysis.

The historical foundation of STAR-CCM+ can be traced to earlier computational fluid dynamics (CFD) solvers such as STAR-CD and CCM+, which were extensively used in the automotive and energy industries. These legacy systems demonstrated strong predictive capabilities, particularly in engine flows and thermal management applications. The development of STAR-CCM+ represented a structural redesign of the software architecture into a modern object-oriented environment, eliminating the need for external meshing tools and disconnected preprocessing stages.

From a numerical perspective, STAR-CCM+ is based on the finite volume method, solving the conservation equations of mass, momentum, and energy over discretized control volumes. This formulation ensures strict conservation properties and robust handling of complex geometries. The platform includes an extensive library of turbulence models, ranging from Reynolds-averaged Navier-Stokes approaches to Large Eddy Simulation. The availability of validated turbulence closures, combined with stable discretization schemes, contributes significantly to its strong correlation with experimental measurements. One of the major strengths of STAR-CCM+ is its advanced automated meshing technology. Polyhedral meshing and surface wrapping techniques allow complex industrial geometries to be prepared with minimal manual intervention. Polyhedral cells improve numerical stability and reduce cell count while maintaining accuracy. This feature is particularly valuable in large-scale simulations such as full vehicle aerodynamics or battery pack cooling systems, where computational efficiency is critical. The software supports a wide range of multiphysics capabilities. These include multiphase flow using Volume of Fluid methods, conjugate heat transfer, fluid structure interaction, radiation modelling, chemical reactions, and electrochemical battery simulation. Such integrated multiphysics modelling is essential in modern research and development, where thermal, mechanical, and fluid phenomena interact simultaneously. For example in electric vehicle battery systems, fluid cooling, heat generation, and structural integrity must be analysed within a unified computational environment.

Slosh Dynamics. DOI: https://doi.org/10.1016/B978-0-443-44330-5.00013-X

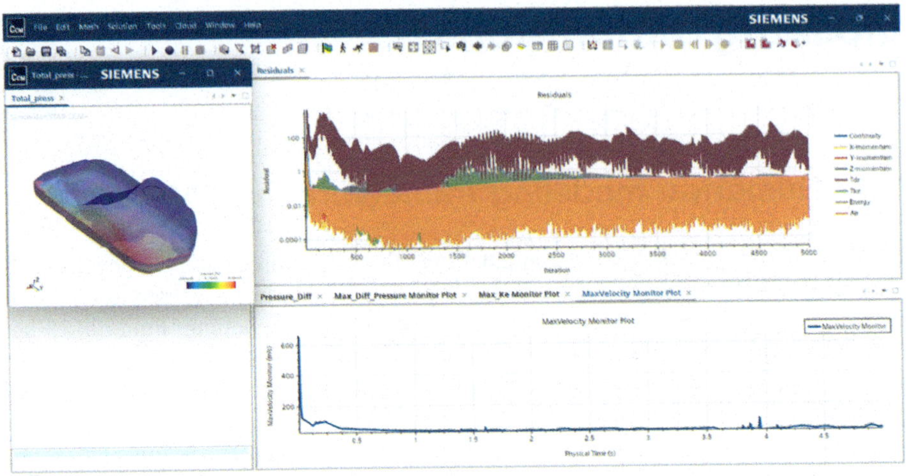

Figure 7.1 STAR CCM+ interface overview. STAR CCM+ interface overview.
No permission required.

After its acquisition by Siemens in 2016, the software expanded further into digital twin integration, battery electrochemistry modelling, and advanced multiphysics simulation.

A fundamental strength of STAR CCM+ lies not only in its numerical solvers but in its programmable architecture and integrated graphical environment. The software combines a comprehensive graphical user interface with an extensive object-oriented application programming interface based on Java. This dual structure enables both interactive modelling and fully script-driven automation within the same computational framework. The Java environment provides significant advantages for advanced users. Because STAR CCM+ is internally structured around a Java object model, every entity in the simulation tree, such as regions, boundaries, physics continua, reports, monitors, and field functions, can be programmatically accessed, modified, and generated. Java ensures platform independence, strong type safety, memory management robustness, and seamless integration with the native solver architecture. This allows reproducible workflows, parametric studies, and scalable batch execution. Importantly, the scalar field functions and logical expressions discussed later in this chapter are formulated within this Java-based logic structure, which guarantees consistency between user-defined automation and internal solver operations.

For users without a formal programming background, STAR CCM+ provides macro recording capabilities. Any sequence of graphical user interface (GUI) actions can be recorded and automatically translated into executable Java macros. These macros can then be edited, parameterised, and reused, effectively bridging the gap between graphical interaction and scripted automation. This feature allows non-programmers to reproduce complex workflows with high reliability. Although direct native support for VBA is not provided, full external automation is still achievable. By adopting a metaprogramming strategy, Excel VBA can be used as a supervisory control layer to orchestrate STAR CCM+ simulations. Through controlled command execution, file manipulation, and batch management, it is possible to automate the complete simulation cycle from geometry import and mesh generation to solution execution and structured report extraction. This level of external automation is feasible because of the stability, transparency, and scripting smoothness of the STAR CCM+ environment, which allows deterministic execution of programmable tasks across large parametric studies. The software supports a wide range of multiphysics capabilities. These include multiphase flow using Volume of Fluid methods, conjugate heat transfer, fluid–structure interaction, radiation modelling, chemical reactions, and electrochemical battery simulation. Such integrated multiphysics modelling is essential in modern research and development, where thermal, mechanical, and fluid phenomena interact simultaneously. For example in electric vehicle battery systems, fluid cooling, heat generation, and structural integrity must be analysed within a unified computational environment.

Industrially, STAR-CCM+ is extensively applied in automotive, aerospace, aviation, and energy sectors. In automotive engineering, it is used for aerodynamic drag reduction, underhood thermal management, electric motor cooling, and fuel cell water management. In aerospace, it supports wing optimisation, rotorcraft aerodynamics, and propulsion system analysis. In battery technology, it enables thermal runaway analysis and coolant channel design optimisation. Across these applications, the reliability of the software is reinforced by decades of validation against experimental campaigns, establishing strong confidence in its predictive performance.

Beyond direct simulation, STAR-CCM+ also functions as a design exploration environment. Parametric studies, automated reporting, and optimisation algorithms allow systematic evaluation of engineering configurations. This transforms the simulation from a verification tool into a driver of conceptual design. Engineers can perform virtual testing campaigns that significantly reduce the need for physical prototypes, thereby accelerating development cycles.

As such, the sloshing simulations conducted in this study were implemented using STAR CCM+, a finite volume-based computational fluid dynamics platform capable of resolving multiphase free surface flows under transient conditions. The modelling strategy was constructed to capture the coupled air-fuel interaction inside a partially filled tank subjected to dynamic excitation.

The physical model selection constitutes the first critical step. For sloshing problems, the flow must be treated as transient, three-dimensional, and incompressible, with gravity explicitly included. Because two immiscible fluids are present, namely liquid and air, a multiphase formulation is required. In this case, an Eulerian multiphase framework with an interface capturing method was employed to resolve the evolving free surface.

Industrially, STAR-CCM+ is extensively applied in automotive, aerospace, aviation, and energy sectors. In automotive engineering, it is used for aerodynamic drag reduction, underhood thermal management, electric motor cooling, and fuel cell water management. In aerospace, it supports wing optimisation, rotorcraft aerodynamics, and propulsion system analysis. In battery technology, it enables thermal runaway analysis and coolant channel design optimisation. Across these applications, the reliability of the software is reinforced by decades of validation against experimental campaigns, establishing strong confidence in its predictive performance.

Beyond direct simulation, STAR-CCM+ also functions as a design exploration environment. Parametric studies, automated reporting, and optimisation algorithms allow systematic evaluation of engineering configurations. This transforms the simulation from a verification tool into a driver of conceptual design. Engineers can perform virtual testing campaigns that significantly reduce the need for physical prototypes, thereby accelerating development cycles.

As such, the sloshing simulations conducted in this study were implemented using STAR CCM+, a finite volume-based computational fluid dynamics platform capable of resolving multiphase free surface flows under transient conditions. The modelling strategy was constructed to capture the coupled air-fuel interaction inside a partially filled tank subjected to dynamic excitation.

The physical model selection constitutes the first critical step. For sloshing problems, the flow must be treated as transient, three-dimensional, and incompressible, with gravity explicitly included. Because two immiscible fluids are present, namely liquid and air, a multiphase formulation is required. In this case, an Eulerian multiphase framework with an interface capturing method was employed to resolve the evolving free surface.

The selected models include:

- Three-dimensional unsteady formulation
- Segregated flow solver
- Multiphase interaction
- Volume of Fluid method for interface tracking
- Gravity body force
- Turbulence modelling using Reynolds-averaged Navier-Stokes equations

This combination allows accurate prediction of the liquid free surface deformation, wave formation, and impact pressures on tank walls during oscillatory motion.

After establishing the governing physical models, the thermophysical properties of each phase must be defined accurately. In sloshing simulations, the density ratio between liquid and gas phases strongly influences interface dynamics, wave amplitude, and impact loads.

For this demonstration, two phases were considered:

1. Water is the primary liquid phase
2. Air is the secondary gaseous phase

The material properties were defined as constant under isothermal conditions, since thermal effects were not the primary focus of this hydrodynamic investigation.

For air, the following properties were specified:

- Density
- Dynamic viscosity

- Molecular weight
- Specific heat
- Thermal conductivity

The liquid phase was defined using density and viscosity values consistent with standard ambient conditions to ensure physical realism in the hydrodynamic response. Because multiphase sloshing behaviour is highly sensitive to density contrast, viscous dissipation, and surface tension effects, it is essential to explicitly document the thermophysical parameters adopted in the simulation.

Although thermal properties such as specific heat and conductivity are not dominant in pure sloshing problems, their inclusion ensures numerical stability and model completeness within the multiphase solver framework.

Accurate specification of viscosity is particularly important because it affects:

- Damping of free surface oscillations
- Boundary layer development near tank walls
- Energy dissipation during impact

The density contrast between liquid and air governs buoyancy-driven interface curvature and determines the inertial response of the fluid mass under acceleration (Fig. 7.2).

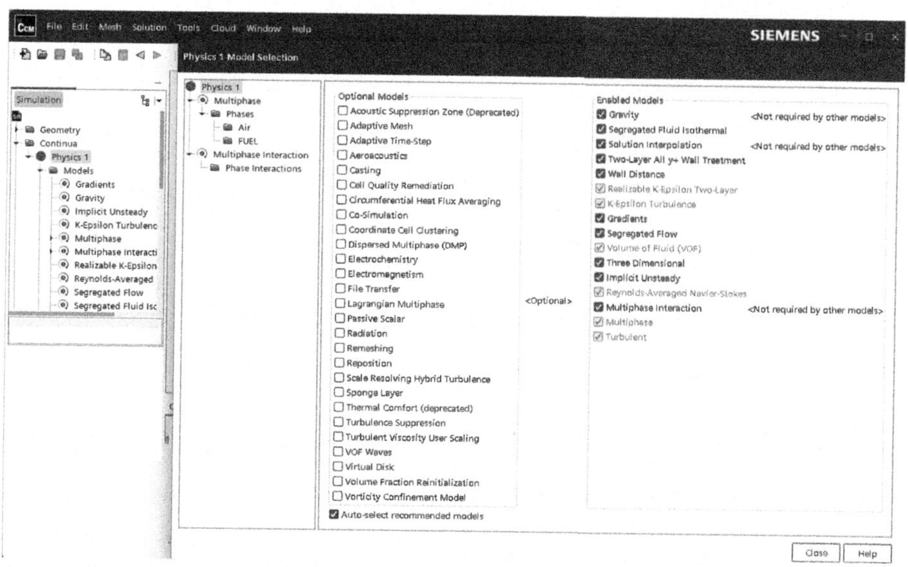

Figure 7.2 Physics model configuration for transient multiphase sloshing simulation in STAR CCM+.
No permission required.

7.2 Solver configuration and thermodynamic assumptions

7.2.1 Segregated isothermal flow formulation

The multiphase sloshing simulation was solved using a segregated flow formulation under isothermal conditions. In this approach, the governing equations for momentum and pressure are solved sequentially rather than simultaneously, which provides numerical stability and computational efficiency for large transient problems.

The simulation was conducted under the assumption of constant temperature, since thermal gradients do not significantly influence the hydrodynamic behaviour of sloshing in the present configuration. The reference temperature was explicitly set to 20°C (as the settings are shown in Fig. 7.3).

It is important not to rely on the software's default value. Instead, the reference temperature should be manually defined to reflect realistic ambient conditions. This ensures consistency in density and viscosity values when temperature-dependent properties are involved.

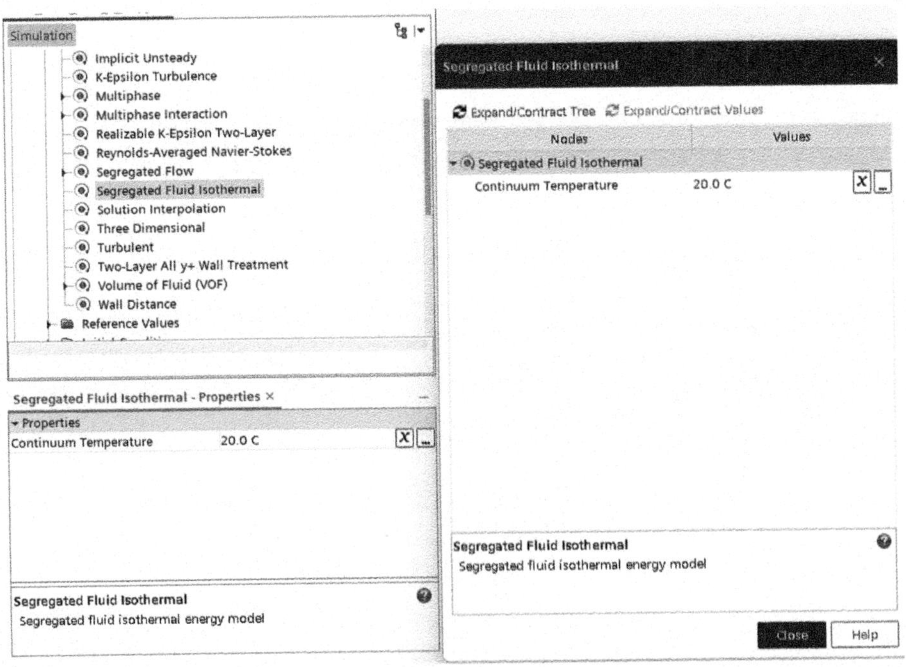

Figure 7.3 Solver configuration showing the segregated isothermal flow model with reference temperature set to 20°C.
No permission required.

Although the flow is modelled as isothermal, the explicit specification of temperature serves two purposes:

1. It guarantees consistency in thermophysical property evaluation.
2. It avoids ambiguity in post-processing and documentation of simulation parameters.

The segregated isothermal model therefore provides an efficient and physically consistent framework for capturing transient multiphase free surface motion without introducing unnecessary thermal complexity.

7.3 Initialisation of multiphase volume fraction

7.3.1 Concept of volume fraction in free surface modelling

In multiphase simulations using the Volume of Fluid method, the interface between liquid and air is represented through a scalar variable known as the volume fraction. This scalar field defines the proportion of each computational cell occupied by a given phase.

For example:

- Volume fraction equal to one indicates a cell fully occupied by liquid.
- Volume fraction equal to zero indicates a cell fully occupied by air.
- Intermediate values represent interface cells containing both phases.

Proper initialisation of the volume fraction field is critical in sloshing simulations because it defines the initial liquid fill level inside the tank.

The initial phase distribution was defined using a scalar field function created within the automation environment of the software. This allows a spatially dependent mathematical expression to describe the initial fluid height.

The procedure involves:

1. Accessing the automation panel.
2. Creating a new scalar field function.
3. Defining the volume fraction expression based on spatial coordinates.

This approach ensures:

- Precise control of the initial fill level.
- Sharp definition of the free surface.
- Reproducibility of the initial condition.

Using field functions (as shown in Fig. 7.4) is superior to manual assignment because it allows parameterisation and systematic variation of liquid height in parametric studies.

7.4 Definition of scalar field functions for parametric control

In computational fluid dynamics, a scalar field is a spatially distributed quantity that assigns a single numerical value to every point in the computational domain. Unlike

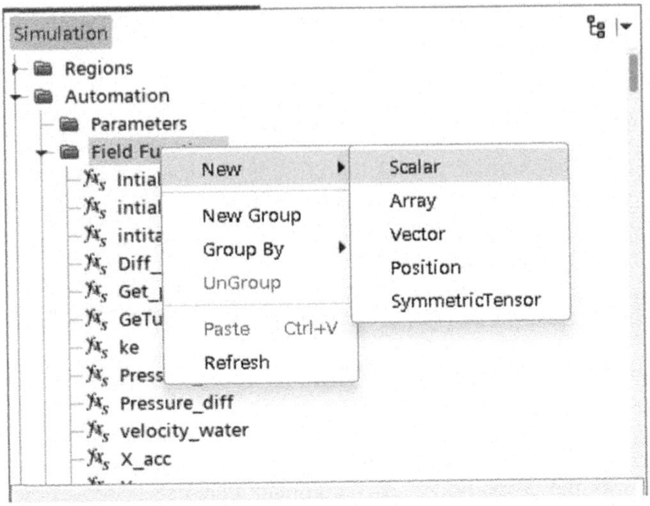

Figure 7.4 Creation of a scalar field function for initialising the volume fraction distribution. No permission required.

vector fields, which possess magnitude and direction, scalar fields represent physical quantities that are fully described by magnitude alone. Examples include pressure, temperature, density, and volume fraction.

Within STAR CCM+, scalar field functions serve as user-defined mathematical expressions that can depend on spatial coordinates, time, or other solution variables. They allow the analyst to construct derived quantities, define parametric variables, impose spatially varying initial conditions, and generate custom post-processing metrics. These functions are evaluated locally within each computational cell and can be updated dynamically during transient simulations.

The utility of scalar fields becomes particularly significant in multiphase sloshing simulations. In Volume of Fluid-based modelling, the interface between liquid and gas phases is represented through a scalar quantity known as the volume fraction. Each computational cell contains a scalar value between zero and one that indicates the proportion of liquid within that cell. The continuous evolution of this scalar field governs the deformation of the free surface and therefore determines wave formation, impact behaviour, and nonlinear oscillatory response.

Beyond interface tracking, scalar field functions enable parametric control of critical quantities such as initial fill height, reference pressure, or derived energy measures. By defining these parameters as scalars rather than embedding fixed constants directly into the model, the simulation becomes flexible and systematically adjustable. This is especially important in sloshing dynamics, where the natural frequency, resonance characteristics, and impact loads are highly sensitive to liquid depth and fill ratio.

Moreover, scalar fields enhance reproducibility and clarity in computational research. They allow transparent documentation of assumptions and facilitate systematic sensitivity analyses. In advanced sloshing studies, scalar field functions may also be used to compute pressure differences between monitoring points, evaluate kinetic energy density, or quantify free surface elevation relative to a reference level.

7.4.1 Preparation of user-defined scalar parameters

To ensure flexibility and reproducibility of the sloshing simulations, a set of user-defined scalar field functions was created. These scalar parameters serve as global variables that control initialisation, post-processing quantities, and derived physical fields.

The defined scalar quantities include:

- Initial air volume fraction
- Initial liquid volume fraction
- Initial liquid height
- Reference pressure
- Pressure difference
- Differentiated pressure term
- Kinetic energy-related parameter
- Liquid velocity magnitude

It is important to note that the order of creation of these scalar functions is not critical (as shown in Fig. 7.5). What is essential is the logical consistency of their definitions and their correct use in initialisation and evaluation expressions.

The introduction of scalar parameters provides several advantages:

1. Centralised control of physical quantities
2. Simplified modification of initial conditions
3. Enhanced clarity in parametric studies
4. Reduced risk of manual errors in repeated simulations

From a computational modelling perspective, defining explicit scalar variables transforms the simulation setup from a static configuration into a parametric framework suitable for systematic investigation.

7.4.2 Definition of initial liquid height parameter

Among the defined scalar variables, the initial liquid height in the tank plays a central role because it determines the fill ratio of the tank. The value of this parameter was obtained from the design criteria of the tank, where the geometric reference level was established.

For example the design of an automotive fuel tank is defined primarily by its nominal capacity, which represents the total internal volume enclosed by the tank structure (as shown in Fig. 7.6). However, only a portion of this volume is usable fuel, because a properly engineered tank also incorporates regions known as dead

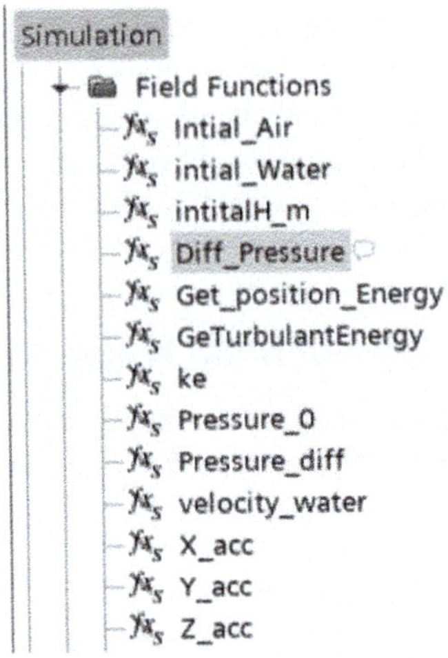

Figure 7.5 User-defined scalar field functions used for parametric control of the sloshing simulation.
No permission required.

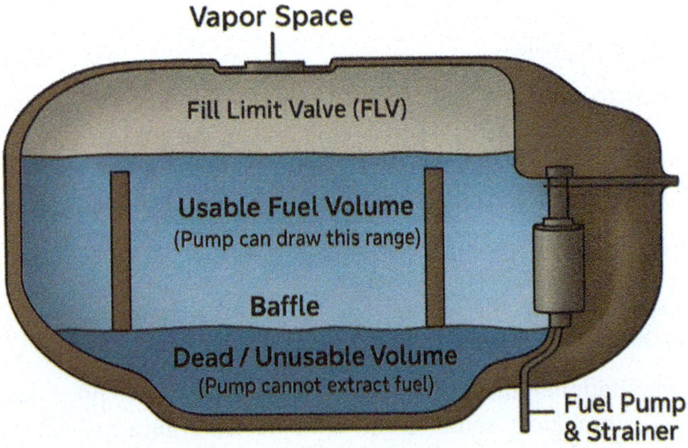

Figure 7.6 Fuel tank schematic showing vapour space and fuel volumes; initial liquid height defines fill ratio based on design reference geometry.
No permission required.

volume or unusable volume. This dead volume is intentionally included to ensure reliable fuel pickup under dynamic driving conditions, prevent pump starvation, and provide compliance with safety and evaporative-emission requirements. In a typical tank, the usable volume extends from the fill-limit valve level down to the lowest point at which the fuel pump module can continuously extract liquid fuel. Below this level, a small layer of fuel remains that cannot be drawn into the pump due to geometric constraints, pump-strainer clearance, and the need to keep the pump submerged for cooling and noise reduction. Additional dead volume is created by internal baffles, structural ribs, and vapour space required for thermal expansion – particularly important to prevent excessive pressure during hot-soak conditions. As a result, the rated capacity printed in vehicle specifications reflects the maximum fillable volume, while the operational capacity available to the driver is slightly lower. A well-designed fuel tank therefore balances packaging constraints, slosh-control geometry, pump-module performance, evaporative-emission requirements, and crash robustness to achieve a stable fuel supply across the entire driving envelope.

The scalar variable representing the initial liquid height was defined as a dimensionless or dimensional parameter, `initialH_m`, depending on the normalisation strategy adopted in the simulation (as shown in Fig. 7.7).

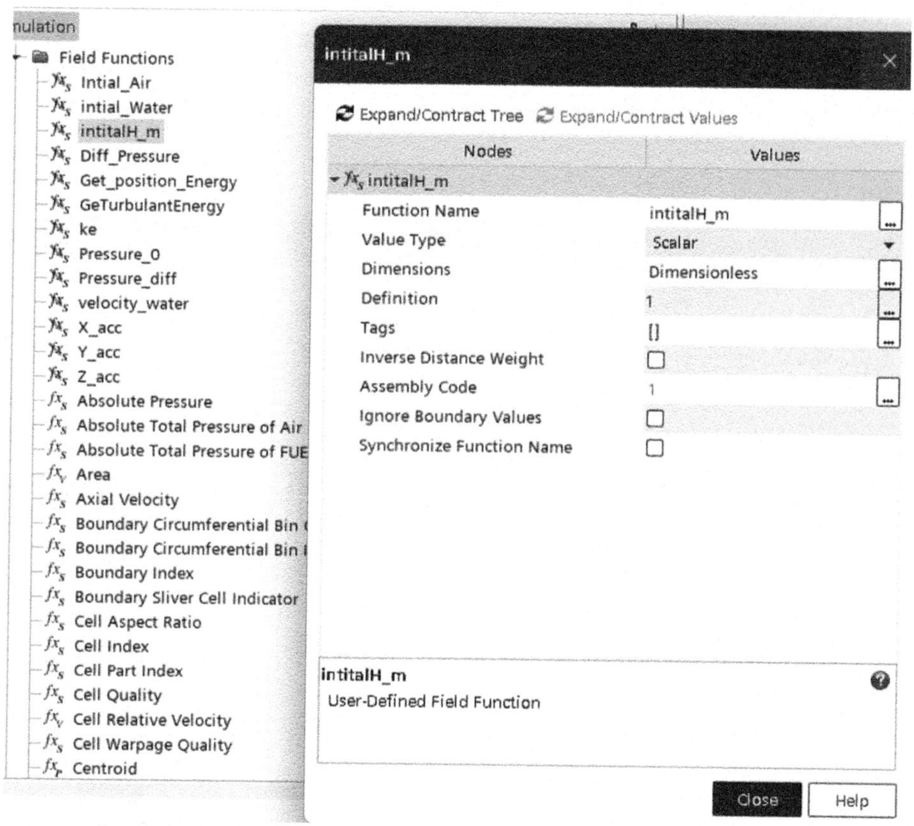

Figure 7.7 Definition of the initial liquid height parameter used for volume fraction initialisation.
No permission required.

This parameter is subsequently used in the volume fraction field function expression to define the spatial distribution of phases. For example the initialisation condition may be expressed conceptually as:

- Liquid phase volume fraction equals one below the specified height
- Liquid phase volume fraction equals zero above that height

By defining the height as a scalar parameter rather than hard-coding it directly into the field function, the model allows straightforward adjustment of fill levels without rewriting expressions.

This approach is particularly useful when:

- Performing sensitivity studies on fill ratio
- Investigating resonance behaviour under varying liquid depths
- Studying nonlinear sloshing regimes

The use of a scalar height parameter therefore enhances both numerical efficiency and experimental comparability.

The initialisation of the multiphase domain is performed using spatially dependent scalar field functions that define the initial distribution of liquid and air within the tank (as shown in Figs. 7.8 and 7.9).

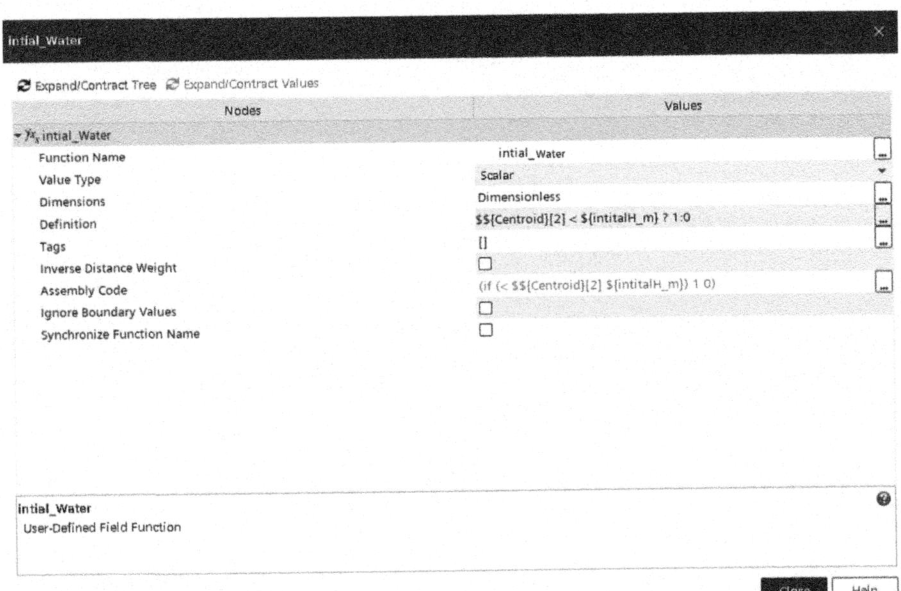

Figure 7.8 Field representation of the liquid phase describing the spatial distribution of liquid regions within the domain.
No permission required.

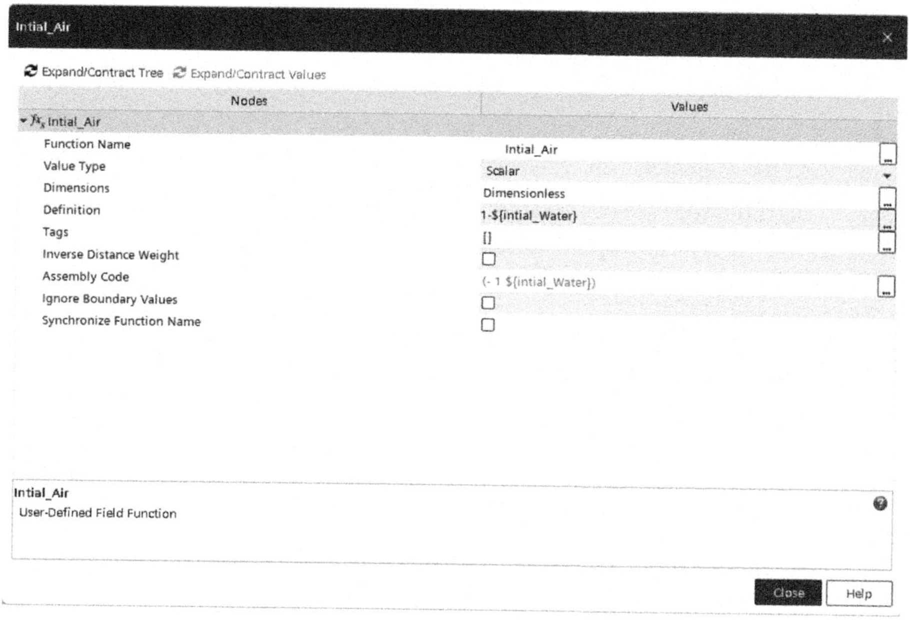

Figure 7.9 User-defined field function in STAR-CCM+ for defining the initial air volume fraction as the complement of the initial water volume fraction, enabling specification of the liquid free-surface level.

The scalar field function representing the initial liquid distribution is defined using a conditional spatial expression of the form:

The centroid coordinate in the vertical direction is less than the prescribed initial height. In STAR CCM+ syntax, this expression is written as:

```
$${Centroid}[2] < ${initialH_m} ? 1 : 0
```

This expression can be interpreted physically as follows:

- The term `Centroid[2]` represents the vertical coordinate of the centre of each computational cell.
- The parameter `InitialH_m` represents the prescribed initial liquid height.
- If the vertical coordinate of the cell is below the specified height, the function assigns a value of 1.
- If the vertical coordinate is above that height, the function assigns a value of 0.

Thus, the scalar field behaves as a Heaviside-type step function that separates the computational domain into two regions:

- The lower region is fully occupied by liquid.
- The upper region is fully occupied by air.

This formulation guarantees a sharp and well-defined initial free surface aligned with the horizontal plane at the specified fill height.

From a mathematical perspective, this represents an indicator function:

Liquid volume fraction equals one for z less than `H0`, Liquid volume fraction equals zero for z greater than or equal to `H0`.

Such an initialisation is essential in sloshing dynamics because the natural frequency and nonlinear response strongly depend on liquid depth.

The air phase initialisation is defined as the complementary field:

```
1 - ${Initial_Water}
```

This ensures strict phase consistency, meaning that:

Liquid volume fraction plus air volume fraction equals one everywhere in the domain.

This constraint is critical in Volume of Fluid simulations to maintain physical consistency and numerical stability. By defining the air fraction as the complement of the liquid fraction, the model avoids gaps or overlaps between phases.

Accurate initialisation of the phase interface is not a minor procedural step. It directly determines:

- Initial hydrostatic pressure distribution
- Gravitational potential energy of the liquid mass
- Onset characteristics of wave motion
- Resonance response under excitation

Any error in initial height specification propagates throughout the transient simulation and can significantly alter predicted impact loads and free surface amplitudes.

The use of coordinate-based conditional scalar functions provides:

- Exact geometric control
- Reproducibility
- Parametric adjustability
- Mathematical transparency

This is particularly important in advanced sloshing studies where the fill ratio is systematically varied to study nonlinear resonance phenomena.

In addition to defining the initial phase distribution, it is essential to initialise the pressure field consistently with the hydrostatic equilibrium condition. Failure to do so may introduce artificial transient oscillations at the beginning of the simulation, which can contaminate the sloshing response.

The initial pressure field was defined using a spatially dependent scalar field function based on the classical hydrostatic relation:

Pressure gradient equals density times gravitational acceleration.

For a fluid at rest under gravity, the pressure varies linearly with depth. Accordingly, the initial pressure was defined using a conditional expression that distinguishes between liquid and air regions.

The implemented functional form is:

If the vertical coordinate is below the initial liquid height, pressure is computed using liquid density. If the vertical coordinate is above the initial liquid height, pressure is computed using air density.

In STAR CCM+ syntax, the expression can be written as:

If z is less than `H0`, Pressure equals minus `rho_liquid` times g times (z minus H0), else Pressure equals minus rho_air times g times (z minus `H0`).

Where:

- z represents the vertical centroid coordinate.
- `H0` is the initial liquid height.
- `rho_liquid` is the liquid density.
- `rho_air` is the air density.
- `g` is gravitational acceleration.

The negative sign is essential because pressure increases downward relative to the free surface reference level.

This initialisation ensures that:

- The pressure at the free surface equals zero gauge pressure.
- The pressure increases linearly with depth in the liquid region.
- The gas phase exhibits its own much smaller hydrostatic gradient due to lower density.

By defining pressure in this way, the simulation begins in a mechanically balanced state. Consequently, any subsequent motion arises only from imposed tank excitation rather than from artificial numerical imbalance.

This is particularly important in sloshing dynamics, where the free surface is sensitive to even small pressure inconsistencies. Improper initialisation may cause:

- Spurious waves at time zero.
- Numerical divergence in the first few time steps.
- Artificial energy injection into the system.

Thus, hydrostatic consistency is not merely a procedural detail but a fundamental requirement for accurate transient multiphase simulation (as shown in Fig. 7.10).

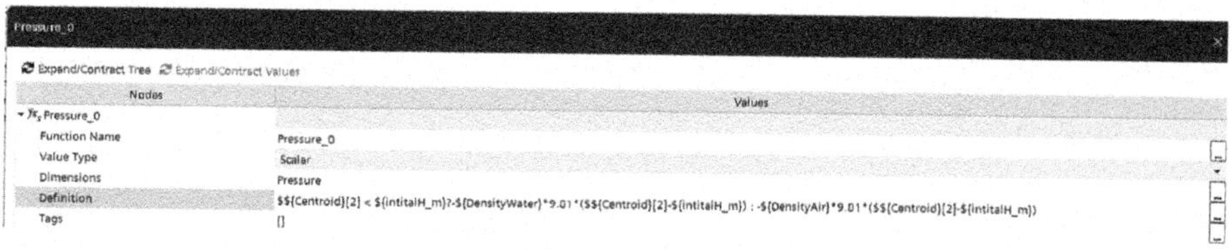

Figure 7.10 Hydrostatic pressure initialisation in multiphase sloshing simulation showing linear pressure variation in liquid and gas regions.
No Permission Required.

Here, important implementation notes are provided to highlight key modelling considerations of the pressure (`pressure_0`) and ensure consistent application of the defined simulation parameters:

1. The negative sign before the density term must not be omitted. Without it, the pressure gradient direction becomes physically incorrect.
2. It is strongly recommended to use explicit numerical density values rather than symbolic references during initialisation. This reduces ambiguity and prevents unintended dependency on dynamically updated material properties.

7.5 Definition of pressure difference as a derived scalar field

To analyse sloshing-induced loads, a derived scalar field representing pressure deviation was introduced. This quantity is defined as:

Pressure difference equals instantaneous pressure minus the initialised hydrostatic pressure (as shown in Fig. 7.11).

In symbolic form:

Pressure difference equals `P` minus `P0`.

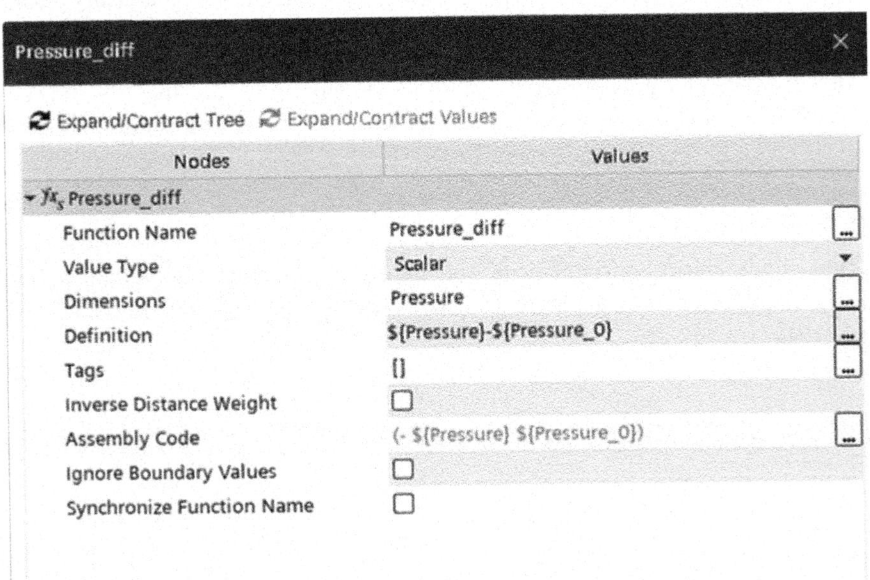

Figure 7.11 Decomposition of total pressure into hydrostatic and dynamic components for sloshing analysis.
No Permission Required.

Where:

- `P` is the instantaneous pressure computed during the transient simulation.
- `P0` is the initial hydrostatic pressure field defined previously.

This derived scalar isolates the dynamic contribution to pressure resulting from tank excitation and fluid motion.

The total pressure field in a sloshing tank consists of two components:

1. Static hydrostatic pressure due to gravity.
2. Dynamic pressure due to fluid acceleration and wave motion.

By subtracting the hydrostatic component, the model extracts the purely dynamic pressure contribution.

This is particularly important when:

- Evaluating impact pressures on tank walls.
- Identifying peak transient loads.
- Comparing simulation results with experimental pressure sensor data.

Using pressure difference instead of total pressure allows a direct interpretation of sloshing-induced forces without gravitational bias.

To analyse the motion of the liquid phase during sloshing, a filtered velocity field was defined specifically for diesel. In a multiphase simulation using the Volume of Fluid method, each computational cell may contain either liquid, gas, or a mixture of both near the interface. If the global velocity magnitude is plotted directly, it may include contributions from both phases, which can make interpretation unclear. To avoid this, a conditional scalar function was introduced so that the velocity magnitude is evaluated only in cells where diesel occupies at least half of the cell volume. In all other cells, the value is set to zero. This ensures that the reported velocity represents the motion of the liquid only, which is physically meaningful when studying sloshing behaviour. Such a filtered velocity field is useful for identifying regions of strong liquid motion, estimating kinetic energy distribution, and analysing resonance effects.

After defining this phase, filtered velocity and the previously established volume fraction and hydrostatic pressure fields, the initial conditions were assigned. The fluid was initialised in a state of hydrostatic equilibrium, meaning that the free surface was flat, the pressure distribution followed the hydrostatic relation, and the velocity of both phases was set to zero. Starting from a physically consistent equilibrium state is essential in sloshing simulations because any artificial velocity or pressure imbalance at the beginning can generate non-physical waves and distort the transient response. By carefully initialising the system in this manner, all subsequent fluid motion arises solely from the imposed tank excitation, ensuring that the computed sloshing dynamics reflect the true physical behaviour of the system.

After defining the scalar field functions and initialising the hydrostatic state, the phase-specific initial conditions were assigned separately for air and diesel. In the multiphase framework, each phase requires its own volume fraction definition under the initial condition settings. The diesel phase was initialised using the previously defined scalar function that assigns a value of one below the specified liquid height and zero above it. The air phase was defined as the complementary fraction, ensuring that the sum of both phases equals unity throughout the domain. This step guarantees a physically consistent interface at the beginning of the simulation and prevents artificial voids or overlapping phase regions. All other quantities, including velocity and pressure, were initialised consistently with the hydrostatic equilibrium state described earlier.

Following the definition of phase settings, the solver criteria were configured to ensure stable and accurate transient computation. An implicit unsteady formulation was used, with a specified time step chosen to capture the dynamic evolution of the free surface while maintaining numerical stability. The temporal discretization scheme was selected to balance robustness and computational efficiency, especially during the early stages of motion when sharp gradients may develop near the interface. Proper solver settings are particularly important in sloshing simulations because the problem involves strong nonlinear free surface deformation and transient pressure fluctuations. An inappropriate time step or discretisation scheme can lead to excessive numerical diffusion or instability. Therefore careful preparation of the solver parameters ensures that the predicted wave motion, impact loads, and resonance behaviour accurately reflect the underlying physical dynamics rather than numerical artifacts.

In STAR CCM+, quantitative analysis of simulation results requires the creation of reports. Reports are essential tools used to extract numerical data from the computational domain during the transient solution. They allow the user to measure specific physical quantities, such as maximum velocity, pressure difference, or total kinetic energy, and to record their evolution over time. Without defining reports, it is not possible to systematically generate time histories or export numerical data for further analysis.

Reports can be configured to compute different types of quantities, including maximum, minimum, average, integral, or surface-based values. In sloshing simulations, typical monitored parameters include maximum liquid velocity, maximum pressure difference, and total kinetic energy of the fluid. These quantities provide direct insight into wave intensity, impact severity, and dynamic energy exchange within the tank.

After defining a report, it is linked to a monitor. The monitor records the report value at each time step, allowing the generation of curves as a function of time. These curves are fundamental for evaluating resonance behaviour, identifying peak loads, and comparing numerical predictions with experimental data.

7.6 Importing external excitation data using file tables

Time-dependent excitation or motion data can be incorporated into the simulation environment through the File Table functionality. This feature enables the direct reading of externally prepared datasets, typically stored in CSV format, that describe quantities varying with time, such as prescribed acceleration profiles or imposed motion histories (as shown in Fig. 7.12). The File Table serves as an interface

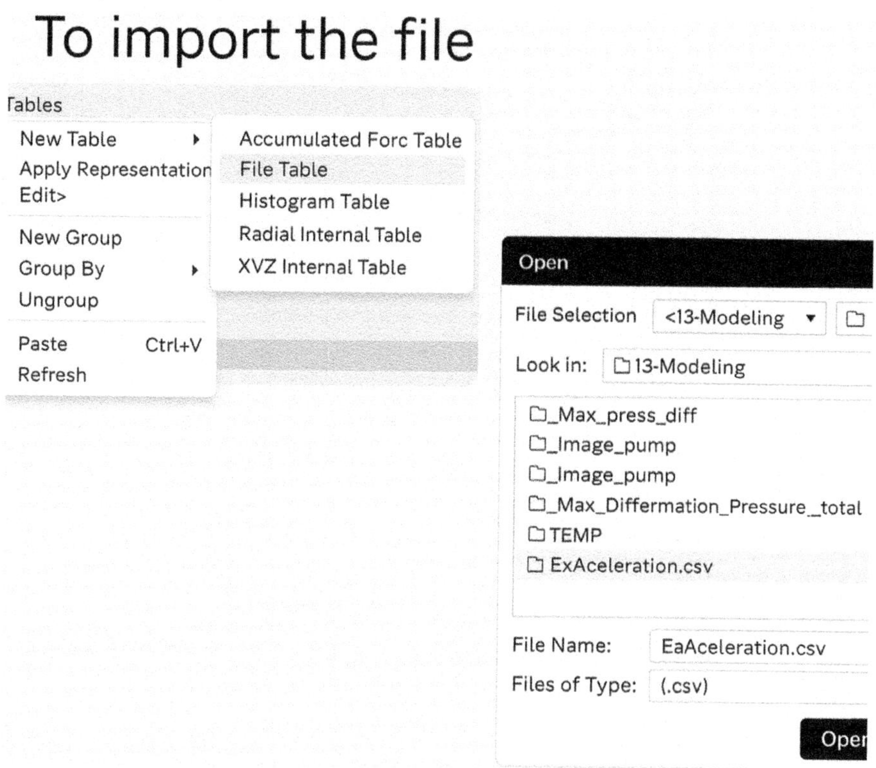

Figure 7.12 Import of time-dependent excitation data using a File Table. No Permission Required.

between measured or experimentally obtained signals and the numerical model. Once imported, the tabulated data can be referenced within boundary condition definitions, body force specifications, or prescribed motion settings, thereby allowing the simulation to reproduce realistic dynamic loading scenarios. For reliable implementation, the structure of the external file must be carefully organised. Clear column definitions, consistent units, and properly formatted time sequences are necessary to ensure correct interpretation by the solver. Any inconsistency in formatting may lead to misalignment between time and excitation values, potentially affecting the physical fidelity of the simulated response.

7.7 Creating and configuring monitoring of reports

After defining the required reports, monitoring must be activated to record their values throughout the transient simulation. Each report can be converted into a monitor, which automatically stores its computed value at every time step. This allows the generation of time-dependent curves for key physical quantities such as

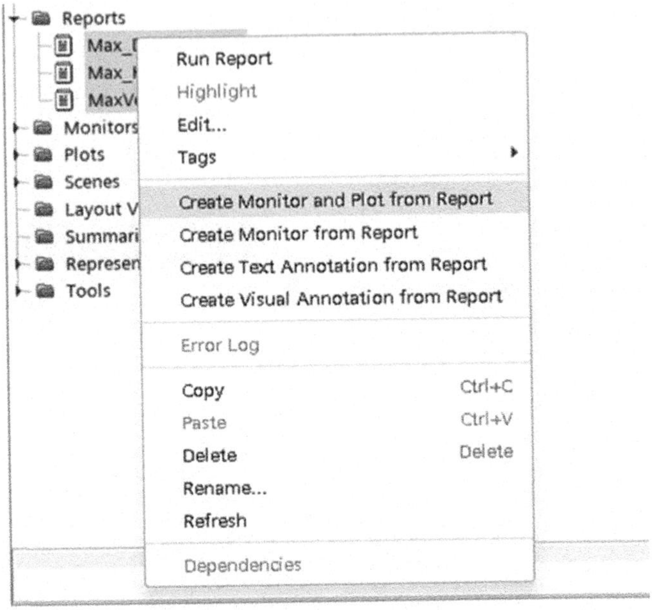

Figure 7.13 Creation of monitors and plots from user-defined reports in STAR CCM+. No Permission Required.

maximum diesel velocity, maximum pressure difference, total kinetic energy, and other response parameters (as shown in Fig. 7.13).

When multiple reports are selected, the software provides the option to create a single combined plot or separate plots for each report. For clarity in sloshing analysis, it is often preferable to generate individual plots so that peak values and oscillatory behaviour can be examined independently.

In addition to creating monitors, it is necessary to configure their properties. By right-clicking on the monitoring objects, specific settings such as sampling frequency, normalisation options, and maximum stored samples can be adjusted. Proper configuration ensures that the recorded data accurately represents the transient behaviour without excessive data storage or loss of resolution.

Monitoring is a critical step in sloshing simulations because it enables quantitative evaluation of resonance effects, impact pressures, and energy evolution over time. Without monitors, the simulation would produce fields but not structured time history data suitable for analysis.

7.8 Sloshing dynamics in a spherical LNG tank

The first case study examines sloshing dynamics within a large-scale spherical storage tank of 10 m internal diameter, representative of liquefied natural gas

Figure 7.14 LNG spherical tank geometry used for sloshing simulation.
No Permission Required.

containment systems used in marine transport vessels (as illustrated in Fig. 7.14). Spherical LNG tanks are widely employed in shipboard applications due to their favourable structural integrity, uniform stress distribution, and geometric efficiency under internal pressure. However, when partially filled and subjected to vessel motion induced by waves, wind loading, or manoeuvreing conditions, the contained cryogenic liquid undergoes sloshing motion that can generate significant transient hydrodynamic loads on the tank walls. Accurate prediction of these loads is essential for structural safety, fatigue life estimation, and operational reliability. The present numerical investigation employs STAR-CCM+ to resolve the transient multiphase flow behaviour within such a tank under prescribed dynamic excitation, thereby demonstrating the solver's capability to capture large-scale free surface motion, nonlinear wave evolution, and energy redistribution phenomena in a geometrically smooth yet dynamically complex domain.

The spherical geometry, with its continuous curvature and absence of sharp discontinuities, provides a well-controlled physical environment for investigating fundamental sloshing mechanisms. For a tank of diameter 10 m, the characteristic length scale is sufficiently large that gravitational forces dominate surface tension effects, particularly for LNG whose density at cryogenic conditions is approximately 450 kg per cubic metre. The dynamic viscosity is relatively low compared to water, which allows wave motion to persist with limited viscous damping under moderate excitation amplitudes. In marine applications, sloshing excitation arises primarily from ship roll and pitch motions. To represent such operating conditions, the tank is subjected to prescribed lateral acceleration profiles corresponding to harmonic roll motion with frequencies in the range typical for large vessels. The acceleration magnitude is selected to reflect realistic sea state conditions while remaining within operational safety margins.

Sloshing is fundamentally governed by the balance between inertia and gravity. When the excitation frequency approaches the natural sloshing frequency of the partially filled sphere, resonance can occur, leading to amplification of free surface oscillations. For a spherical tank, natural modes depend strongly on fill level and gravitational acceleration. At moderate filling ratios, the dominant mode typically corresponds to a single large standing wave oscillating between opposite sides of the tank. Because the domain is symmetric, the modal structure is spatially coherent and can be clearly identified through both interface shape evolution and velocity field patterns. This clarity makes the spherical configuration particularly suitable for validating interface tracking accuracy and the modal resolution capability of the numerical framework.

The tank is considered partially filled at representative operational levels, for example between 30% and 70% of total volume. At lower fill levels, the free surface has sufficient vertical clearance to develop pronounced wave amplitudes. Under near resonant excitation, the wave crest can rise significantly along the tank wall, leading to elevated impact pressures in localised regions. At higher fill ratios, the free surface motion becomes more constrained, and the liquid mass behaves increasingly like a bulk moving body with reduced deformation amplitude but increased inertial loading. By varying the fill ratio systematically, the sensitivity of sloshing dynamics to geometric confinement can be investigated quantitatively.

The computational setup begins with a precise geometry definition to ensure a smooth spherical boundary without surface artifacts. A high-quality polyhedral volume mesh is generated throughout the domain. Polyhedral cells offer improved numerical stability and reduced numerical diffusion compared to tetrahedral discretisations, particularly in multiphase flows involving complex interface evolution. Near the tank wall, prism layer refinement is introduced to resolve boundary-layer velocity gradients and accurately capture shear stress distribution. This near-wall resolution becomes especially important during high velocity phases when the fluid accelerates upward along the curved surface.

The multiphase flow is modelled using the Volume of Fluid method, which tracks the liquid-gas interface through a phase fraction field. This method ensures conservation of mass for each phase while maintaining a sharp interface representation during strong deformation. Gravitational acceleration is included explicitly, and no-slip boundary conditions are imposed at the tank wall. The excitation is implemented as a time-dependent body force representing ship roll acceleration. Adaptive time stepping is employed to maintain numerical stability during periods of rapid interface deformation, such as crest formation or near-impact events. Convergence is monitored through residual reduction and global mass conservation cheques to ensure physical credibility of the solution.

At low excitation amplitudes, the simulation reveals periodic oscillatory motion characterised by smooth standing waves. The free surface elevation varies sinusoidally in time, and the pressure field evolves in a predictable manner. Kinetic energy and potential energy exchange cyclically throughout each oscillation period. The velocity magnitude is highest near the mid-height region of the wall, where the fluid accelerates upward during crest formation. Because the geometry is smooth and symmetric, the flow field remains organised and coherent, with minimal generation of secondary vortical structures.

As excitation amplitude increases or as excitation frequency approaches resonance, nonlinear effects become increasingly pronounced. Wave steepening develops near the wall, leading to increased curvature at the crest and localised intensification of velocity. The interface may exhibit asymmetric deformation even within the symmetric geometry, particularly when transient effects dominate. Localised pressure peaks arise at the contact region between the liquid crest and the wall. These pressure transients are of primary engineering interest because they contribute to structural loading. The solver captures these transient events with high temporal resolution, demonstrating stability even during rapid acceleration phases.

Under sufficiently strong forcing, the simulations indicate the possibility of partial wave breaking. Although spherical geometry does not contain internal obstacles that promote chaotic flow development, strong resonance can still produce localised splashing and transient jet formation near the wall. The ability of the numerical scheme to maintain interface sharpness under these conditions highlights the robustness of the multiphase formulation. Importantly, energy redistribution can be quantified through temporal monitoring of total kinetic energy within the liquid phase. In the absence of internal damping structures, energy decay following excitation cessation occurs gradually, indicating relatively weak intrinsic damping dominated by viscous dissipation at the boundary.

The 10 m diameter scale introduces additional physical considerations compared to small laboratory tanks. The characteristic sloshing period increases with tank dimension, and the associated fluid mass is substantial. Consequently, the hydrodynamic loads generated during large amplitude motion can be significant. Even moderate free surface elevations correspond to large displaced volumes and therefore large inertial forces. From a ship design perspective, understanding these forces is essential for assessing structural reinforcement requirements and evaluating the need for sloshing mitigation measures such as internal baffles or operational filling constraints.

Another important aspect investigated in this case study is the spatial distribution of shear stress along the tank wall. During crest formation, the upward-moving fluid generates elevated shear near the wall surface. Mapping of velocity gradients reveals zones of intensified wall interaction, particularly near the intersection of the free surface with the boundary. These regions may experience repeated cyclic loading during prolonged sea states, potentially contributing to fatigue effects. By resolving these gradients accurately, the simulation provides valuable insight into structural design considerations.

As such, using the same numerical framework established earlier in this chapter, the spherical LNG tank case was implemented without modification of the fundamental modelling structure. The same multiphase formulation based on the Volume of Fluid method was adopted, together with the segregated unsteady solver, hydrostatic pressure initialisation, and scalar field-based phase definition.

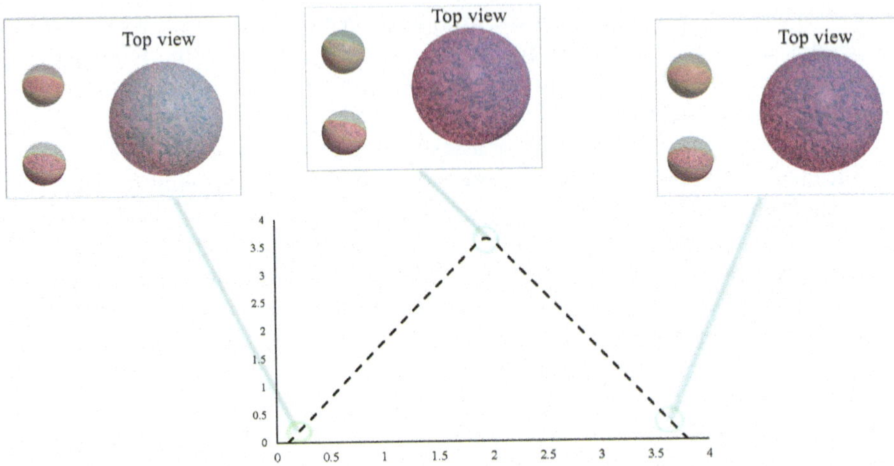

Figure 7.15 Simulation results of sloshing in a spherical LNG tank.
No Permission Required.

The initialisation strategy ensured mechanical equilibrium at the start of the computation, while the prescribed harmonic acceleration profile was introduced through the previously described external excitation approach. Mesh topology, near-wall refinement strategy, turbulence treatment, and monitoring configuration were applied consistently with the validated workflow to maintain methodological uniformity (Fig. 7.15).

7.9 Sloshing dynamics automotive fuel tank under road excitation

The second case study investigates transient multiphase flow behaviour within an automotive fuel tank of characteristic longitudinal length 0.7 m, partially filled to 50% capacity with gasoline, subjected to dynamic excitation representative of real road conditions (As shown in Fig. 7.16). Unlike the idealised spherical LNG tank discussed previously, the automotive fuel tank represents a geometrically complex, structurally integrated component embedded within a vehicle platform. Its design is constrained by packaging limitations, chassis geometry, safety regulations, structural reinforcement requirements, and fuel delivery system integration. The objective of this case study is to demonstrate the capability of STAR CCM+ to resolve sloshing phenomena within a highly irregular industrial geometry subjected to realistic acceleration profiles associated with vehicle acceleration, braking, cornering, and road surface irregularities. Due to the geometric complexity of the tank, including internal ribs, local depressions, asymmetric cavities, and small-scale structural features, surface preparation plays a critical role prior to volume discretisation. The computational domain was therefore generated using the surface wrapping

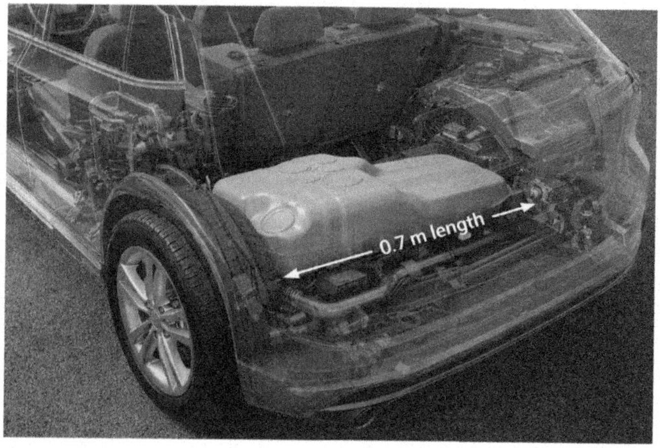

Figure 7.16 Automotive fuel tank geometry for sloshing analysis.
No Permission Required.

technique discussed in Chapter 6. This approach reconstructs a clean, watertight representation of the internal fluid volume by enveloping the original CAD geometry within a controlled background discretization. The wrapping process eliminates minor gaps, overlaps, and non-manifold features while preserving geometrically and physically relevant details that influence fuel motion. By employing the wrapping strategy, a numerically robust and mesh-consistent surface was obtained, enabling stable polyhedral volume meshing and reliable multiphase interface tracking. This preprocessing step is essential when transitioning from idealised geometries to real industrial components, where raw CAD data are rarely suitable for direct simulation without controlled geometric abstraction.

The automotive fuel tank considered here possesses a longitudinal dimension of approximately 0.7 m, a transverse width of approximately 0.5 m, and a variable height distribution shaped to accommodate underbody packaging constraints. The internal volume includes geometric discontinuities such as internal ribs, localised depressions, asymmetric cavities, and structural stiffeners. These features are not merely geometric artifacts; they play an important role in controlling fuel motion, mitigating excessive sloshing, and ensuring consistent fuel pickup under transient manoeuvres. At a 50% fill ratio, the free surface occupies a central region of the tank, leaving significant headspace above and allowing substantial fluid displacement under acceleration.

Gasoline is modelled as an incompressible Newtonian liquid with a density of approximately 720–750 kg per cubic metre and dynamic viscosity near 0.6–0.8 mPa·s depending on temperature. Compared to LNG, gasoline exhibits higher density and slightly greater viscous dissipation under ambient conditions. However, the relatively small tank dimension implies shorter natural sloshing periods and a more rapid response to excitation. In automotive operation, sloshing excitation arises from multiple sources, including longitudinal acceleration during braking and acceleration, lateral acceleration during cornering, and vertical oscillations induced by road surface irregularities. These excitations are inherently transient and broadband in frequency content rather than purely harmonic.

To replicate realistic operating conditions, the tank is subjected to prescribed acceleration profiles derived from representative vehicle manoeuvre scenarios. A typical braking event may involve deceleration on the order of 0.6–0.8 g sustained for several seconds, while cornering events may generate lateral acceleration up to 0.5 g. Additionally, vertical accelerations of smaller magnitude but higher frequency may arise due to road roughness. In the present simulation, combined longitudinal and lateral acceleration components are applied sequentially to examine the directional sensitivity of the sloshing response. The excitation is implemented as time-dependent body forces acting on the liquid phase within the non-inertial reference frame of the vehicle.

The numerical workflow begins with detailed geometry preparation. Automotive fuel tank CAD models frequently contain small fillets, minor surface gaps, and embedded hardware features that must be treated appropriately before meshing. Automated surface wrapping is employed to produce a watertight computational domain that preserves essential geometric features while eliminating unnecessary small-scale artifacts that would otherwise degrade mesh quality. This preprocessing step is critical in industrial CFD applications where geometry complexity can compromise numerical stability.

Following surface preparation, a polyhedral volume mesh is generated throughout the tank interior. Polyhedral cells are particularly advantageous for irregular geometries because they conform more naturally to complex surfaces and require fewer cells for equivalent accuracy compared to tetrahedral meshes. Prism layer refinement is introduced near solid boundaries to resolve boundary layer flow and capture wall shear stress accurately. Although the Reynolds number within the tank may vary depending on acceleration intensity, transient inertial effects dominate during rapid manoeuvres. Therefore adequate near-wall resolution is essential for predicting velocity gradients and dissipation rates.

The air-gasoline interface is captured using the Volume of Fluid approach. This method solves a transport equation for the volume fraction of gasoline while enforcing mass conservation. Interface compression schemes are utilised to minimise numerical diffusion and maintain a sharp phase boundary during large deformation events. Gravitational acceleration is included, and the applied vehicle acceleration is superimposed as an additional body force. No slip boundary conditions are imposed at the tank walls, and surface tension effects are included, although they are secondary relative to inertial forces at this scale.

At moderate longitudinal acceleration, such as steady braking, the free surface tilts in the direction opposite to motion due to inertial redistribution of the liquid mass. Initially, the interface deforms smoothly, forming a quasi-planar inclined surface. As deceleration continues, fluid accumulates towards the front wall of the tank. Depending on the internal geometry, this accumulation may lead to localised impact events where the rising liquid contacts structural ribs or localided protrusions. The velocity field reveals strong forward-directed flow near the tank floor, accompanied by recirculation zones forming behind internal baffles. These vortical structures enhance mixing and contribute to energy dissipation.

When braking is released abruptly, the stored potential energy within the tilted free surface converts into kinetic energy, generating backward sloshing motion. The interface oscillates longitudinally, and multiple wave fronts may develop due to reflection from irregular internal surfaces. Unlike the spherical case, where symmetry promotes coherent standing waves, the automotive tank exhibits spatially fragmented motion. The internal ribs break the flow into localised compartments, causing partial phase lag between different regions of the liquid. This spatial heterogeneity significantly modifies pressure distribution patterns.

Under lateral acceleration corresponding to cornering manoeuvres, the free surface tilts transversely. Because the tank geometry is asymmetric in many designs, lateral sloshing may be more severe in one direction than the other. Simulation results show that fuel can migrate preferentially towards certain cavities, temporarily isolating regions of low fluid depth elsewhere in the tank (As illustrated in Fig. 7.17). This behaviour is critical for fuel pickup reliability. In real vehicles, insufficient fuel coverage near the pump inlet can lead to transient fuel starvation. By resolving the transient interface evolution, the numerical model provides insight into the likelihood and duration of such exposure events.

Combined longitudinal and lateral excitations produce even more complex flow patterns. The interface may twist and deform non uniformly, and secondary wave systems may emerge. Velocity magnitude contours demonstrate the development of high-speed streaks along the tank floor where fluid is forced through narrow passages between internal ribs. These localised jets increase shear stress and enhance viscous

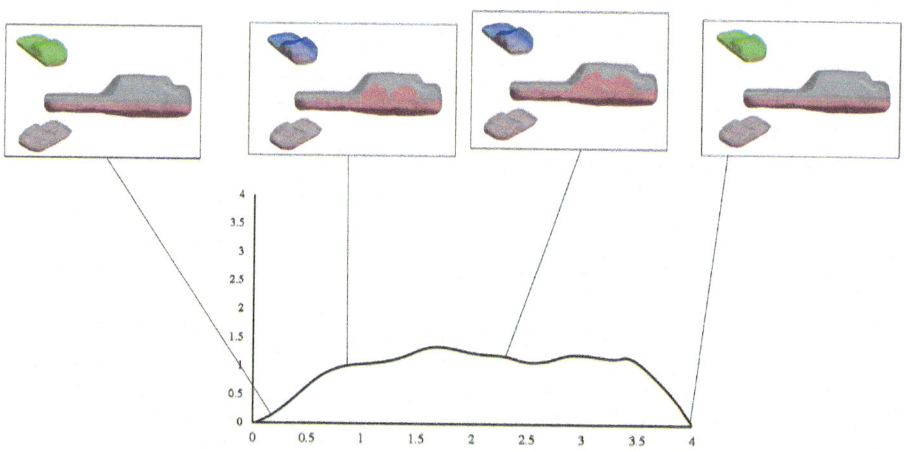

Figure 7.17 Simulation results of sloshing in an automotive fuel tank under dynamic excitation.
No Permission Required.

dissipation. Recirculation zones behind structural elements persist even after primary excitation diminishes, illustrating the role of geometry in controlling damping characteristics.

An important aspect of this case study is the analysis of pressure distribution along the tank walls. During sudden deceleration, peak pressures are observed near the forward wall where accumulated fluid mass impacts the boundary. These pressures are transient but can contribute to structural loading and acoustic excitation. Similarly, during cornering, elevated pressures occur along lateral walls. By monitoring wall pressure time histories at selected locations, the severity and frequency of impact events can be quantified. Such information is valuable for structural durability analysis and noise vibration assessment.

Energy analysis provides additional insight into damping mechanisms. The total kinetic energy of the liquid phase is computed as a function of time. Following an impulsive acceleration event, kinetic energy rises sharply and then decays due to viscous dissipation and interaction with internal structures. Compared to smooth geometries, the presence of ribs and cavities increases turbulence generation and enhances energy dissipation. This geometric damping is beneficial from a vehicle dynamics perspective because it reduces sustained oscillation and minimises secondary sloshing cycles that could affect handling stability.

Fill level sensitivity is also investigated. At 50% fill, the free surface has sufficient mobility to develop significant amplitude motion. If the fill level decreases further, larger free surface excursions become possible, potentially increasing impact severity. Conversely, at higher fill levels, the liquid mass behaves more rigidly, reducing surface deformation but increasing overall inertial load. Parametric variation of fill ratio reveals nonlinear dependence of peak pressure and velocity magnitude on fill percentage, emphasising the importance of considering operational fuel levels in design evaluation. Time step sensitivity analysis is performed to ensure accurate resolution of transient impact peaks. Because automotive sloshing events may occur over fractions of a second, sufficient temporal resolution is required. Adaptive time stepping allows refinement during high acceleration phases while maintaining computational efficiency during quiescent periods. Mesh independence studies confirm that predicted free surface shape and peak pressure values converge with refinement, reinforcing confidence in the solution.

Acknowledgements

The author sincerely thanks Siemens for its generous support, technical guidance, and valuable advice throughout this work. Its assistance was indispensable to the successful completion of this research.The author also wishes to acknowledge Toyota Technological Institute, Japan, for its support, facilities, and stimulating research environment, which greatly aided the development of this study.

Advanced practical sloshing dynamics

8

8.1 Introduction to advanced practice in sloshing dynamics

Advanced practice in sloshing dynamics extends far beyond classical analytical descriptions of free surface motion. Traditionally, sloshing studies focused on predicting wave heights, estimating hydrodynamic pressures, and evaluating load effects under simplified assumptions. While such analyses remain fundamental, modern engineering applications demand a far more comprehensive and integrated approach. Today, sloshing must be understood within the broader context of system-level dynamics, where fluid motion interacts continuously with structural flexibility, operational vibration environments, thermal effects, and manufacturing constraints.

In contemporary engineering systems, tanks are rarely isolated components. They are embedded within complex mechanical assemblies that operate under diverse and often severe dynamic conditions. Aerospace propellant tanks experience launch vibrations, attitude control manoeuvres, and prolonged microgravity operation. Automotive fuel systems are subjected to engine excitation, road-induced vibration, acceleration, and braking cycles. Marine storage tanks encounter wave-induced motion and variable filling conditions. In all these environments, the dynamic behaviour of the contained liquid cannot be treated independently of the supporting structure. Instead, the tank–fluid system behaves as a coupled dynamic entity whose response depends on both hydrodynamic and structural characteristics.

One of the most critical phenomena in such systems is resonance. A partially filled tank behaves like a distributed mass oscillator. The liquid inside possesses natural sloshing frequencies that depend primarily on geometry and fill level. At the same time, the tank structure and its supporting components exhibit their own natural vibration modes governed by stiffness and mass distribution. When the natural sloshing frequencies of the fluid align with structural natural frequencies or with dominant frequencies in the operational excitation spectrum, dynamic amplification occurs. This amplification can lead to excessive stress, fatigue crack initiation, wall impact pressures, noise, and degradation of overall system performance.

In early design practice, engineers typically addressed resonance issues through conservative reinforcement strategies. Increasing wall thickness, adding baffles and ribs, or introducing mass dampers were common solutions. Although these approaches may reduce vibration amplitudes, they often introduce weight penalties and fail to address the root cause of the spectral interaction. In addition, they are typically reactive measures implemented after resonance problems are identified during testing.

Slosh Dynamics. DOI: https://doi.org/10.1016/B978-0-443-44330-5.00005-0

Advanced practice in sloshing dynamics adopts a fundamentally different philosophy. Instead of reacting to resonance after it appears, the objective is to design the tank–fluid system such that resonance conditions are inherently avoided. This requires deliberate spectral management. The natural sloshing frequencies of the liquid must be positioned outside the operational vibration range of the system. Simultaneously, structural natural frequencies must be separated from both sloshing modes and external excitation bands. The problem thus becomes one of coordinated frequency design.

In realistic operating environments, excitation is rarely characterised by a single dominant frequency. Automotive systems, for example experience broadband vibration spanning a wide frequency range associated with engine firing harmonics, drivetrain dynamics, and road roughness. Aerospace structures are exposed to launch vibration spectra that include both low-frequency and high-frequency components. Marine systems are influenced by complex wave spectra that vary with sea state. Consequently, the design objective is not simply to minimise sloshing amplitude at one frequency, but to ensure that the entire spectrum of fluid and structural natural frequencies lies outside critical operational bands.

This requirement introduces the concept of spectral separation. The tank–fluid system must be engineered so that sufficient frequency margins exist between sloshing modes, structural modes, and excitation frequencies. Adequate separation reduces the likelihood of dynamic amplification and enhances long-term durability. Achieving this objective demands a predictive and computationally rigorous framework capable of modifying the intrinsic dynamic characteristics of the system.

An additional layer of complexity arises from structural compliance. In advanced systems, the tank cannot be assumed rigid. Wall flexibility modifies boundary conditions for the fluid and leads to coupled fluid–structure modes. The global dynamic behaviour is therefore determined by the interaction between fluid inertia and structural elasticity. Changes in stiffness distribution or internal geometry influence not only structural frequencies but also hydrodynamic modes. The tank must be treated as an integrated multiphysics system rather than as separate fluid and structural components.

Within this integrated framework, geometry plays a central role. The natural sloshing frequency depends strongly on tank dimensions and liquid depth. Alterations in internal configuration, such as the introduction of partitions or shaping of boundaries, modify effective wave propagation characteristics. Structural modifications influence stiffness and mass distribution, which in turn alter vibration modes. Thus frequency control is fundamentally a geometric and material distribution problem.

Modern computational tools enable engineers to analyse these interactions with high fidelity. Fluid simulations capture sloshing modes and pressure fields. Finite element models evaluate structural dynamics. Coupled analyses quantify fluid–structure interaction effects. However, analysis alone is insufficient. To truly practice advanced sloshing engineering, one must move from analysis to synthesis – from predicting dynamic response to intentionally shaping it.

This transition sets the stage for topology optimisation as a design methodology (Shimoda, Hikasa, et al., 2023). Rather than manually adjusting ribs, thicknesses, or internal partitions based on intuition, topology optimisation provides a systematic approach to redistributing material and geometry within a defined design space. Through iterative computational procedures, the material layout evolves until spectral separation objectives are satisfied under volume and manufacturability constraints.

Before exploring topology optimisation in detail, it is essential to recognise the conceptual shift represented by advanced sloshing practice. The tank is no longer viewed as a passive container resisting unpredictable fluid motion. Instead, it becomes a dynamically engineered component whose natural frequencies are intentionally positioned relative to its operational vibration environment. Sloshing behaviour is not merely mitigated; it is managed through deliberate spectral design.

In this context, topology optimisation emerges as a powerful tool to achieve frequency-tailored tank configurations. By enabling controlled redistribution of stiffness and modification of internal geometry, it allows engineers to design tanks whose sloshing natural frequencies do not resonate with structural modes or with the operational vibration range. The following section introduces topology optimisation as the computational framework that makes such spectral coordination possible.

8.2 Topology optimisation for sloshing resistant structures

Topology optimisation is a design methodology that determines the optimal distribution of material within a prescribed design domain in order to achieve a specified performance objective under given constraints. Unlike traditional size or shape optimisation, which modifies dimensions or boundary contours of a predefined configuration, topology optimisation allows the *connectivity* of the structure itself to evolve. Holes may appear or disappear, internal members may emerge automatically, and load paths are discovered rather than imposed. In this sense, topology optimisation transforms structural design from an intuition-driven process into a rigorous computational synthesis problem grounded in variational mechanics and numerical optimisation theory.

At its core, topology optimisation is formulated as a constrained minimisation problem. Let the structural displacement field be governed by the equilibrium equation

$$K(\rho)u = f \tag{8.1}$$

where $K(\rho)$ is the stiffness matrix dependent on the spatial material distribution $\rho(x)$, u is the displacement vector, and f represents external loading. The general topology optimisation problem may be written as

$$\min_{\rho}: J(u(\rho), \rho) \tag{8.2}$$

subject to

$$\sum_e \rho_e v_e \leq V^*, \degree 0 \leq \rho_e \leq 1 \quad (8.3)$$

Here, J is the objective function, often structural compliance, eigenfrequency, stress, or dynamic response; v_e is the element volume; and V^* is the allowable material fraction. The design variable ρ_e represents the relative density of each finite element, interpolating between void and solid.

The historical development of topology optimisation can be traced back to Michell's 1904 work on optimal truss structures (Hegemier & Prager, 1969; Michell, 1904). Michell derived theoretical layouts for minimum weight trusses under given loading, demonstrating that optimal material distribution follows principal stress trajectories. Although purely analytical and restricted to truss continua, Michell's theory established the principle that structural efficiency arises from aligning material with load paths. However, the absence of computational tools limited practical implementation for many decades.

The modern era of topology optimisation began in the late 20th century with the advent of finite element analysis and nonlinear programming methods. In the 1980s, the homogenisation method emerged as a foundational breakthrough (Al Ali & Shimoda, 2024; Sigmund, 1994). Researchers proposed modelling material as a periodic microstructure whose effective properties could vary continuously between solid and void (Bendsøe & Kikuchi, 1988; Suzuki & Kikuchi, 1991). By solving cell problems at the microscale, designers could approximate intermediate material states and determine optimal macroscale distributions. Although mathematically elegant, homogenisation required complex microstructural parameterisation and was computationally demanding.

A major practical milestone occurred in the early 1990s with the introduction of the density-based Solid Isotropic Material with Penalisation method (Bendsøe, 1989). This formulation simplified implementation by interpolating material stiffness directly as

$$E(\rho) = E_0 \rho^p \quad (8.4)$$

where p is a penalisation factor that drives the design towards discrete solid–void solutions. The density approach eliminated the need for explicit microstructure modelling and enabled large-scale finite element optimisation. Its simplicity, numerical robustness, and compatibility with gradient-based solvers contributed to widespread adoption across structural mechanics.

Parallel to density methods, level set approaches emerged as an alternative representation of topology. Instead of assigning density values to elements, the boundary of the structure is implicitly described by a signed distance function $\phi(x)$. The evolution of the boundary is governed by the Hamilton–Jacobi equation,

$$\frac{\partial \phi}{\partial t} + V_n |\nabla \phi| = 0 \quad (8.5)$$

where V_n is the normal velocity derived from shape sensitivities. Level set methods provide crisp interfaces and avoid grey regions, making them attractive for manufacturing-oriented design (Abass et al., 2019). However, they require more sophisticated numerical treatment and initialisation strategies.

During the 2000s, topology optimisation expanded beyond static compliance (Liu & Shimoda, 2014) minimisation into dynamic and multiphysics domains (Shimoda et al., 2024). Eigenfrequency maximisation (Shimoda et al., 2024), buckling resistance (Neves et al., 2002), thermal conduction (Al Ali et al., 2023), fluid flow, and acoustic optimisation were incorporated into unified frameworks. The extension to fluid–structure interaction and Navier-Stokes-based optimisation opened new frontiers (Yaji et al., 2015; Zeng et al., 2020), allowing engineers to design channels, heat exchangers, and flow control devices.

Simultaneously, filtering techniques and projection schemes were developed to address numerical instabilities such as checkerboarding and mesh dependency. Density filtering introduced spatial averaging of design variables, while Heaviside projection sharpened intermediate densities. These advancements improved convergence behaviour and manufacturability.

The integration of additive manufacturing in the 2010s marked another transformative phase (Al Ali & Shimoda, 2022a). Previously, topology-optimised geometries were often difficult to fabricate due to geometric complexity. With the maturation of selective laser melting, electron beam melting, and advanced polymer printing, highly intricate lattice and organic forms became manufacturable. Topology optimisation shifted from theoretical exploration to industrial deployment, particularly in aerospace, automotive, and biomedical sectors.

More recently, machine learning and reduced-order modelling have been incorporated to accelerate optimisation cycles. Surrogate models approximate system responses, enabling rapid iteration. Concurrent multiscale optimisation integrates microstructural design with macroscale topology, producing architected materials tailored for stiffness, damping, or thermal performance.

In the context of tank systems subjected to sloshing, topology optimisation represents the convergence of structural mechanics, fluid dynamics, and spectral analysis. Historically, tank design relied on empirical reinforcement and conservative safety margins. The evolution of topology optimisation now permits systematic codesign of structural supports, internal partitions, and damping architectures. Instead of reacting to dynamic issues during testing, engineers can proactively engineer the spectral characteristics of the tank–fluid system.

Thus, the historical progression of topology optimisation reflects a broader evolution in engineering philosophy: from deterministic geometry specification towards algorithmic material distribution guided by physics and computation. What began as a theoretical study of optimal trusses has matured into a multiphysics design paradigm capable of reshaping fluid–structure systems at both macro and microscales. This conceptual and historical foundation provides the necessary framework to understand its application to sloshing-resistant tank design and to the intentional shifting of natural sloshing frequencies discussed in the subsequent section.

Sloshing-resistant structural design represents a paradigm shift from passive strengthening towards physics-guided material distribution. Traditional design methodologies rely on empirical safety factors, local stiffening, or heuristic reinforcement of tank walls and supports. While such approaches may increase global stiffness, they do not explicitly account for the dynamic interaction between the liquid free surface and the structural system. As a result, resonance conditions can emerge when dominant sloshing frequencies coincide with structural eigenfrequencies, leading to amplified dynamic stresses, fatigue accumulation, and even catastrophic failure. Topology optimisation provides a systematic framework to address this challenge by redistributing material within a prescribed design domain so that dynamic performance criteria are satisfied under strict volume and manufacturability constraints.

The fundamental idea is to treat the support structure, internal baffles, or external stiffening system as a continuum whose material density can vary spatially between solid and void. Rather than predefining ribs or beams, the optimisation process determines where material is most effective in altering modal characteristics, increasing damping capacity, or minimising dynamic amplification factors. This approach is particularly powerful in sloshing problems because the excitation frequencies are not arbitrary. They are determined by fluid depth, tank geometry, and gravitational acceleration. Consequently, structural design can be tailored to avoid frequency coincidence with dominant sloshing modes.

The dynamic behaviour of a partially filled tank can be modelled as a coupled fluid–structure system. The structural domain obeys the classical elastodynamic equation

$$M_s \ddot{u} + C_s \dot{u} + K_s u = f_{fs} \tag{8.6}$$

where M_s, C_s, and K_s denote structural mass, damping, and stiffness matrices, respectively, and f_{fs} represents the fluid–structure interaction force. The fluid domain, when described in potential form for moderate sloshing amplitudes, yields an equivalent dynamic system that can be expressed in discretised form as

$$M_f \ddot{\phi} + K_f \phi = f_{sf} \tag{8.7}$$

Here ϕ represents the velocity potential at the free surface nodes, and f_{sf} is the coupling force induced by structural acceleration. In a fully coupled framework, these two subsystems are assembled into a global eigenvalue problem,

$$\begin{bmatrix} K_s & C_{fs} \\ C_{sf} & K_f \end{bmatrix} \begin{bmatrix} u \\ \phi \end{bmatrix} = \omega^2 \begin{bmatrix} M_s & 0 \\ 0 & M_f \end{bmatrix} \begin{bmatrix} u \\ \phi \end{bmatrix} \tag{8.8}$$

The eigenvalues represent the coupled sloshing–structure frequencies. Resonance occurs when these frequencies align with external excitation frequencies such as seismic loading, vehicle acceleration, or marine wave forcing. The objective of topology optimisation is therefore to control these eigenvalues by modifying the structural stiffness distribution.

In the density-based method, also known as the Solid Isotropic Material with Penalisation approach, a scalar design variable (ρ_e^p in [0,1]) is assigned to each finite element e within the structural design domain. The effective Young's modulus of each element is interpolated as

$$E_e(\rho_e) = E_0 \rho_e^p \tag{8.9}$$

where E_0 is the material modulus and p is a penalisation exponent typically between three and five. This interpolation encourages the solution to converge towards a black and white distribution, reducing intermediate density regions. The structural stiffness matrix becomes a function of the density field,

$$K_s(\rho) = \sum_e \rho_e^p K_e^0 \tag{8.10}$$

with K_e^0 representing the elemental stiffness matrix for solid material.

For sloshing-resistant design, several objective formulations are possible. A direct strategy is frequency separation. Let ω_{sl} denote a dominant sloshing frequency determined from fluid analysis. The optimisation problem can be formulated as

$$\min_\rho \frac{1}{|\omega_1(\rho) - \omega_{sl}|}, \tag{8.11}$$

subject to

$$\sum_e \rho_e v_e \le V^*, \ {}^\circ 0 \le \rho_e \le 1 \tag{8.12}$$

Here, $\omega_1(\rho)$ is the first structural eigenfrequency, v_e is the element volume, and V^* is the prescribed material fraction. This objective penalises frequency proximity, thereby forcing the optimised design to shift its natural frequency away from the sloshing mode.

Alternatively, the objective may focus on minimising dynamic compliance under harmonic excitation at the sloshing frequency,

$$J(\rho) = u^T K_s(\rho) u \tag{8.13}$$

where u is the displacement response to harmonic loading. In this formulation, the structure is optimised to reduce vibration amplitude at critical frequencies.

Sensitivity analysis plays a central role in the optimisation process. For eigenvalue-based objectives, the derivative of the eigenfrequency with respect to the density variable is derived from perturbation theory,

$$\frac{\partial \omega_i}{\partial \rho_e} = \frac{1}{2\omega_i} \frac{\phi_i^T \left(\frac{\partial K_s}{\partial \rho_e} - \omega_i^2 \frac{\partial M_s}{\partial \rho_e} \right) \phi_i}{\phi_i^T M_s \phi_i}. \tag{8.14}$$

This expression quantifies how a small increase in material density in element e influences the eigenfrequency ω_i. Regions where the derivative magnitude is high indicate a strong influence on dynamic behaviour. Gradient-based optimisers, such as the Method of Moving Asymptotes, update the density field iteratively using these sensitivities.

When fluid–structure coupling is explicitly considered, the optimisation must account for how structural modifications alter the coupled eigenvalues. In such cases, the sensitivity expression includes contributions from both structural and fluid matrices. Although computationally intensive, this approach ensures that resonance avoidance is addressed at the system level rather than the structural level alone.

One of the most powerful aspects of topology optimisation in sloshing applications is its ability to create nonintuitive load paths. Instead of thickening walls uniformly, the method may generate diagonal struts, lattice-like support networks, or curved reinforcement patterns that align with principal dynamic stress trajectories. Such geometries often resemble biological structures optimised for vibration resistance. These configurations can significantly reduce mass while achieving substantial shifts in eigenfrequencies.

Manufacturability constraints are essential for practical implementation. For additive manufacturing, minimum length scale constraints ensure that optimised members exceed printable thickness limits. Density filtering techniques introduce a convolution operation,

$$\tilde{\rho} e = \frac{\Sigma j \in N_e w_{ej} \rho_j}{\Sigma_{j \in N_e} w_{ej}}, \tag{8.15}$$

where w_{ej} are spatial weights and N_e denotes neighbouring elements within a filter radius. This process smooths the design and eliminates checkerboard instabilities. Projection methods may further sharpen the filtered densities to approach binary distributions while maintaining numerical stability.

In applications such as automotive fuel tanks, resonance conditions may occur during acceleration or cornering. By embedding optimised internal ribs or support frames derived from topology optimisation, the dynamic amplification factor can be significantly reduced. In marine liquefied natural gas (LNG) tanks, where sloshing-induced impact pressures can be severe, optimised external support systems can shift structural frequencies away from dominant sloshing harmonics, thereby mitigating stress concentration.

An advanced extension of this framework involves concurrent optimisation of macroscale support structures and microscale porous inserts. Porous or lattice-filled regions can enhance energy dissipation through viscous effects. In such multiphysics optimisation, additional constraints are introduced to control permeability or damping characteristics. The objective may then combine frequency separation and energy dissipation maximisation, yielding hybrid designs that resist resonance while absorbing vibrational energy.

Uncertainty quantification can also be incorporated. Sloshing frequencies depend on fill level, which may vary during operation. Let the fill ratio be treated as a random variable with probability distribution $P(h)$. The expected objective becomes

$$\mathbb{E}\Big[J(\rho)] = \int J(\rho, h) P(h), dh. \tag{8.16}$$

Optimisation under this stochastic framework produces designs robust to variations in liquid height. Polynomial chaos expansion or Monte Carlo sampling can approximate the expectation efficiently.

Computational cost remains a challenge. Coupled eigenvalue analysis with thousands of design variables requires high-performance computing resources. Model order reduction techniques, such as Proper Orthogonal Decomposition, can accelerate repeated eigenvalue evaluations within the optimisation loop. Reduced models preserve dominant dynamic characteristics while reducing system size.

Experimental validation is critical to confirm the predicted performance improvements. Scaled tank experiments with accelerometers and pressure sensors can measure frequency response functions before and after structural optimisation. Typically, optimised designs demonstrate measurable shifts in peak response frequencies and reductions in amplitude at resonance.

From a conceptual standpoint, topology optimisation transforms sloshing-resistant design from a reactive process into a predictive and proactive one. Instead of waiting for vibration issues to appear in testing, engineers can integrate resonance avoidance directly into early-stage conceptual design. This is particularly important in aerospace and marine applications, where weight penalties are costly and structural failure risks are unacceptable.

Furthermore, the methodology integrates naturally with digital workflows. Starting from a CAD model, the design domain is discretised into finite elements. Sloshing frequencies are computed via fluid analysis, structural modes are identified, and optimisation iteratively refines the material layout. The final density field is converted back into manufacturable geometry using iso surface extraction techniques. The entire process can be embedded within a multiphysics simulation environment, ensuring reproducibility and traceability.

8.3 Shifting the natural sloshing frequency through geometry and structural–fluid codesign

While the previous discussion focused primarily on shifting *structural* eigenfrequencies away from dominant sloshing modes, an equally powerful and less explored strategy consists of directly shifting the natural sloshing frequency itself through topology optimisation of the tank geometry and its internal configuration. Rather than modifying only the stiffness distribution of the support system, the design domain can be extended to include internal fluid boundaries, free surface shaping elements, or compliant partitions that alter the hydrodynamic spectrum of the contained liquid.

For a rigid rectangular tank of length L and liquid depth h, the fundamental sloshing frequency under small amplitude assumptions is approximately

$$\omega_{sl}^2 = gk{\cdot}\tanh\,(kh),\ where{:}\ {}^{\circ}k = \frac{\pi}{L} \tag{8.17}$$

This expression clearly shows that the sloshing frequency depends on geometric parameters, gravity, and fill height. In conventional design, these parameters are fixed by functional requirements. However, topology optimisation allows controlled redistribution of internal geometry in a way that modifies the effective hydrodynamic length scale, alters boundary conditions, and perturbs modal shapes.

If internal partitions or porous inserts are introduced, the effective wavenumber becomes geometry dependent:

$$k_{\text{eff}} = k\,(\rho) \tag{8.18}$$

where the density field ρ\rhoρ defines the topology of internal obstacles. Consequently, the sloshing frequency becomes a design-dependent quantity,

$$\omega_{sl}^2(\rho) = g k_{\mathrm{eff}}(\rho) \tanh\left(k_{\mathrm{eff}}(\rho) h\right) \tag{8.19}$$

The optimisation objective may therefore be formulated as

$$\min_{\rho} \left| \omega_{sl}(\rho) - \omega_{\mathrm{target}} \right| \tag{8.20}$$

where ωtarget\omega_{target}ωtarget is chosen to avoid resonance with external excitation frequencies or structural modes. Unlike structural frequency tuning, this approach reshapes the fluid modal spectrum itself.

In practice, internal topology modifies sloshing dynamics through three primary mechanisms.

First, geometric partitioning reduces coherent wavelength. When longitudinal obstacles subdivide the tank into smaller effective compartments, the fundamental mode shifts towards higher frequencies due to the reduced effective wavelength. This mechanism resembles the behaviour of a string constrained at intermediate points, increasing stiffness and frequency.

Second, compliant internal structures introduce additional coupling between fluid inertia and structural flexibility. If lightweight partitions are optimised for tailored compliance, they may act as secondary oscillators. The coupled system then exhibits mode splitting, where the original sloshing modes divide into multiple hybrid modes. Controlled mode splitting can be exploited to spread energy across several frequencies rather than concentrating it at a single resonance peak.

Third, porous topology inserts alter the effective boundary condition from a no-slip rigid wall to a distributed damping interface. The modified dispersion relation in the presence of porous media introduces additional dissipative terms, effectively reducing peak response amplitude and slightly shifting resonant frequency due to altered inertial distribution.

To integrate these mechanisms into topology optimisation, the fluid domain must be treated as an active design region. A common approach employs a Brinkman penalisation model in which the momentum equation is augmented by a resistance term,

$$\rho_f \left(\frac{\partial \mathbf{v}}{\partial t} + \mathbf{v} \cdot \nabla \mathbf{v} \right) = -\nabla p + \mu \nabla^2 \mathbf{v} - \alpha(\rho) \mathbf{v} \tag{8.21}$$

Here, α(ρ) interpolates between zero in fluid regions and a large value in solid regions. The sloshing eigenvalue problem derived from the linearised equations becomes dependent on ρ, and eigenfrequency sensitivities can be computed through adjoint methods.

The derivative of the sloshing eigenfrequency with respect to the design variable can be expressed as

$$\frac{\partial \omega_{sl}}{\partial \rho_e} = \frac{\psi^T \left(\frac{\partial K_f}{\partial \rho_e} - \omega_{sl}^2 \frac{\partial M_f}{\partial \rho_e} \right) \psi}{2 \omega_{sl} \psi^T M_f \psi} \tag{8.22}$$

where ψ\psiψ is the fluid eigenmode. This gradient identifies fluid regions where material insertion most effectively shifts frequency. Typically, high sensitivity zones appear near free surface antinodes and along dominant flow paths.

An advanced codesign strategy considers both structural stiffness and fluid topology simultaneously. The global objective becomes

$$\min_{\rho_s,\rho_f}[w_1\,|\omega_{sl}(\rho_f)-\omega_{\text{target}}|+w_2\,|\omega_{str}(\rho_s)-\omega_{sl}(\rho_f)|] \tag{8.23}$$

where ρ_f and ρ_s represent structural and fluid design variables, respectively, and w_1, and w_2 are weighting factors. The first term tunes the sloshing frequency to a safe band, while the second ensures separation from structural resonance. This dual control framework achieves complete spectral coordination between fluid and structure.

From a physical interpretation standpoint, shifting the sloshing frequency is fundamentally equivalent to redistributing effective fluid inertia. Internal topology can redirect flow patterns, shorten wave travel distance, or introduce phase lag between different fluid compartments. The resulting modal structure often exhibits locased oscillation patterns rather than global coherent waves. Such localised modes are less effective in generating large wall pressures and therefore reduce structural demand.

In aerospace fuel tanks operating under microgravity, the situation becomes even more complex. In reduced gravity environments, surface tension replaces gravity as the dominant restoring force. The natural frequency then follows a capillary dispersion relation,

$$\omega^2=\frac{\sigma}{\rho_f}k^3 \tag{8.24}$$

where σ\sigmaσ is surface tension. Topology optimisation of internal vanes or wicking structures can modify effective curvature and interfacial area, thereby shifting capillary sloshing frequencies. This is particularly relevant for propellant management systems where resonance can disrupt attitude control.

Another promising avenue involves geometric tailoring of the tank shell itself. Instead of flat walls, topology optimisation can produce smoothly curved or corrugated boundaries that alter wave reflection behaviour. Nonuniform wall curvature disrupts standing wave formation by introducing phase shifts at reflection points. Optimisation may aim to minimise constructive interference by shaping boundaries that break symmetry and suppress global mode amplification.

When applied to automotive applications, frequency shifting is crucial during transient manoeuvres. Vehicle acceleration spectra are not monochromatic but contain broadband components. A tank whose sloshing frequency falls within dominant vehicle excitation bands will experience amplified oscillations. By shifting the fluid frequency slightly outside this band through internal topology design, dynamic load transfer to the chassis can be reduced significantly.

Computational implementation of frequency shifting optimisation typically requires repeated solution of fluid eigenvalue problems. To reduce computational cost, reduced-order models can approximate sloshing dynamics using modal truncation. The fluid displacement field is expressed as

$$\eta(x, t) = \sum_{i=1}^{N} q_i(t)\Phi_i(x) \tag{8.25}$$

where $\Phi_i(x)$ are basis modes. Optimisation then modifies coefficients within a reduced stiffness and mass representation. This dramatically accelerates convergence while retaining sufficient accuracy.

Experimental validation of frequency shifting strategies involves modal testing using accelerometers and high-speed imaging. By exciting the tank at low amplitude and measuring free decay oscillations, the natural frequency can be extracted. Optimised designs typically exhibit measurable upward or downward shifts compared to baseline geometry, confirming predictive accuracy.

It is important to emphasise that frequency shifting does not necessarily aim to increase frequency universally. In some cases, lowering the sloshing frequency away from structural modes may be more effective. The direction of shift depends on external excitation characteristics and system-level design constraints. Optimisation therefore identifies the most beneficial spectral relocation rather than imposing a predetermined trend.

From a theoretical standpoint, shifting the sloshing frequency represents a direct manipulation of the system's dispersion relation. Rather than reacting to resonance after it emerges, engineers reshape the spectral landscape itself. This proactive control aligns with modern performance-based design philosophy in which dynamic behaviour is deliberately engineered at early design stages.

8.4 Introduction: sloshing as a critical challenge in structural design

Liquid sloshing is an inherently complex dynamic phenomenon that occurs when a partially filled container is subjected to external excitation. Such excitation may arise from seismic loading, vehicle acceleration and deceleration, wave action in marine environments, or operational disturbances in industrial systems. Under these conditions, the free surface of the liquid undergoes oscillatory motion, leading to time-varying hydrodynamic pressures and inertial forces acting on the container walls and its supporting structure.

From a mechanical engineering perspective, the most critical consequence of sloshing is not merely the presence of additional loads, but the dynamic coupling between the fluid motion and the supporting structure. Sloshing introduces dominant excitation frequencies that depend on the container geometry, fill level, gravitational acceleration, and boundary conditions. When these excitation frequencies approach the natural frequencies of the structural support system, resonance may occur. Resonance amplification can dramatically increase displacement amplitudes, stress levels, and fatigue damage, ultimately compromising structural integrity and service life.

Traditional structural design approaches often treat sloshing-induced loads using simplified equivalent static forces or conservative amplification factors. While such approaches may ensure safety, they tend to obscure the underlying dynamic mechanisms and often lead to excessive material usage. More importantly, they do not directly address the fundamental issue of frequency interaction between the sloshing fluid and the supporting structure.

The challenge is therefore not only to design a structure that is strong enough, but to design one that is dynamically compatible with the fluid it supports. This requirement motivates the adoption of design methodologies that explicitly incorporate dynamic behaviour into the early stages of structural synthesis.

8.5 Limitations of experience-based structural design for sloshing problems

In conventional engineering practice, structural layouts are typically proposed based on prior experience, empirical guidelines, and incremental modification of known design archetypes. While this approach has proven effective for many standard applications, it becomes increasingly inadequate for systems governed by complex multiphysics interactions such as fluid–structure coupling.

Sloshing-induced dynamics present several features that challenge intuition-based design. First, the spatial distribution of hydrodynamic pressure is highly nonuniform and time-dependent. Second, the dominant sloshing modes are often low-frequency and global in nature, making them particularly effective at exciting the fundamental modes of the support structure. Third, small geometric or stiffness changes can lead to significant shifts in natural frequencies, making manual tuning unreliable.

As a result, experience-based design often leads to overstiffened or overly massive structures, implemented as a precautionary measure rather than as an optimised solution. This not only increases material and manufacturing costs, but may also introduce new dynamic issues by shifting resonance into another operational regime.

To overcome these limitations, a design framework is required that relies on physical laws and computational optimisation rather than heuristic judgement. Topology optimisation provides such a framework by allowing structural configurations to emerge naturally from mathematically defined objectives and constraints.

Topology optimisation is a nonparametric optimisation technique that seeks the optimal material distribution within a prescribed design domain. Unlike traditional design methods, which assume a fixed structural layout, topology optimisation allows both geometry and connectivity to evolve freely during the optimisation process.

A defining characteristic of topology optimisation is that it is data-driven (2025 Najmon and Tovar, 2025; Hirotani et al., 2025; Yamasaki et al., 2021). The optimisation process relies directly on governing equations derived from mechanics, such as equilibrium, compatibility, and constitutive relationships. The final design is therefore not the result of engineering intuition, but the outcome of a systematic search for optimality within a mathematically defined design space.

This characteristic is particularly advantageous for sloshing-sensitive structures. Because sloshing–structure interaction is governed by well-defined physical principles, it can be incorporated into the optimisation problem through appropriate objective functions and constraints. The resulting designs inherently reflect the true dynamic behaviour of the system rather than simplified assumptions.

Furthermore, topology optimisation is especially valuable during the conceptual design stage, where design freedom is greatest, and performance improvements are most impactful. At this stage, decisions regarding stiffness distribution, load paths, and structural topology have long-lasting consequences that cannot easily be corrected later.

8.6 Design objective: avoiding resonance with sloshing motion

In this section, the primary design objective is to ensure that the tank support structure does not resonate with the dominant sloshing modes of the contained liquid. Rather than directly modelling the full fluid–structure interaction problem within the optimisation loop, a frequency-based strategy is adopted.

The central idea is to maximise the natural frequency of a selected structural mode, typically the first eigenfrequency, while limiting material usage through a volume constraint. Since sloshing excitation is generally dominated by low-frequency modes, increasing the fundamental natural frequency of the support structure effectively separates the structural response from the sloshing excitation spectrum.

This approach is physically intuitive and computationally efficient. By focusing on eigenfrequency optimisation, the essential dynamic characteristics of the structure are captured without the need for time-domain fluid simulations at every optimisation iteration.

The dynamic behaviour of the tank support structure is governed by the classical equation of motion for linear elastic systems:

$$\boldsymbol{M}\dot{\boldsymbol{U}} + \boldsymbol{K}\boldsymbol{U} = \boldsymbol{F} \tag{8.26}$$

where $\boldsymbol{M}$ is the global mass matrix, $\boldsymbol{K}$ is the global stiffness matrix, $\boldsymbol{U}$ is the displacement vector, and $\boldsymbol{F}$ represents external forces.

For free vibration analysis, the forcing term vanishes, leading to the generalised eigenvalue problem:

$$(\boldsymbol{K} - \omega^2\boldsymbol{M})\boldsymbol{U} = 0 \tag{8.27}$$

The eigenvalues ω represent the natural circular frequencies of the structure, while the corresponding eigenvectors are corresponding to the displacement and is define the mode shapes.

Both the stiffness and mass matrices depend on the material distribution within the design domain. This dependency forms the basis for topology optimisation of dynamic properties.

The design domain Ω_d represents the region in which material may be distributed to form the support structure. The domain is discretised into finite elements, and each element is assigned a continuous design variable ρ, representing the local material density.

$$0 < \rho \leq 1, ^\circ \forall \rho \in \Omega_d \tag{8.28}$$

A value of $\rho = 1$ corresponds to solid material, while values approaching zero represent void regions (Al Ali & Shimoda, 2022b; Shimoda, Umemura, et al., 2023).

The objective of the optimisation is to maximise the eigenfrequency of a specified mode (i), expressed as:

$$\max_{\rho} \omega_i \tag{8.29}$$

In the context of the present study, emphasis is placed on maximising the first eigenfrequency $\omega_{i=1}$, as it governs the global dynamic response and is most susceptible to excitation by sloshing motion.

Material usage is controlled through a global volume constraint:

$$\int_{\Omega_d} \rho, \, d\Omega \leq v \tag{8.30}$$

where v is the prescribed maximum volume fraction relative to the fully solid domain.

The topology optimisation problem can thus be formulated as:

$$\begin{array}{ll} & \omega_i \\ \max\limits_{\rho} & (\boldsymbol{K}(\rho) - \omega^2 \boldsymbol{M}(\rho))\boldsymbol{U} = 0 \\ \text{subject to} & \int_{\Omega_d} \rho, \, d\Omega \leq v \\ & 0 < \rho \leq 1^\circ \, \forall \rho \in \Omega_d \end{array} \tag{8.31}$$

This formulation directly links material distribution to dynamic performance.

The density-based approach, commonly known as the solid isotropic material with penalization (SIMP) method, interpolates material properties as functions of the density variable. The elastic modulus of each element is expressed as:

$$E(\rho) = \rho^p E_0 \tag{8.32}$$

where E_0 is the Young's modulus of the solid material and p is the penalisation exponent. Penalisation discourages intermediate densities by making them mechanically inefficient, thereby promoting clear solid–void designs.

The mass density is typically interpolated linearly:

$$\rho_m(\rho) = \rho \rho_0 \tag{8.33}$$

where ρ_0 is the mass density of the solid material.

Eigenfrequency is governed by the ratio of stiffness to mass. Increasing stiffness raises the eigenfrequency, while increasing mass lowers it. The density-based formulation explicitly captures this competition. During optimisation, material is therefore placed only where it contributes more to stiffness than to mass for the selected mode.

This balance is particularly important for sloshing-sensitive structures, where indiscriminate mass addition may reduce dynamic performance.

Efficient optimisation requires the sensitivity of the objective function with respect to the design variables. For eigenfrequency optimisation, the sensitivity of the eigenvalue with respect to density is given by:

$$\frac{\partial \omega_i}{\partial \rho} = \frac{1}{2\omega_i} \boldsymbol{U}_i^T \left(\frac{\partial \boldsymbol{K}}{\partial \rho} \omega_i^2 \frac{\partial \boldsymbol{M}}{\partial \rho} \right) \boldsymbol{U}_i \tag{8.34}$$

This expression reveals the physical mechanism of the optimisation process. Material is favoured in regions where it contributes strongly to modal stiffness while adding relatively little modal mass.

The optimisation problem is typically solved using gradient-based algorithms such as the Method of Moving Asymptotes. To ensure numerical stability and mesh-independent results, density or sensitivity filtering techniques are applied. These filters impose a minimum length scale and prevent numerical artifacts such as checkerboard patterns.

From a physical standpoint, the optimised support structure naturally aligns material along efficient load paths associated with the selected vibration mode. Regions experiencing high modal strain energy are reinforced, while material is removed from dynamically inactive regions. For tank support structures, this often results in truss-like or arching configurations that efficiently transmit inertial forces to the foundation. Importantly, these configurations are not prescribed in advance but emerge organically from the optimisation process. By maximising the first eigen-frequency, the resulting structure exhibits increased resistance to low-frequency excitation, which is characteristic of sloshing motion. This frequency separation strategy provides a direct and physically meaningful approach to mitigating sloshing-induced resonance.

8.7 Study of dynamic stability of a four-leg supported automotive fuel tank

This case study investigates the structural and dynamic behaviour of an automotive fuel tank supported by four vertical steel legs with square cross sections. The tank, representative of a realistic automotive geometry with a characteristic longitudinal length of approximately 0.7 m, is elevated on four rigid supports, each having a height of 0.5 m and a square cross section of 0.1 m × 0.1 m. The legs are bolted to a rigid base plate at the bottom and connected to the tank structure at the top. The tank is considered half-filled with gasoline, creating a free surface that is capable of undergoing significant sloshing motion under dynamic excitation. The overall system therefore consists of a coupled fluid–structure assembly in which liquid motion and structural vibration may interact.

The motivation behind this investigation arises from the possibility of resonance coupling between the sloshing frequency of the fuel inside the tank and the natural frequency of the supporting legs. When a partially filled tank is subjected to dynamic loading such as road irregularities, braking, cornering, or external harmonic excitation, the liquid develops gravity-driven free surface oscillations. At the same time, the structural support system behaves as an elastic framework with its own characteristic vibration modes. If the dominant sloshing frequency approaches the first bending mode of the support legs, resonance amplification may occur. Such coupling can lead to increased vibration amplitudes, elevated stress levels at mounting interfaces, accelerated fatigue accumulation, and reduced structural reliability. Therefore controlling the dynamic separation between fluid resonance and structural natural frequencies is a critical design requirement.

For a tank of this size and fill level, the fundamental sloshing frequency typically lies in the low-frequency range, often between one and a few hertz depending on geometry and excitation direction. The vertical support legs, acting mechanically as cantilever-like columns fixed at the base, exhibit bending-dominated vibration modes. Their natural frequency depends primarily on the elastic modulus of the material, the effective mass of the tank–fluid system, the second moment of area of the square cross-section, and the boundary conditions. With a leg height of 0.5 m and a solid square section of 0.1 m width, the stiffness may not automatically guarantee sufficient frequency separation from the sloshing response. Consequently, a detailed dynamic analysis is required to ensure that the structural eigenfrequency does not coincide with the dominant fluid mode.

The structural model is developed by representing each leg as a vertical elastic member rigidly fixed at the base and coupled to the tank mass at the top interface. Modal analysis is performed to extract the fundamental bending frequency and higher vibration modes. Particular attention is given to the first bending mode, as it typically governs resonance interaction. The total effective mass includes both the tank structure and the contained liquid, as fluid inertia contributes to the global dynamic behaviour. By evaluating the modal characteristics of the complete assembly, the potential for resonance overlap can be identified.

In parallel, the sloshing behaviour of the half-filled tank is evaluated using multiphase computational fluid dynamics. The free surface evolution underprescribed dynamic excitation allows identification of the dominant sloshing frequency. Analytical estimation of sloshing modes provides an initial approximation, while numerical simulation captures nonlinear effects and geometric influences. Comparing the fluid natural frequency with the structural eigenfrequency establishes the resonance margin of the system.

To achieve adequate frequency separation, strategies focus on increasing the stiffness-to-mass ratio of the support legs rather than merely increasing mass. The bending stiffness of each leg is governed by the second moment of area, which scales strongly with cross-sectional dimensions. However, increasing the solid cross-section increases weight and may not be structurally efficient. Therefore alternative geometric modifications such as hollow square sections, internal reinforcement ribs, tapered profiles, or topology optimisation can be employed to enhance stiffness while controlling mass. The design objective is to shift the structural natural frequency sufficiently away from the sloshing frequency band, thereby minimising the risk of dynamic amplification.

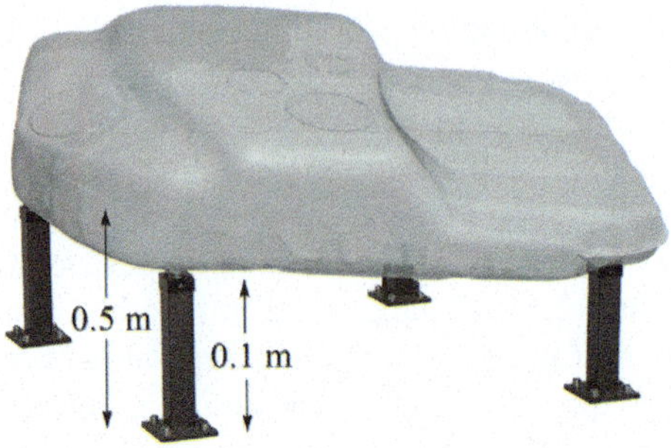

Figure 8.1 Four-legged supported automotive fuel tank configuration.

The computational modelling framework used for topology optimisation of the tank support legs is illustrated in Fig. 8.1.

The design domain corresponds to one representative leg of the four-leg support system that carries the tank mass and transfers dynamic loads to the foundation. Two loading scenarios were considered in order to account for realistic operational conditions. In the first configuration, the tank mass is assumed to act concentrically relative to the leg axis, representing an idealised symmetric load transfer condition. In the second configuration, the mass is shifted towards one side of the support, simulating eccentric loading conditions that may arise during sloshing events when liquid mass redistribution induces transient shifts in the centre of gravity. These two cases allow the optimisation algorithm to account for both symmetric and asymmetric dynamic loading states.

The primary objective of the topology optimisation is to minimise structural volume while simultaneously increasing the natural frequency of the support leg. The fundamental design requirement is to shift the structural eigenfrequency of the legs away from the dominant sloshing resonance frequency of the tank. When the natural frequency of the support structure coincides with the sloshing frequency of the contained fluid, dynamic amplification may occur, leading to excessive vibration amplitudes and potential structural fatigue. Therefore the optimisation problem is formulated to increase stiffness-to-mass ratio while reducing overall material usage.

To ensure consistency in the parametric study and isolate geometric influence from material variability, the Young modulus and density are normalised to unity. The Poisson ratio is fixed at 0.3, representing a typical isotropic structural material. This nondimensional formulation allows the optimisation process to focus exclusively on the geometric distribution of material within the design domain rather than absolute material scaling effects.

The resulting optimised geometry is shown in Fig. 8.2. The algorithm converges towards an elongated hollow pillar configuration characterised by strategically distributed material along principal stress paths. Instead of a solid column, the optimised structure exhibits internal cavities that remove material from regions of low strain energy while preserving load-bearing pathways aligned with the dominant bending and axial stress directions. The final geometry reveals pronounced vertical

Figure 8.2 Topology optimised support leg geometry.

stiffening members oriented along the principal loading axis, accompanied by horizontal reinforcement features that enhance torsional rigidity under eccentric loading. The cross-sectional profile indicates a hollow core that significantly reduces mass without compromising structural stiffness. This internal void formation is not arbitrary; it emerges as a consequence of minimising compliance under dynamic constraints while satisfying volume reduction targets.

The geometry also exhibits a gradual tapering towards the lower boundary where the leg interfaces with the foundation. This narrowing reflects the redistribution of bending moments along the leg height. The base region experiences high constraint reactions, and therefore material concentration remains sufficient to maintain local stability. Towards the mid-height region, where bending curvature is dominant, material is arranged in a tubular-like pattern that maximises the second moment of area. This geometric configuration enhances flexural stiffness and contributes directly to the upward shift of natural frequency. The upper region of the leg, interfacing with the tank support frame, shows symmetric reinforcement to maintain alignment under combined axial and lateral dynamic forces.

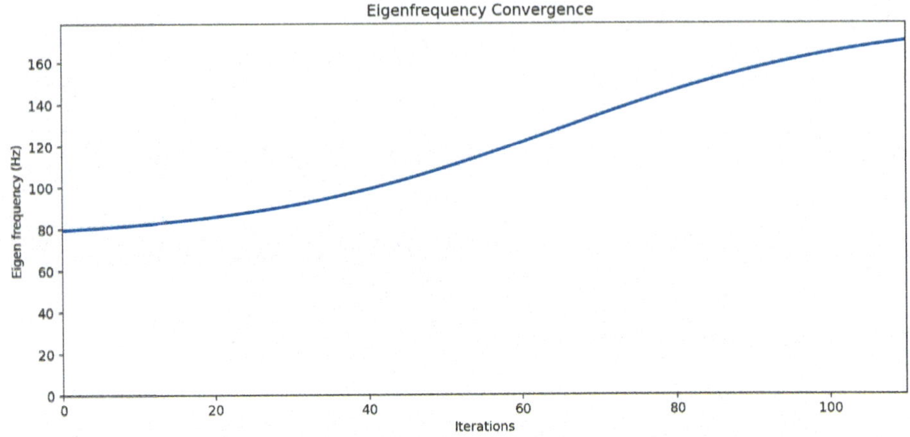

Figure 8.3 Convergence history of the fundamental eigenfrequency during topology optimisation.

Fig. 8.3 presents the eigenfrequency evolution throughout the optimisation process. The horizontal axis represents the iteration number, while the vertical axis denotes the fundamental natural frequency of the leg structure. At the beginning of the optimisation, the frequency is relatively low due to inefficient material distribution in the initial solid design. As iterations progress, the frequency increases rapidly, indicating improved stiffness-to-mass efficiency. Around the mid-stage of the optimisation cycle, the rate of improvement decreases, and the curve gradually approaches a plateau. This plateau signifies convergence of the topology optimisation algorithm towards a near-optimal structural configuration.

The convergence behaviour demonstrates two important characteristics. First, the steep initial increase in eigenfrequency indicates effective redistribution of material from low-strain regions to high-strain regions. Second, the smooth stabilisation of the curve confirms numerical robustness and stability of the optimisation process. The absence of oscillatory behaviour in the convergence history suggests that the solution space is well-conditioned and that the algorithm effectively balances volume constraint and frequency maximisation.

Quantitatively, the optimised leg design achieves a significant upward shift in the fundamental eigenfrequency relative to the nonoptimised baseline configuration. The increase in natural frequency ensures that the structural resonance of the support system is separated from the dominant sloshing frequency range of the tank. This frequency separation reduces the likelihood of dynamic coupling between fluid motion and structural vibration. At the same time, the optimised configuration achieves substantial material reduction, resulting in a lighter support structure without compromising stiffness.

The spatial distribution of volume reduction follows a physically meaningful pattern. Material removal is concentrated in regions where strain energy density is minimal, typically near the central core of the original solid column. Conversely, material is retained and even redistributed along load paths corresponding to bending and axial stress trajectories. This principle aligns with classical structural optimisation theory and resembles strategies used in large-scale civil engineering structures, where hollow or tubular sections provide high stiffness with reduced weight.

From a dynamic systems perspective, the optimisation of the four legs collectively enhances the global vibration characteristics of the tank-support assembly. Because the legs function as boundary constraints for the tank, their stiffness directly influences the coupled structural-fluid dynamic response. By increasing the eigenfrequency of the legs beyond the sloshing resonance frequency, the structure avoids resonance amplification during transient fluid excitation. This decoupling mechanism improves structural durability and reduces cyclic stress accumulation at the tank-support interface.

Furthermore, the inclusion of eccentric loading conditions in the optimisation ensures robustness against asymmetric sloshing events. During intense sloshing, transient shifts in liquid mass can induce lateral and torsional forces on the support structure. The optimised geometry demonstrates improved torsional stiffness due to its distributed material layout, thereby enhancing resistance to off-axis excitation.

As illustrated in Fig. 8.4, the topology optimised configuration of the tank support leg embodies a structurally efficient geometry that simultaneously reduces structural mass and increases the fundamental natural frequency of the support system.

The resulting design evolves from a conventional prismatic column into a hollow, load-path-oriented structure in which material is redistributed along principal stress trajectories. Regions subjected to low strain energy are systematically removed, while highly stressed zones are reinforced, producing a tapered, internally relieved pillar with improved stiffness to mass performance. This redistribution enhances bending rigidity without proportional mass increase, thereby shifting the eigenfrequency of the leg upward and creating a clear separation from the dominant sloshing frequencies of the partially filled tank. The dynamic decoupling between structural resonance and fluid oscillation is therefore achieved not by overdesign, but through intelligent material placement.

It is important to recognise that topology optimisation often generates geometries that are difficult to manufacture using conventional subtractive or formative

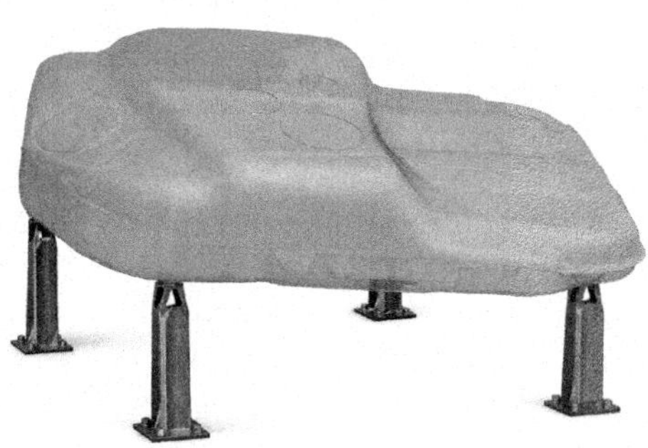

Figure 8.4 Final additively manufacturable topology optimised support leg design for a sloshing-resistant fuel tank application.

processes (Al Ali et al., 2024; Al Ali, 2025; Miki & Yamada, 2021; Shu et al., 2025; Tian et al., 2024). The algorithm does not constrain itself to prismatic beams, constant-thickness plates, or easily machined cavities. Instead, it produces organically evolving shapes with smooth transitions, internal voids, nonuniform cross sections, and curvature-driven load paths. In many engineering applications, these optimised geometries remain theoretical because casting, milling, or extrusion processes impose geometric restrictions such as draft angles, tool accessibility, or uniform wall thickness. However, in the present case, the optimised leg configuration remains manufacturable due to the absence of enclosed internal cavities and the presence of accessible hollow regions. Even if the geometry were more complex, modern additive manufacturing technologies would enable its realisation without significant geometric compromise.

Additive manufacturing, also known as layer-wise fabrication, is a production methodology in which components are built directly from digital three-dimensional models by sequentially depositing or consolidating material layer by layer (Hikasa et al., 2024; Ishimoto et al., 2017; Shimoda, Hikasa, et al., 2023). Unlike subtractive manufacturing, where material is removed from a bulk solid, additive manufacturing constructs the geometry only where material is structurally required. This approach inherently aligns with topology optimisation, since both methodologies are based on material efficiency and geometric freedom.

In metal additive manufacturing, several technologies are commonly employed. One of the most established methods is powder bed fusion, which includes selective laser melting and electron beam melting. In these processes, a thin layer of metallic powder, such as titanium alloy, stainless steel, or aluminium alloy, is spread across a build platform. A high-energy laser or electron beam selectively melts the powder according to the cross-sectional slice of the digital model. Once the layer solidifies, the platform lowers, a new powder layer is deposited, and the process repeats until the full three-dimensional component is completed. This technique enables the production of intricate internal channels, lattice structures, variable wall thickness regions, and smooth curvature transitions that are nearly impossible to fabricate using conventional machining.

Another metal additive technique is directed energy deposition, where metallic powder or wire is fed into a focused energy source and deposited onto a substrate in a controlled manner. This method is particularly suitable for larger structural components and repair applications. Binder jetting and metal extrusion processes also exist, each offering different advantages in terms of cost, surface finish, mechanical properties, and scalability.

A key advantage of additive manufacturing is geometric complexity without a cost penalty. In traditional processes, complexity increases manufacturing time and expense. In additive systems, the cost is primarily related to material volume and build time, rather than geometric intricacy. Therefore topology optimised structures with curved ribs, tapered profiles, or nonuniform cross sections can be produced without additional tooling or mould design. Furthermore, additive manufacturing allows consolidation of multiple parts into a single monolithic structure, reducing assembly interfaces and potential stress concentrations.

From a mechanical perspective, additively manufactured metals exhibit microstructures influenced by rapid solidification and thermal gradients. Postprocessing treatments such as heat treatment, hot isostatic pressing, or surface finishing may be applied to enhance mechanical strength, reduce porosity, and improve fatigue performance. For structural support legs subjected to dynamic loading, fatigue resistance and surface integrity are critical considerations. Proper parameter control and postprocessing ensure that the mechanical performance meets engineering requirements.

In the context of the optimised tank support legs, additive manufacturing provides a direct pathway from numerical optimisation to physical realisation. The tapered, hollow, and locally thickened geometry can be fabricated as a single integrated component with base plates and mounting interfaces included in the digital model. The resulting structure preserves the stiffness-to-mass benefits predicted by simulation, enabling effective upward shifting of structural eigenfrequency and reliable separation from sloshing-induced excitation frequencies.

References

Abass, R. S., Al Ali, M., & Al Ali, M. (2019). Shape and topology optimization design for total hip joint implant. In *Proceedings of the World Congress on Engineering 2019, WCE 2019, July 3-5, 2019, London, UK*. Newswood Limited, 1–6. https://www.iaeng.org/publication/WCE2019/WCE2019_pp559-564.pdf.

Al Ali, M. (2025). *Advanced Techniques in Porous Structure Design for Additive Manufacturing*. Wiley. 10.1002/9781394312719.

Al Ali, M., & Shimoda, M. (2022a). Investigation of concurrent multiscale topology optimization for designing lightweight macrostructure with high thermal conductivity. *International Journal of Thermal Sciences, 179*, 107653. https://doi.org/10.1016/j.ijthermalsci.2022.107653.

Al Ali, M., & Shimoda, M. (2022). Toward multiphysics multiscale concurrent topology optimization for lightweight structures with high heat conductivity and high stiffness using MATLAB. *Structural and Multidisciplinary Optimization, 65*, 207. https://doi.org/10.1007/s00158-022-03291-0.

Al Ali, M., & Shimoda, M. (2024). Exploring the influence of initial design domain dependencies in concurrent multiscale topology optimization for heat conductivity maximization. *Computer Physics Communications, 295*, 108968. https://doi.org/10.1016/j.cpc.2023.108968.

Al Ali, M., Shimoda, M., & Naguib, M. (2024). Optimizing Additive Manufacturable Structures with Computer Vision to Enhance Material Efficiency and Structural Stability. *Modelling, 5*(3), 1286–1297. https://doi.org/10.3390/modelling5030066.

Al Ali, M., Shimoda, M., Benaissa, B., & Kobayashi, M. (2023). Non-parametric optimization for lightweight and high heat conductive structures under convection using metaheuristic structure binary-distribution method. *Applied Thermal Engineering, 233*, 121124. https://doi.org/10.1016/j.applthermaleng.2023.121124.

Bendsøe, M. P. (1989). Optimal shape design as a material distribution problem. *Structural Optimization, 1*(4), 193–202. https://doi.org/10.1007/bf01650949.

Bendsøe, M. P., & Kikuchi, N. (1988). Generating optimal topologies in structural design using a homogenization method. *Computer Methods in Applied Mechanics and Engineering, 71*(2), 197–224. 10.1016/0045-7825(88)90086-2.

Hegemier, G. A., & Prager, W. (1969). On Michell trusses. *International Journal of Mechanical Sciences, 11*(2), 209–215. https://doi.org/10.1016/0020-7403(69)90006-X.

Hikasa, M., Shimoda, M., & Musaddiq, A. A. (2024). Concurrent multi-scale optimization of macro- and micro-shapes of laminated porous shell structure. *Structural and Multidisciplinary Optimization, 67*, 57. https://doi.org/10.1007/s00158-023-03726-2.

Hirotani, S., Yaji, K., Makihara, K., & Otsuka, K. (2025). Data-Driven Real-Time Topology Optimization Using Consistent Rotation-Based Moving Morphable Components. *AIAA International, Japan AIAA Journal, 63*(10), 4491–4497. https://doi.org/10.2514/1.J065458.

Ishimoto, T., Hagihara, K., Hisamoto, K., Sun, S. H., & Nakano, T. (2017). Crystallographic texture control of beta-type Ti–15Mo–5Zr–3Al alloy by selective laser melting for the development of novel implants with a biocompatible low Young's modulus. *Scripta Materialia, 132*, 34–38. https://doi.org/10.1016/j.scriptamat.2016.12.038.

Liu, Y., & Shimoda, M. (2014). A non-parametric solution to shape identification problem of free-form shells for desired deformation mode. *Computers and Structures, 441*, 1–11. 10.1016/j.compstruc.2014.07.020.

Michell, A. G. M. (1904). LVIII. The limits of economy of material in frame-structures. *The London, Edinburgh, and Dublin Philosophical Magazine and Journal of Science, 8*(47), 589–597. https://doi.org/10.1080/14786440409463229.

Miki, T., & Yamada, T. (2021). Topology optimization considering the distortion in additive manufacturing. *Finite Elements in Analysis and Design, 193*, 103558. https://doi.org/10.1016/j.finel.2021.103558.

Najmon, J. C., & Tovar, A. (2025). Multiscale topology optimization using machine learning and displacement-driven microstructures. *Structural and Multidisciplinary Optimization, 68*, 172. https://doi.org/10.1007/s00158-025-04115-7.

Neves, M. M., Sigmund, O., & Bendsøe, M. P. (2002). Topology optimization of periodic microstructures with a penalization of highly localized buckling modes. *International journal for numerical methods in engineering, 54*(6), 809–834. https://doi.org/10.1002/nme.449.

Shimoda, M., Hikasa, M., & Al Ali, M. (2023). Micropore shape optimization of porous laminated shell structures. *Additive Manufacturing, 69*, 103530. https://doi.org/10.1016/j.addma.2023.103530.

Shimoda, M., Umemura, M., Al Ali, M., & Tsukihara, R. (2023). Shape and topology optimization method for fiber placement design of CFRP plate and shell structures. *Composite Structures, 309*, 116729. https://doi.org/10.1016/j.compstruct.2023.116729.

Shimoda, M., Fujita, J., Al Ali, M., & Kamiya, A. (2024). Concurrent multiscale and multi-material optimization method for natural vibration design of porous structures. *International journal for numerical methods in engineering, 125*(7). https://doi.org/10.1002/nme.7424.

Shu, Z., Gao, L., & Li, H. (2025). Data-driven multiscale topology optimization of solid-lattice hybrid structures for additive manufacturing. *Additive Manufacturing, 110*, 104920. https://doi.org/10.1016/j.addma.2025.104920.

Sigmund, O. (1994). Materials with prescribed constitutive parameters: An inverse homogenization problem. *International Journal of Solids and Structures, 31*(17), 2313–2329. https://doi.org/10.1016/0020–7683(94)90154-6.

Suzuki, K., & Kikuchi, N. (1991). A homogenization method for shape and topology optimization. *Computer Methods in Applied Mechanics and Engineering, 93*(3), 291–318. https://doi.org/10.1016/0045–7825(91)90245-2.

Tian, W., Li, Q., Wang, Q., Chen, D., & Gao, W. (2024). Additive manufacturing error quantification on stability of composite sandwich plates with lattice-cores through machine learning technique. *Composite Structures, 327*, 117645. https://doi.org/10.1016/j.compstruct.2023.117645.

Yaji, K., Yamada, T., Kubo, S., Izui, K., & Nishiwaki, S. (2015). A topology optimization method for a coupled thermal-fluid problem using level set boundary expressions. *International Journal of Heat and Mass Transfer, 81*, 878–888. https://doi.org/10.1016/j.ijheatmasstransfer.2014.11.005.

Yamasaki, S., Yaji, K., & Fujita, K. (2021). Data-driven topology design using a deep generative model. *Structural and Multidisciplinary Optimization, 64*(3), 1401–1420. https://doi.org/10.1007/s00158-021-02926-y.

Zeng, T., Wang, H., Yang, M., & Alexandersen, J. (2020). Topology optimization of heat sinks for instantaneous chip cooling using a transient pseudo-3D thermofluid model. *International Journal of Heat and Mass Transfer, 154*, 119681. https://doi.org/10.1016/j.ijheatmasstransfer.2020.119681.

Future outlook

9

9.1 Introduction

In practical engineering systems, sloshing cannot be treated as a simple free surface oscillation problem governed solely by gravity waves inside an idealised rigid container. Real-world sloshing is inherently multiphysics in nature. It involves strong nonlinear coupling between fluid inertia, structural flexibility, excitation dynamics, and often thermal or gas phase interactions. When a partially filled container is subjected to external acceleration, whether from vehicle braking, seismic excitation, marine motion, or rocket thrust variation, the liquid does not merely oscillate harmonically. Instead, it redistributes mass, modifies pressure fields, alters the structural load path, and interacts dynamically with the support structure. The system therefore behaves as a coupled oscillator composed of a distributed fluid mass and an elastic structural framework. Understanding this coupling is the foundation of advanced practical sloshing dynamics.

At the core of the phenomenon lies the nonlinear free surface. Under small amplitude excitation, linear potential theory may provide approximate predictions of natural sloshing frequencies. However, in practical conditions, excitation amplitudes are rarely infinitesimal. As wave elevation increases, geometric nonlinearities become dominant. The free surface curvature evolves in time, leading to energy transfer between modes, frequency shift, and potential wave breaking. This nonlinear behaviour modifies the apparent stiffness of the fluid system. Consequently, the sloshing frequency is not constant but amplitude dependent. In severe cases, hydraulic jumps and impact loads occur, generating localised pressure peaks that far exceed predictions from linear models. These impact loads are particularly critical in aerospace propellant tanks and automotive fuel systems, where structural fatigue and vibration transmission must be carefully controlled. The second fundamental component is fluid–structure interaction. In many simplified analyses, the tank wall is assumed rigid. In reality, structural flexibility alters the dynamic response significantly. The liquid exerts time-varying hydrodynamic pressure on the wall, and the wall deformation feeds back into the fluid domain by modifying boundary conditions. This bidirectional coupling forms a fully coupled eigenvalue problem. The combined system has natural modes that differ from those of the isolated fluid or structure alone. In slender supports or lightweight frames, structural modes may approach the sloshing frequency range. When spectral proximity occurs, resonance amplification can arise. Advanced analysis, therefore requires solving the coupled fluid momentum equations and structural elasticity equations simultaneously, ensuring consistency of kinematic and dynamic interface conditions.

Damping mechanisms in real systems are also multiphysical. Pure inviscid theory predicts perpetual oscillation. In practice, energy dissipates through several channels.

Slosh Dynamics. DOI: https://doi.org/10.1016/B978-0-443-44330-5.00004-9

Viscous shear within boundary layers converts kinetic energy into heat. Turbulent eddies form during violent sloshing, increasing dissipation. Structural material damping absorbs part of the vibrational energy transmitted from the fluid. Additionally, if the tank contains a gas phase, compressibility effects and interfacial interactions contribute further damping. In cryogenic storage applications, thermodynamic coupling between liquid and vapour phases introduces additional complexity, since evaporation and condensation modify mass distribution and pressure evolution. Accurately capturing these damping contributions is essential for realistic amplitude prediction. External excitation characteristics strongly influence system behaviour. Practical tanks experience broadband and transient inputs rather than single-frequency sinusoidal forcing. Vehicle acceleration profiles include stochastic components generated by road irregularities. Seismic events involve impulsive loading with broad frequency content. In aerospace systems, thrust oscillations and attitude control manoeuvres produce multidirectional acceleration fields. Because sloshing behaves as a frequency-selective amplifier, the overlap between the excitation spectrum and system eigenfrequencies determines response severity. Therefore spectral analysis becomes a key tool in practical design. The objective is not only to compute natural frequencies but also to understand modal participation factors and frequency bandwidth under operational loading. From a computational perspective, the governing equations are the Navier-Stokes equations with free surface tracking, coupled with structural elasticity equations. Interface capturing techniques such as the Volume of Fluid method have become standard in industrial solvers. However, high-fidelity simulation of violent sloshing remains numerically challenging. Interface smearing, artificial numerical diffusion, and time step restrictions can compromise accuracy. Moreover, when coupled with structural solvers, stability of partitioned algorithms must be carefully managed to avoid divergence due to added mass effects. Recent developments emphasise strongly coupled schemes and adaptive mesh refinement near the free surface to preserve sharp interface representation while maintaining computational efficiency.

An additional layer of complexity arises from geometric irregularities. Real tanks rarely possess perfect symmetry. Automotive fuel tanks incorporate internal baffles, curved walls, and packaging constraints. Marine liquefied natural gas (LNG) tanks may have membrane containment systems. The geometry directly influences modal shape and frequency. Even a small geometric asymmetry can split degenerate modes and alter wave patterns. Therefore realistic modelling requires precise geometric representation rather than idealised shapes. This trend reflects the broader movement in 2025 and 2026 towards digital twin methodology, where the computational model mirrors the actual manufactured system, including structural supports and mounting conditions. Thermal coupling is increasingly relevant in advanced applications. In electric vehicles and hydrogen storage systems, tanks operate in thermally varying environments. Temperature gradients modify fluid density and viscosity, thereby influencing sloshing dynamics. In cryogenic propellant tanks, stratification may develop due to thermal layering, creating density gradients that alter natural frequencies. In such cases, the problem becomes thermo-fluid structural interaction rather than purely hydrodynamic. Multiphysics solvers now integrate heat transfer and phase change modelling into sloshing simulations to account for these effects.

Another important aspect is scale interaction. Sloshing involves macroscopic wave motion but may also generate small-scale turbulence and vortical structures. Multiscale modelling approaches aim to resolve large-scale dynamics accurately while representing small-scale turbulence through advanced models or subgrid techniques. In recent research trends, reduced-order models trained using high-fidelity simulation data provide efficient prediction capability for control and optimisation studies. Machine learning algorithms classify dominant sloshing modes and predict amplitude growth under varying excitation conditions, enabling rapid design iteration without full computational fluid dynamics (CFD) re-computation. Uncertainty quantification has also gained importance. Manufacturing tolerances, fill level variation, and material property variability introduce uncertainty in system response. Probabilistic analysis frameworks evaluate how these uncertainties affect resonance risk and pressure peak distribution. This approach moves sloshing design from deterministic evaluation toward reliability-based engineering.

9.2 Frequency tailoring and resonance management

In advanced practical sloshing dynamics, the objective is no longer confined to calculating natural frequencies and reporting potential resonance conditions. The paradigm has shifted towards frequency engineering, where the designer intentionally manipulates the dynamic characteristics of both the fluid domain and the supporting structure in order to prevent spectral overlap under realistic operational excitations. In modern lightweight systems such as electric vehicles, aerospace propellant tanks, and marine storage units, structural mass is minimised for efficiency. This reduction in stiffness-to-mass ratio often brings structural eigenfrequencies closer to the fundamental sloshing modes. If untreated, such proximity can trigger resonance amplification, leading to excessive vibration, structural fatigue, noise transmission, and even loss of stability. Therefore frequency tailoring has become a deliberate design strategy rather than a passive analytical step. Once the fluid eigenfrequencies are characterised, attention shifts towards structural eigenmode control. Structural frequency shifting is achieved through the redistribution of stiffness and mass within the supporting framework. Increasing stiffness locally may elevate a critical structural mode, while selective mass redistribution may reduce modal participation in the frequency range of interest. The design objective is to create sufficient spectral separation between dominant sloshing modes and structural modes so that resonance interaction is minimised. Importantly, this process must preserve overall mechanical performance, manufacturability, and weight constraints. Therefore structural modification is not arbitrary but guided by sensitivity analysis and modal participation factors that quantify the contribution of each mode to system response.

Topology optimisation has emerged as a powerful tool for structural eigenmode tailoring. Rather than relying solely on intuitive reinforcement strategies, topology optimisation distributes material within a prescribed design domain to achieve targeted dynamic objectives. In the context of sloshing, the optimisation problem may include eigenfrequency constraints that push structural modes away from fluid modes, or maximise frequency separation while minimising mass. Because topology optimisation is data-driven and does not depend purely on prior engineering intuition, it enables exploration of nonconventional structural layouts that achieve superior stiffness-to-weight ratios. With advances in additive manufacturing, these optimised geometries can now be fabricated with high precision, allowing complex load paths and internal lattice structures that were previously impractical. As a result, frequency engineering has transitioned from a theoretical possibility to a manufacturable reality in 2025 and beyond. Sensitivity analysis plays a crucial role in understanding how variations in liquid depth and geometry influence resonance behaviour. Sloshing frequency is strongly dependent on fill level, and operational tanks rarely operate at a single constant volume. During vehicle motion or fuel consumption, liquid depth changes continuously, shifting the natural frequency of the fluid domain. Similarly, geometric parameters such as tank curvature, aspect ratio, and internal partitions alter modal distribution. Sensitivity studies quantify how small changes in these parameters influence eigenvalues and modal shapes. This information guides robust design strategies that maintain adequate spectral separation across the entire operational envelope rather than at a single nominal condition. Such robustness is particularly important in aerospace and hydrogen storage systems where safety margins must be guaranteed. Experimental modal validation remains an indispensable component of frequency tailoring. Numerical predictions must be confirmed through controlled testing to ensure physical reliability. Modal testing of the empty structure provides baseline structural frequencies, while partially filled tank experiments reveal coupled fluid–structure modes. Techniques such as impact hammer testing, shaker excitation, and high-speed-free-surface visualisation allow measurement of modal shapes and damping ratios. By comparing experimental data with simulation results, model parameters can be calibrated, improving predictive capability.

9.3 Industrial computational fluid dynamics and high fidelity simulation strategies

Accurate prediction of practical sloshing phenomena requires more than solving the governing equations of fluid motion. It demands a carefully designed computational workflow that balances physical fidelity, numerical stability, and computational efficiency. Industrial computational fluid dynamics has evolved into a mature discipline capable of resolving complex free surface flows under transient and nonlinear excitation. However, sloshing remains one of the most demanding applications because it combines moving interfaces, strong inertial forces, turbulence generation, impact loading, and sometimes fluid–structure coupling. For this reason, high fidelity simulation strategies must be constructed systematically, beginning with interface capturing methodology and extending to mesh management, turbulence modelling, time integration control, and large-scale parallel computation. Mesh strategy represents the second pillar of reliable simulation. Sloshing flows exhibit localised gradients near the free surface, especially during impact events against tank walls. Uniform coarse meshes fail to capture these gradients accurately, whereas uniformly fine meshes become computationally prohibitive. Adaptive mesh refinement offers an effective compromise. By dynamically refining cells in regions of high interface curvature or pressure gradient, the solver concentrates resolution where it is physically required. This adaptive capability enhances the prediction of peak loads and wave steepness while maintaining manageable cell counts elsewhere in the domain. In addition, boundary layer refinement near solid walls improves representation of viscous shear and damping effects. Mesh independence studies remain mandatory, as sloshing predictions can be highly sensitive to spatial resolution, particularly in nonlinear regimes. Turbulence modelling introduces another layer of complexity. While low-amplitude sloshing may remain laminar, violent excitation generates vortical structures and turbulent mixing within the liquid. Standard turbulence models developed for steady channel flow are not always directly applicable. The transient and highly anisotropic nature of sloshing demands careful model selection. Reynolds-averaged models provide computational efficiency but may underestimate transient vortical energy. Large eddy simulation resolves large-scale turbulent structures explicitly, offering improved physical realism at the cost of increased computational demand. In practice, model selection depends on the intensity of sloshing and the required level of detail. For impact load prediction, capturing large-scale vortices and coherent structures becomes important because they influence pressure distribution and energy dissipation. Therefore turbulence modelling must be aligned with the specific engineering objective of the study.

Temporal discretisation is equally critical. Sloshing involves rapid changes in velocity and pressure, especially during wall impact and wave overturning. An excessively large time step introduces numerical damping and phase error, distorting predicted frequency and amplitude. Conversely, an extremely small time step increases computational cost without proportional benefit if spatial resolution is insufficient. Stability constraints such as the Courant number provide guidance, but physical time scale resolution must also be considered. The natural sloshing period should be resolved by multiple time steps to capture the dynamic evolution accurately. Implicit time integration schemes often enhance stability for stiff multiphase problems, yet they must be combined with appropriate convergence criteria to prevent artificial dissipation. Careful monitoring of energy balance and wave amplitude evolution helps detect excessive numerical damping.

Management of numerical dissipation represents a subtle but important challenge. All discretization schemes introduce some level of artificial diffusion. In sloshing simulations, excessive numerical dissipation may suppress wave amplitude growth and underestimate resonance effects. Conversely, insufficient stabilisation may generate unphysical oscillations near the interface. Achieving the correct balance requires the selection of higher-order spatial schemes and controlled interface compression. Validation against experimental data is essential to ensure that predicted damping arises from physical viscosity rather than numerical artefact. Grid convergence and time step refinement studies provide quantitative measures of numerical error and support confidence in simulation reliability. High-performance computing infrastructure enables large-scale sloshing simulations with millions of computational cells and long transient durations. Parallel scaling across multiple processors reduces wall clock time, making parametric studies feasible. Efficient domain decomposition ensures balanced workload distribution and minimises communication overhead between processors. In addition to traditional central processing units, graphics processing units (GPUs) offer substantial acceleration for multiphase flow calculations. GPU-based solvers exploit massive parallelism to accelerate matrix operations and interface advection algorithms. Such acceleration is particularly valuable for design optimisation workflows where repeated simulations are required. Nonetheless, scalability must be evaluated carefully, since communication latency and memory bandwidth limitations can reduce parallel efficiency if not properly managed. An integrated computational workflow for sloshing, therefore begins with accurate geometric modelling, proceeds through careful mesh generation with adaptive refinement capability, applies robust Volume of Fluid interface tracking, selects turbulence models consistent with flow regime, and implements time stepping strategies that preserve physical fidelity. Throughout the process, validation against experimental measurements and sensitivity analysis ensures reliability. The ultimate objective is not merely to visualise free surface motion, but to quantify hydrodynamic pressure, modal interaction, damping characteristics, and structural loading under realistic excitation. High-fidelity simulation becomes a decision-making tool for structural design, frequency tailoring, and safety evaluation.

9.4 Emerging frontiers: artificial intelligence in sloshing dynamics

Artificial intelligence is redefining how sloshing dynamics are analysed, predicted, and controlled. Traditionally, sloshing assessment relied on analytical approximations, experimental testing, and high-fidelity computational fluid dynamics. While these approaches remain fundamental, they are computationally intensive and often limited in real-time applicability. Artificial intelligence introduces a complementary paradigm in which data-driven models learn the underlying dynamic behaviour of fluid motion and provide rapid predictive capability across a wide range of operating conditions. In practical engineering systems, this shift enables faster design iteration, improved robustness assessment, and adaptive control strategies.

One of the most significant developments is the construction of reduced-order models using machine learning. High fidelity simulations generate large datasets describing free surface elevation, pressure distribution, and velocity fields under varying fill levels and excitation spectra. Neural networks can be trained on this data to approximate the nonlinear mapping between input excitation and sloshing response. Once trained, these surrogate models evaluate system behaviour in fractions of a second compared to hours or days required for full computational fluid dynamics. This capability is particularly valuable in optimisation workflows, where repeated simulations are needed to adjust geometry or structural stiffness.

Rather than replacing physics-based solvers, artificial intelligence augments them by accelerating parametric exploration (Huo & Lyu, 2025; Ma et al., 2024; Moya et al., 2019; Wu et al., 2025).

Mode classification and resonance prediction represent another powerful application. Sloshing systems exhibit multiple eigenmodes whose interaction with structural modes determines resonance risk. Machine learning algorithms can analyse time series data from simulation or experiment and automatically identify dominant modal contributions. Pattern recognition techniques classify wave behaviour into regimes such as linear oscillation, nonlinear amplification, or impact-dominated motion. By correlating excitation characteristics with modal response, predictive models can estimate the likelihood of resonance without performing full transient analysis. This approach enhances safety assessment in automotive, aerospace, and marine applications where operational conditions vary continuously (illustrated in Fig. 9.1).

Artificial intelligence also supports uncertainty quantification. Real systems experience variability in fill level, temperature, material properties, and manufacturing tolerances. Traditional deterministic simulation evaluates only nominal conditions. Probabilistic frameworks combined with machine learning enable rapid evaluation of thousands of parameter combinations. Bayesian inference methods update predictive models based on new experimental measurements, refining confidence intervals of pressure peaks and frequency shifts. This integration transforms sloshing analysis into a reliability-based engineering discipline rather than a single scenario evaluation (Filo & Lempa Paweland Wisowski, 2025; Wei et al., 2021).

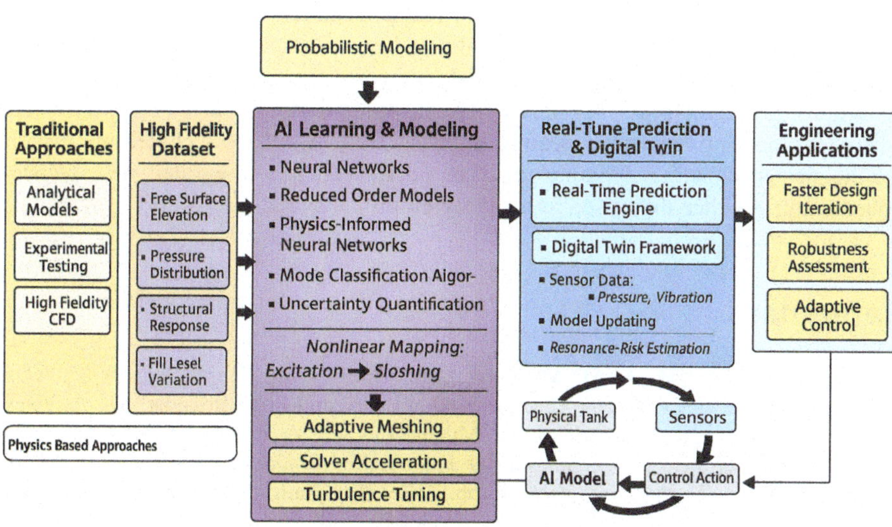

Figure 9.1 Machine learning enhanced sloshing design framework integrating high fidelity simulation, surrogate modelling, and resonance prediction for rapid multiphysics analysis and frequency engineering.

In the context of digital twins, artificial intelligence enables continuous synchronisation between computational models and physical systems. Sensors installed on tanks measure acceleration, pressure, and structural vibration in real time. Data streams are processed through trained models that estimate internal fluid motion and predict potential resonance conditions. When deviations from expected behaviour are detected, control systems can adjust operational parameters to mitigate dynamic amplification. For example, vehicle suspension tuning or thrust modulation in aerospace systems can be informed by real-time sloshing prediction. Such adaptive control frameworks represent a significant advancement beyond static design-based mitigation. Physics-informed neural networks provide an additional frontier. Unlike purely data-driven models, these networks incorporate governing equations of fluid motion directly into the learning process. By embedding conservation laws and boundary conditions within the training algorithm, the resulting models respect physical constraints while benefiting from data adaptability. This hybrid approach reduces the amount of required training data and enhances extrapolation reliability. For sloshing dynamics, physics-informed networks can approximate free surface evolution while ensuring mass and momentum conservation, thereby maintaining physical consistency.

Artificial intelligence further assists in computational efficiency. Adaptive meshing strategies can be guided by learning algorithms that identify regions requiring refinement based on evolving flow features. Turbulence model parameters may be tuned dynamically according to flow regime classification. Solver acceleration techniques use machine learning to predict optimal relaxation factors and convergence strategies, reducing computational time in multiphase simulations. Despite its transformative potential, artificial intelligence must be applied with methodological rigour (Benaissa et al., 2021; Amoura et al., 2023). Model training requires high-quality datasets generated through validated experiments or trusted numerical simulations. Overfitting must be avoided to ensure generalisation beyond training conditions. Interpretability remains essential, particularly in safety-critical systems where engineering decisions require transparent justification. Therefore artificial intelligence in sloshing dynamics is most powerful when integrated with physical understanding rather than used as a purely empirical black box.

References

Amoura, N., Benaissa, B., Al Ali, M., & Khatir, S. (2023). Deep Neural Network and YUKI Algorithm for Inner Damage Characterization Based on Elastic Boundary Displacement. In R. Capozucca, S. Khatir, & G. Milani (Eds.). *Proceedings of the International Conference of Steel and Composite for Engineering Structures. ICSCES 2022. Lecture Notes in Civil Engineering, vol 317* (pp. 220–233). Cham: Springer. https://doi.org/10.1007/978-3-031-24041-6_18.

Benaissa, B., Hocine, N. A., Khatir, S., Riahi, M. K., & Mirjalili, S. (2021). YUKI Algorithm and POD-RBF for Elastostatic and dynamic crack identification. *Journal of Computational Science, 55*, 101451. https://doi.org/10.1016/j.jocs.2021.101451.

Filo, G., & Lempa Pawełand Wisowski, K. (2025). Review of deterministic and AI-based methods for fluid motion modelling and sloshing analysis. *Energies, 18*(5), 1263.

Huo, J., & Lyu, J. (2025). A data-driven equivalent modeling approach for large-amplitude liquid sloshing under microgravity environment. *Advances in Astronautics, 8*(1), 47–59.

Ma, B., Yan, S., Upham, M. P., & Yue, B. (2024). Data-driven parameter identification of an equivalent mechanical model for large amplitude liquid sloshing. *Physics of Fluids, 36*, 112124. https://doi.org/10.1063/5.0241049.

Moya, B., González, D., Alfaro, I., Chinesta, F., & Cueto, E. (2019). Learning slosh dynamics by means of data. *Computational Mechanics, 64*(2), 511–523.

Wei, Z., Zhai, G., Wang, Z., Wang, W., & Ji, S. (2021). An artificial intelligence segmentation method for recognizing the free surface in a sloshing tank. *Ocean Engineering, 220*, 108488.

Wu, J., Zhang, J., Yu, Y., Shentu, Z., & Xue, M. (2025). Research on sloshing characteristics for liquefied iron ore and size optimization of anti-sloshing baffle based on data-driven models. *Applied Ocean Research, 158*, 104604.

Index

Note: Page numbers followed by "*f*" and "*t*" refer to figures and tables, respectively.

F

Printed and bound by CPI Group (UK) Ltd, Croydon, CR0 4YY
22/09/2026
02220473-0004